Advances in Neurosurgery 18

K.-A. Bushe · M. Brock · M. Klinger (Eds.)

Stabilizing Craniocervical Operations

Calcium Antagonists in SAH

Current Legal Issues

With 143 Figures

Springer-Verlag Berlin Heidelberg New York
London Paris Tokyo Hong Kong

Proceedings of the 40th Annual Meeting of the
Deutsche Gesellschaft für Neurochirurgie
Würzburg, May 7–10, 1989

Prof. Dr. K.-A. Bushe
Neurochirurgische Universitätsklinik
Josef-Schneider-Straße 11, D-8700 Würzburg

Prof. Dr. Mario Brock
Neurochirurgische Klinik und Poliklinik
Universitätsklinikum Steglitz, Freie Universität Berlin
Hindenburgdamm 30, D-1000 Berlin 45

Prof. Dr. Margareta Klinger
Neurochirurgische Klinik der Universität Erlangen-Nürnberg
Schwabachanlage 6 (Kopfklinikum), D-8520 Erlangen

ISBN 3-540-51967-X Springer-Verlag Berlin Heidelberg New York
ISBN 0-387-51967-X Springer-Verlag New York Berlin Heidelberg

Library of Congress Cataloging-in-Publication Data. Deutsche Gesellschaft für Neurochirurgie. Tagung (40th : 1989 : Würzburg, Germany) Stabilizing craniocervical operations : calcium antagonists in SAH : curent legal issues / K.-A. Bushe, M. Brock, M. Klinger (eds.). (Advances in neurosurgery ; 18) "Proceedings of the 40th Annual Meeting of the Deutsche Gesellschaft für Neurochirurgie, Würzburg, May 7–10, 1989"–T.p. verso. English and German. Includes bibliographical references. ISBN 0-387-51967-X (U.S. : alk. paper) 1. Vertebrae, Cervical–Surgery–Congresses. 2. Syringomyelia–Congresses. 3. Nimodipine–Congresses. I. Bushe, K.-A. (Karl-August) II. Brock, M. (Mario), 1983– . III. Klinger, M. (Margareta), 1943– . IV. Title. V. Series. [DNLM: 1. Cervical Vertebrae–surgery–congresses. 2. Fracture Fixation, Intramedullary–methods–congresses. 3. Neurosurgery–Germany, West–legislation. 4. Nimodipine–therapeutic use–congresses. 5. Syringomyelia–therapy–congresses. W1 AD684N v. 18 / WE 725 D486s 1989] RD531.D48 1989 617.4'8–dc20 90-9635 CIP

Offsetprinting and Binding: Druckhaus Beltz, Hemsbach/Bergstr.
2114/3130-543210 – Printed on acid-free paper

Preface

Demands on a Neurosurgeon Under Routine Clinical Conditions*

"Hardly any other field of surgery requires such meticulous asepsis. No other field requires such a protection of the tissue and such reliable hemostasis. No one will question that neurosurgery makes exceedingly high demands. However, the degree of the demands on the personality of the surgeon who has to carry out very serious operations to the exclusion of all others and has little opportuinity to recover psychologically by working on simpler cases is concealed from the outer world. The mental strain which the coworkers and staff have to tolerate may not be denied".

So wrote Wilhelm Tönnis in 1939. Although fifty years have now elapsed, his appraisal is still relevant today – hence my intention to discuss the everyday demands placed on neurosurgeons. My aim is not self-glorification to engender sympathy; rather it is to highlight the effects and repercussions of such demands for the well-being of the patients entrusted to us. The major onerous demands include:

- those in the operating theater
- those resulting from staff problems in nursing
- those resulting from legal developments
- those resulting from the increasing administrative tasks that cost valuable time and energy which are lost to our actual work in looking after patients: science and research also suffer from this.

I shall not go into the latter point within this preface.

Let me first turn to the strains occurring in the operating theater. In 1983 Ogilvie convincingly described the nature of stress in the life of the surgeon. He stated in conclusion that surgeons are confronted with demands from all sides which they have to master with decisive-

* Paper presented on the occasion of the 40th Anniversary of the German Society of Neurosurgery in Würzburg, 7 May 1989

ness, self-confidence, and considerable occupational skill and artistry. Of course, they bear the sole responsibility for the consequences and repercussions of their actions and those of their staff. Ogilvie discussed the stress in the life of a surgeon with all its resultant manifestations, but on the other hand assumed a very critical posture with regard to the behavior of surgeons who act very carelessly under stress and yet are by no means prepared to accept any help, in particular not from psychologists and behavioral researchers. Of course, this resistance has a traditional background.

In former times, surgical operations took place in lecture theaters in front of admiring and mixed audiences. The surgeon had to be bold: manual dexterity and quickness were the prerequisites. He thus had to be rigorous or at least appear to be rigorous. With the further development of surgical medicine, quickness in the operation became completely unimportant, and tranquility pervaded the operating theater. Even at that time Harvey Cushing, the creator of modern neurosurgery who experienced the beginning of this transformation, insisted on complete calm during the operation and prohibited any conversation, in part to avoid droplet infection but above all in order to be able to concentrate without distractions and with undivided attention to his work. In the epoch of microsurgery, calmness on the part of the operating surgeon has become a matter of course. Operations proceed in an oderly way in accordance with a strategic plan, naturally excepting emergency operations.

Eperience and observation in the operating theater and a series of multivarious occupational physiological and psychological investigations on surgeons have shown that despite the progress, operations still constitute a strain (apart from the tolerable physical effort). There is no question that neurosurgeons are individuals with a high and long-lasting performance strain. They are usually already reactive in the operating theater and have a high catecholamine level in their blood. They thus begin their surgical work under a raised anticipatory tension.

In six published investigations, pulse rate, ECG, and rate of breathing have been registered continuously in 6–30 surgeons during entire operations. Furthermore, the excretion of catecholamines such as adrenaline and noradrenaline and of adrenocortical hormones in the urine, and the secretion of adrenaline and noradrenaline in the blood, were measured in some investigations.

Foster reports that the pulse rate in eight surgeons averaged 121 during the operations, with maximum peaks above 150 in some cases. If these average pulse rates were attained by test subjects engaged in physical work, they could not maintain this for longer than 10–

15 min. It is to be borne in mind that these elevations of pulse rate are tolerated as a matter of course over many hours by the surgeon, especially in our specialty. In a comparative study, it could be shown in the same surgeons that this increase in pulse rate could be abolished by administration of 40 mg of a beta-blocker, so that the rate reached values such as those measured in surgeons during their patient rounds.

Timio et al. measured the excretion of catecholamines and of adrenocortical hormones in 30 surgeons before and after administration of beta-blockers. At the same time, a cross-over experiment was carried out in which one group received placebo and the other group, 40 mg oxyprenolol. During the operation, the excretion of adrenaline and adrenocortical hormones rose significantly in the placebo group, but the rise was suppressed by administration of beta-blockers. There was also a distinct coincidence between the increased excretion of the hormones mentioned and the rise in the pulse rate.

Even years ago, Weis and Lazarus reported on the pulse rate of anesthetists performing anesthesia. They also mentioned in this connection that a rise in the pulse rate in normal life is the correlate of increased aerobic muscular work. However, the rise in pulse rate under psychological strain is a manifestation of an ergotropic reaction which is mediated by catecholamines and is not accompanied by a raised oxygen requirement of the entire body but by an increase in the oxygen requirement of the heart muscle.

In my hospital, Poimann et al. have investigated stress in 19 surgeons during operations with varying degrees of difficulty in the context of a field study. The pulse and breathing rates were recorded. In addition, the surgeons specified the mental and physical strain they experienced during certain operations on the basis of an interval scale. The operations were recorded by video in order to be able to appraise the movement intensity of the surgeon. As comparison values for the raised parameters, the basic values during ward duty or during assistance at an operation were registered.

The pulse rate is a good parameter of strain in neurosurgical operations owing to the relatively low physical activity. It shows a high correlation with the subjectively experienced mental strain. In all operations investigated, the pulse rate in the surgeons was higher (about 20%–25% over the basic values on ward duty, the resting value, or values during assistance at operations).

Various surgeons show individual pulse rate curves independent of the kind of operation, which can be regarded as an indication of the individual coping structure of the surgeon in question. Experienced surgeons with more than 400 operations of the same kind display a significantly lower pulse rate during the entire operation than surgeons with less experience. However, there are also differences in this regard. The outer appearance and behavior of the surgeon may be misleading. A surgeon known to me who appeared to be calm and relaxed (a former member of the staff of my hospital) and had many years of experience in the same operations showed extremely high pulse values even in the preparation for the operation while he was washing his hands and disinfecting the site of surgery. These raised pulse values were lowered only in the course of the actual operation when the situation had been mastered.

Another experienced surgeon showed a high elevation of the pulse rate in a difficult situation involving hemostasis. The pulse rate immediately fell to initial values when he was supported in the difficult situation by an additional surgeon.

The above two examples prove that it is the anticipatory tension and the uncertainty with regard to the success or failure of the operation which lead to the excessive circulatory situation. Such examples contradict the general view according to which there is an emotional distance between the experienced surgeon and his patients. Every surgeon confronts his patients every day and in every operation in a state of anticipatory tension and preoccupied with care and solicitude. It is correct that the requirement to commiserate, and to feel the ups and downs of the patient, far exceeds what a person is able to achieve day in and day out as part of his job. Nevertheless, the thoughts of the surgeon before and during the operation in vital and functiona important regions are centred on questions as to the success of treatment.

We know from the results reported and from other investigati that excessive secretion of catecholamines (adrenaline) results in de rioration in all activities secretion and their implementation, especia in complex activities carried out under stress. This has also been fou in appearances of artists. The excessive secretion of adrenaline cau ventricular tachycardia and muscle tremor, and leads to anxiety as w as to gaps in memory in instrumental virtuosi and actors who have know their part by heart. In turn, anxiety leads to raised secretion adrenaline which increases the fear, i.e., a vicious circle is forme Experience shows that extreme stage fright can be alleviated by bet blockers but not by sedatives. Naturally, such stress factors are al of particular importance in other professions involving the associate experiences of risk, responsibility, decisional conflicts, and pressu of time.

Our knowledge of these new results constrains us to reflect on the stress on the surgeon and the stress in the operating theater, since there are numerous ways of alleviating it, e.g., by psychological training and testing of colleagues in advanced training and by ascertaining the feasibility of preoperative administration of beta-blockers, which have proved effective in artists.

The success of a surgical operation can only be ensured by qualified preoperative and postoperative care of the patiens. For this, we require suitable and motivated personnel at the doctor's side as responsible partners in the care of the patient. Officially, it was reported from Bonn by a specialist journal in February 1989 that the shortage of nursing staff reported in the preceding weeks (especially by the lay press) definitely did not exist. In the further course of the article, staff problems on a regional level were confirmed, but the seriousness of the acute situation was not generally recognized.

The increase in the number of patients owing to more sophisticated diagnostics and surgical methods, with a consequent increase in the frequency of operations, and the drastic shortening of hospital stays make hospitals a specialized high-performance enterprise. The duration of patient stays in the Neurosurgical Division in Würzburg was 20 days in 1969 but only 9.6 days in 1989, with a bed utilization of 94%. Although posts have been assigned to compensate for locum service, shortening of working time, protection of mothers-to-be and nursing mothers, etc., as is now usual elsewhere in the economy, these are by no means sufficient to compensate for the high stress owing to greater nursing intensity, etc. The high demands on each individual nurse owing to inadequate staff allocations have led to an increased loss of nursing personnel (especially recently), necessitating amendments of job allocation and duty plans. Despite their commitment, these experienced and qualified staff no longer see any future in their profession owing to the excessive stress, and opt out. This results in beds in important functional areas being placed out of service. Already, bed cancellations owing to the shortage of nursing staff have led to intolerably long waiting periods for inpatient admission (at present, more than 3 months). Long-postponed admission dates understandably encounter lack of understanding on the part of the waiting patients and lead to a burdensome tension between the doctor and patient and to an unfavorable press. Furthermore, integration into the world of work is delayed. In addition, we are exposed to the danger that we have to make a selection of patiens for the operation. This is a conflict situation entailing an additional burden for the physician, apart from the legal problems which are yet to be discussed. Naturally, acute cases are always admitted immediately; the nursing problems mentioned arise from this. Although the German Hospital Association has recommended an extrapolatory evaluation to determine staff requirements, this is not being carried out in our hospitals at present. In an international comparison, West Germany takes 16th place with regard to the employment of nursing staff among the 21 OECD countries.

The percentage of hospital costs in the gross national product in West Germany is 2.8%, as compared to 9.4% for the entire health service. The corresponding figures in the Netherlands are 5.7% for hospital costs and 8.5% for the entire health service. From the preceding figures it can be seen that the share of hospital costs in the total costs of the health service in West Germany is about 30%, which is far less than in all other comparable countries. The costs per case are twice as high in the USA as in West Germany, despite an extremely low average stay in hospital of 7 days in the USA. On the other hand, the costs of the other medical care sectors, outpatient care, dental care, drugs, remedies, and auxiliary aids are very much higher in West Germany than in comparable countries.

The main cause of the low hospital costs is the staff number per bed, which in West Germany is the lowest of all comparable countries. These observations indicate the necessity for measures to alleviate the shortage of nursing staff. These are:

1. Clear definition of nursing work and dispensation from administrative tasks.
2. Adapted staff alocations enabling regular duty times for nursing staff.
3. Adequate performance-related pay.
4. Re-recruitment and training of nursing staff who have left the profession, and the creation of part-time nursing posts.
5. Promotion of job careers, state promotion, and financing of postgraduate and vocational training measures.
6. In competition with other service sectors, nursing professions must be rendered more attractive and represented as being attractive in a deliberate image promotion campaign.

Everyday clinical routine is further aggravated by the forensic risks, which have become greater. These include conflict situations in the area of provision of medical information to patients, which constitute an increasingly intolerable burden for the doctor-patient relationship. This is threatening to become a question of medical survival, especially in the high-risk surgical disciplines (including neurosurgery).

Medical science, with its present-day possibilities, has made more progress in the last 30 years than in the centuries which have elapsed since Paracelsus. These successes are imparted to medical laymen almost daily, especially by television. On the basis of this information, the relationship of trust between the patient and the doctor has been transformed into an expectational attitude. The predominant feature is faith in human omnipotence, in the power of medical progress, in technological perfection, and in routine controllability of medical measures, with resulting excessive expectations on the part of the patient. Unavoidable fateful event sequences are not accepted. Under the influence of medical progress, the patient expects an adequate result of treatment with restoration of full functional integrity.

One-sided antimedical reporting in the press and media has given rise to a distorted, negative picture of the doctor which is persistently cultivated by "patient protection associations." In the April edition of a well-known economics magazine this year, new rights of patients for compensation were reported on over 19 pages, and indeed a manual has been published with a table listing the amounts of money which patients can claim in the event of liability. However, it has to be added that lack of sensitivity and inept behavior on the part of the physician following a complication can also give rise to desires for retaliation and similar feelings among patients and/or their relatives.

As long as I have worked as a physician, I have always regarded the informational talk as an essential part of our attention to the patient. It is a measure to build a relationship of trust on the basis of which we can save the patient from later disappointments and share responsiblity with him or her. We should not regard the duty of providing information as a formality forced on us by court verdicts and should also not carry it out primarily with the consequences for legal liability in mind.

On the one hand, the Chairman of the 6th Senate of the Federal High Court responsible for medical liability stated in an article in a specialist journal that information provided by the doctor must be restricted to important points. On the other hand, the extent of the duty to inform about risks has been incomprehensibly extended and tightened up by the Federal High Court in that it requires that patients be informed about not recognizable typical risks even when these are very rare.

Given their knowledge of the court verdicts, the hospital finance bodies insist on backing by information in juridical terms. If the heads of hospitals do not take care that all the doctors under them inform in accordance with the principles of legal verdicts, they are threatened by recourse because of contravention of professional duty. If he does not wish to lose his job and endanger his reputation, the individual physician has no alternative but to conform to legal decisions, even if his medical conscience violently objects to this because of the burden placed on the patient before consent to the operation is obtained by providing him or her with knowledge of extensive risks whose significance in the case in question *cannot be known with certainty*. It is intolerable that anxiety about legal consequences influences the behavior of doctors. We are in danger of practicing defensive medicine which ultimately harms the patient looking for help, i.e., by abandoning a human life which it might have been possible to save by taking risks.

This development must not be allowed to continue. However, this presupposes a change in the legal system which enables partnership and avoids antagonistic positions. The principle of liability for damage caused by intentional and negligent acts has to be abandoned and replaced by an insurance solution oriented to the principle of causation. Patient insurance has proved effective in Sweden over the past 10 years. The study group "Doctors and Lawyers" of the working group "Medical-Scientific Specialist Associations" introduced this possibility into the discussion years ago. It helps the doctor and the patient to the same extent. The claims of the patient are not aimed against the doctor, but directly against the insurer, and are thus in principle based not on a negligent act on the part of the doctor, but on the damage which has arisen owing to treatment. The system has the advantage of managing without the legislature; only a contract between the medical profession and the insurer is required. Such a transition from liability insurance to full comprehensive insurance would promote the relationship of trust between the doctor and patient, and such an insurance system has already proved effective for 100 years in respect of the liability of employers to pay damages with regard to employees' accidents.

However, besides medical care, major tasks of university hospitals are teaching, science, and research. The current strain referred to allows little time and space for this at present; it is the task of the ministries responsible to provide the necessary staffing and the appropriate scope for this. Up to now, neurosurgeons have been leisure time and weekend researchers. Those of us who took part in our annual meeting for the 40th time have participated in the developmental phase of our specialty with commitment and enthusiasm, have foregone free time, and have suffered further restrictions. Since 15 professorial chairs in our specialty were filled last year, and are to be filled this year and next year, we hope and wish that the new generation representing neurosurgery will attain better conditions than those described above and can continue working on the development of our specialty.

The *Advances in Neurosurgery* have always been published in English and this tradition will continue. However, the Current Legal Issues Section deals with legal matters that will serve as reference material for medical expertise reports in German courts and for German insurance companies. Since the legal system is specific to each country, the editors feel that it is important to present these specialized legal problems in the German language. Furthermore, the exact terminology in German is invaluable for assessing employee compensation reports.

K.-A. Bushe

Contents

Syringomyelia

List of Contributors

Reflections on the Future of German Neurosurgery: Further Development on the Foundations Laid by Tönnis

F. Loew[1]

Wilhelm Tönnis Memorial Lecture[2]

Wilhelm Tönnis was the father of German neurosurgery. Every neurosurgical "infant" in Germany knows this. However, knowledge of how he achieved this status is becoming lost in the mists of the past with the passing away of those who witnessed it, were involved in it, or at least heard vivid first-hand anecdotes about it.

No doubt, there were pioneers of neurosurgery in Germany before and alongside Tönnis. Let me remind you of Fedor Krause and Otfried Foerster. Their memory is honored by our Society through the awarding of the Otfried Foerster and Fedor Krause medals. However, they were neurologists or surgeons who *also* carried out neurosurgical operations. Foerster simultaneously utilized any operation on the nervous system as a neurophysiological experiment and thus expanded our knowledge of the function of the human nervous system. Krause developed new methods of cranial surgery. However, neither had the aim of establishing neurosurgery as an independent discipline. Nor did they aspire to extend our specialty with the objective of providing comprehensive care of the population. One of the important merits of Wilhelm Tönnis is that he not only had this objective but also succeeded in attaining it.

In 1929 Tönnis was accorded the right to lecture on general surgery at the University of Würzburg. Thanks to a Rockefeller Stipendium, he was able to familiarize himself with neurosurgery in 1932 under Olivecrona in Stockholm after working for 6 months under Professors Nonne and Pette in the University Neurological Hospital in Hamburg in accordance with a stipulation of Olivecrona, and after simultaneously having learnt Swedish (Fig. 1). He related how on his first day in the Stockholm hospital, he assisted in an operation on a meningioma which lasted more than 8 hours and a little later participated in just as long an operation for the postoperative hemorrhage that occurred. This impressively demonstrates the stamina which a neurosurgeon had to have at that time.

---

[1] Neurochirurgische Universitätsklinik, Postfach, D–6650 Homburg/Saar

[2] This lecture was presented on the occasion of the awarding of the Wilhelm Tönnis Medal on 8 May 1989 at the Annual Congress of the Deutsche Gesellschaft für Neurochirurgie in Würzburg.

Advances in Neurosurgery, Vol. 18
K.-A. Bushe, M. Brock, and M. Klinger (Eds.)

Fig. 1. Wilhelm Tönnis in 1932, just before his postgraduate scholarship in Stockholm

Upon his return from Stockholm he achieved a crucial change in procedure in Würzburg: His superior at that time, Professor Fritz König, had initially assumed as a matter of course that he himself would continue to operate on the nervous system and that Tönnis would merely assist him. This was the usual practice at German universities. It is proof of a felicitous combination of persuasive power, diplomacy, and willpower that Tönnis succeeded in convincing Professor König of the necessity of breaking with the tradition of the omnipotent hospital boss, so that he gave Tönnis a free hand in neurosurgery. This combination of persuasive power and diplomacy, stamina, and, if necessary, obduracy, the high demands he made of himself and his staff (Fig. 2), and finally, in the subsequent period, the tail wind of excellent surgical successes and scientific results quickly brought the ship of neurosurgery to full speed under its captain Tönnis:

1934 Lectureship in neurosurgery, assistant professorship, and founding of a neurosurgery department.

1937 Tönnis was asked to accept the first German Chair of Neurosurgery at the University of Berlin and at the same time to become Head of the University Neurosurgical Hospital and of the Department of Tumor Research and Experimental Pathology of the Brain at the Kaiser Wilhelm Institute of Brain Research in Berlin-Buch (Fig. 3).
Inauguration of the *Zentralblatt für Neurochirurgie*.
International recognition was demonstrated symbolically by the Meeting of the British Society of Neurological Surgeons in Berlin and Breslau in 1937. Basically, this was a first European congress of neurosurgeons; apart from the

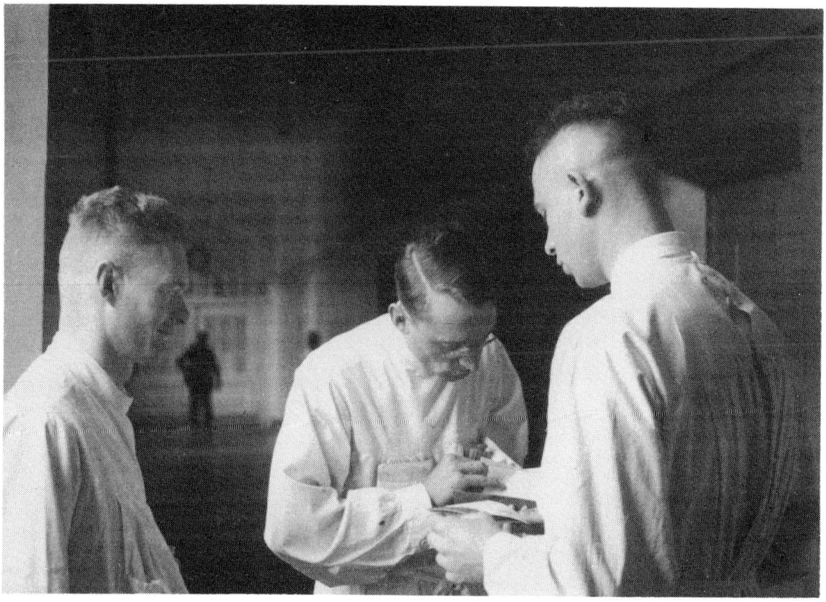

Fig. 2. Pioneer staff at the Neurosurgery Department in Würzburg. *Left to right:* Riechert, Fischer, Röttgen

Fig. 3. Hugo Spatz in 1937. The scientific collaboration with this prominent brain researcher in the context of the Kaiser Wilhelm (later Max Planck) Institute of Brain Research developed into a lifelong friendship

Fig. 4. Sir Geoffrey Jefferson during the joint congress with the Society of British Neurological Surgeons in 1937 in Berlin

British – above all mention should be made of their president, Sir Geoffrey Jefferson (Fig. 4), and Norman Dott (Fig. 5) – the congress was attended by Olivecrona and other Scandinavians, de Martel from France, Martin from Belgium, Fasiani from Italy, and Krayenbühl from Switzerland.

The war brought this development to an abrupt halt, but at the same time also showed how Tönnis mastered new tasks and found excellent solutions. Whereas tumors of the central nervous system had been at the center of interest until that time, brain injuries now became more important. As a consultant neurosurgeon to the air force, he created an organization for taking care of patients with head injuries which the Americans considered, alongside rocket technology, to be one of the few German achievements worthy of imitation after the collapse of the Nazi regime. The principles were as follows:

- At the front, emergency care only
- Air transport for final surgical treatment in specialized military hospitals (Figs. 6, 7)
- Regular training of all physicians working there
- Rehabilitation to the level of occupational reintegration in military rehabilitation centers created especially for this purpose

Fig. 5. Herbert Olivecrona, Norman Dott, and Wilhelm Tönnis during the 1937 congress in Berlin

– All of the above under a uniform management

Improvement of operative methods and classification of brain injuries were, of course, other features of this creative period. Thus countless brain-injured soldiers have Wilhelm Tönnis to thank, directly or indirectly, for their survival and restored health.

The end of the war and the collapse of the Nazi regime forced our people and thus also Wilhelm Tönnis to make a completely new start. In accordance with his realistic appraisal of the situation, he did not return to Berlin, but assumed the position of head of surgical department in the Knappschaftskrankenhaus in the Ruhr. Bochum-Langendreer achieved its reputation thanks to him.

I myself joined Tönnis in June 1946 at the suggestion of his former pupil Sorgo. Tönnis was head of a department comprising 250 beds for general surgery and 100 beds for neurosurgery. Under the senior residents Klug and Hering, who, like Tönnis, were responsible both for general surgery and for neurosurgery, only Heinz Bormann and myself were initially responsible for the 100 neurosurgical beds. One month later, Kurt Schürmann joined us. Subsequently Hans-Werner Pia, Frank Marguth, Reinhold Frowein, Wolfgang Schiefer, Werner Krenkel, and Wilhelm

Fig. 6. Preparation of the air transport of patients with head injuries during the war. The picture shows, among others, the nurses Alice and Gertraud (second and fourth from the right), who remained important members of Wilhelm Tönnis's staff even after the war

Driesen also came, to name but a few (Fig. 8). Unfortunately, some of these have left us before their time.

Only after I myself had become head of a neurosurgical department could I fully appreciate what Tönnis had achieved at that time. He had a huge daily program of operations, including practically all neurosurgical and the more difficult general surgical operations, not to mention outpatient duties, clinical visits, administrative burdens, and problems of procurement of medical equipment. However, over and above this he revived the old Kaiser Wilhelm Institute together with Professor Zülch in the Knappschaftskrankenhaus, thus creating the basis for the later Max Planck Institute of Brain Research. He organized scientific programs on which we, grudgingly, worked at night, presented papers, and reestablished contacts which were necessary in order to be able to return to a university (Fig. 9).

Among other things the classification of closed brain injuries, on which I worked as well, dates from this time. Moving away from the static terms of concussion and cerebral contusion, which were arbitrarily defined according to the duration of primary unconsciousness, the classification was based on the duration of normalization of all disorders of cerebral function. It thus introduced dynamic clinical

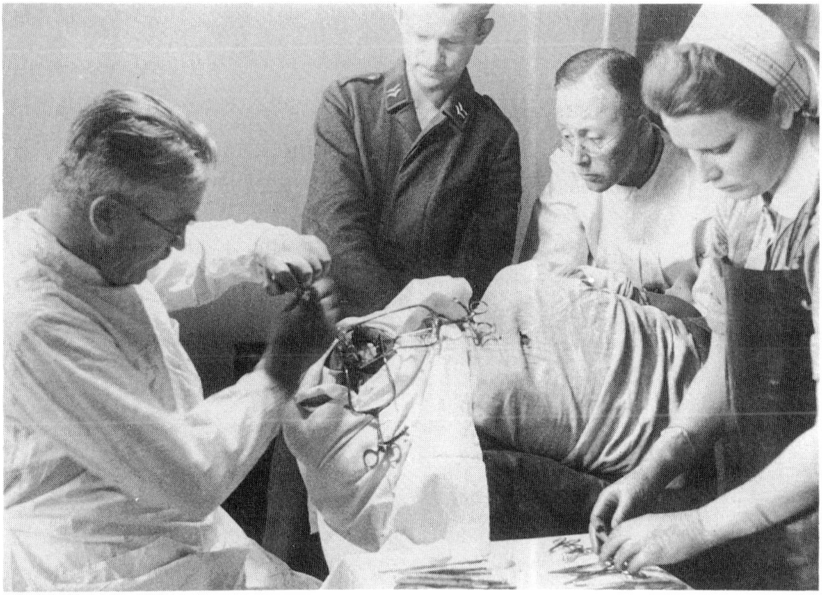

Fig. 7. Brain operation under war conditions

criteria as a new feature and also took into consideration posttraumatic vegetative dysregulations, in particular those of the circulation.

Looking back, it is readily understandable why he did not always react with patience to the permanent stress to which we were exposed at the time, but rather could explode in anger and demand the ultimate in performance not only from himself but also from us.

He succeeded in returning to a university. In 1951 work began both at the newly built Neurochirurgischen Universitätsklinik in Cologne and at the Department of Tumor Research and Experimental Pathology of the Brain at the Max Planck Institute of Brain Research under the same roof (Figs. 10, 11). An animal experimental laboratory was added. The development of clinical neurosurgery over its entire spectrum was accompanied by very many scientific studies which led, among other things, to the development of modern neuroradiology. It should not be forgotten that radiology owes the methods of contrast imaging of hollow systems of the body (especially of the blood vessels and the CSF spaces) to neurosurgeons. Within the context of this lecture it is not possible to recount all the clinical and scientific contributions made during the Cologne period by Tönnis and also by his staff, thanks to his initiative, suggestions, and active participation.

International contacts were also reestablished during this period. This was difficult, since he was unjustly regarded as a militarist and Nazi because of his suc-

Fig. 8. Staff at the Knappschaftskrankenhaus in Bochum-Langendreer

Fig. 9. Wilhelm Tönnis in Bochum-Langendreer

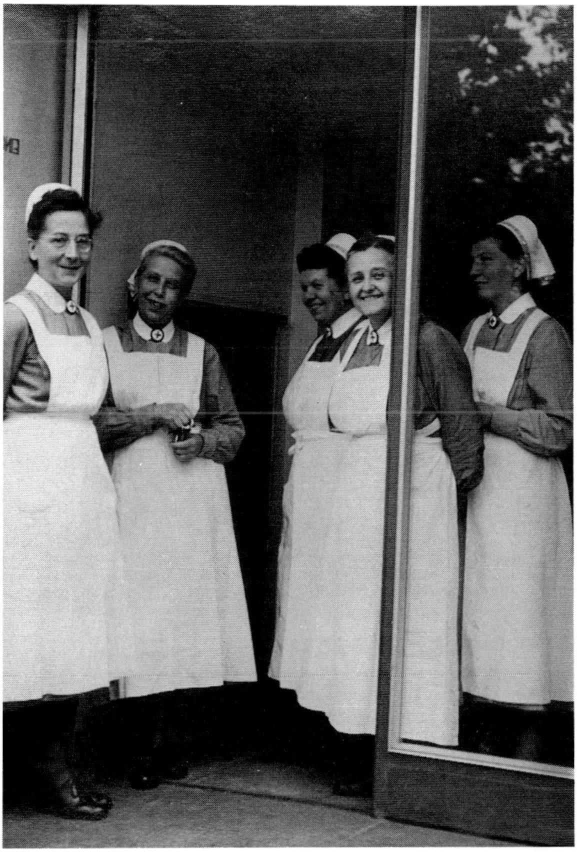

Fig. 10. Important mainstays of the Cologne hospital: the nurses Hiltraud, Gertraud, Claudia, Alice, and Dorothee

cessful work for our wounded soldiers, for which he had been praised by the Nazi regime. Landmarks in regaining recognition and in reconciliation were visits from such prominent foreign colleagues as Earl Walker, Paul Bucy, and Percival Bailey and the First International Congress of Neurological Sciences held in Brussels in 1957 (Fig. 12).

The recognition of neurosurgery as an independent discipline in West Germany, overcoming the supremacy of general surgeons and neurologists, the introduction of a separate "specialty" designation (*Facharztbezeichnung*), and the subsequent logical creation of neurosurgical chairs at an increasing number of West German universities also took place during this period.

Visible evidence of Tönnis's achievements, and the culmination of his successful development of our discipline, is provided by his presidencies of the Gesamtverband Deutscher Nervenärzte and the Deutsche Gesellschaft für Chirurgie, numerous honorary memberships, honorary doctorates, and other distinctions, and his position

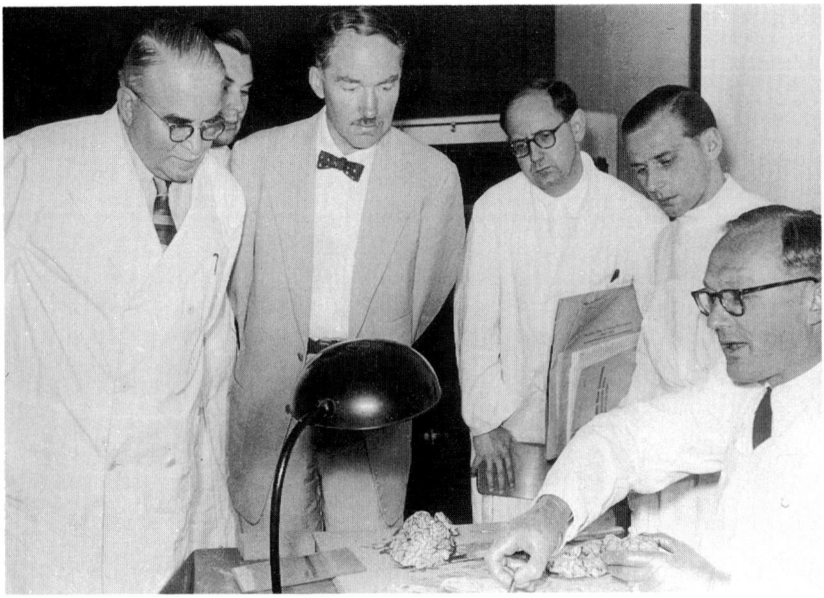

Fig. 11. Neuropathological demonstration by K.J. Zülch

as Rector of Cologne University (Fig. 13). Up to his death, shortly after his 80th year of life (Fig. 14), he was able to follow with pride and satisfaction how the work he had done in developing and expanding neurosurgery was continued and consolidated by the next and subsequent generations.

What is the position today and what is the outlook for the future?

In my opinion, we are at the end of the phase of expansion and consoldiation and at the beginning of a phase of defence, in which we are confronted by the necessity of some reorientation. Let me state the reasons for this.

Comprehensive neurosurgical care of our population has been attained. In the 1960s it was still necessary to increase the number of neurosurgical departments in the interest of patients with head injuries due to traffic accidents. Today, we are confronted by the problem of limiting the number of neurosurgical departments. The head injury argument is no longer valid: thanks to obligatory seatbelts, their number has declined. Moreover, the ever-increasing density of the helicopter transport network has shortened the distance to the neurosurgeon. If there are too many departments, the opportunities for the individual neurosurgeon to gain experience with rarer and more difficult clinical pictures will be reduced, i.e., standards will be lowered.

Limitation of the number of neurosurgery departments and simultaneous restriction of the number of postgraduate training posts for neurosurgeons, as implemented

Fig. 12. At the executive table during the International Congress of Neurological Sciences in 1957 in Brussels: Paul Bucy, Geoffrey Jefferson, Herbert Olivecrona, and Wilhelm Tönnis

in an exemplary way a long time ago in the Netherlands and later also in Spain, are impeded by various conflicting individual interests of which I shall name only some:

- In the course of the reform of the health service, hospital institutions threatened by closure of poorly filled wards are discovering their affection for neurosurgery as a hope of survival.
- Staff members of our university hospitals with limited-period contracts are looking for jobs outside of university. This goes as far as positions in general practitioner wards or practices almost without surgical facilities.

The situation is likely to become more acute when advances in European integration allow the influx of colleagues who have received neurosurgical training in other countries of the European Community. It must also be taken into account that on the one hand a neurosurgical department cannot maintain and raise its standard with constantly changing staff, while on the other hand, new, younger staff may introduce fresh dynamism.

We therefore require solutions within our traditional hierarchical hospital system which enable especially qualified colleagues to be retained in the departments for specific tasks without our specialty being disintegrated by the creation of depart-

Fig. 13. Wilhelm Tönnis as Rector of the University of Cologne

Fig. 14. Wilhelm Tönnis during the last period of his life

ments, and without blocking all posts with staff members who are slowly becoming rather old.

We need not only a West German but also a European solution to this problem. I regard it as one of the most important tasks for the officers of our Society that simultaneous approaches be made to the Länder Governments and the Federal Government, to the Scientific Council (Wissenschaftsrat), and to the Chamber of Physicians, within the context of the European Association of Neurosurgical Societies and the administration of the European Community in Brussels, in order to achieve a reasonable solution. This presupposes the elaboration of an overall approach and a long-term strategy.

Influence within international organizations does not fall from heaven like manna. Without commitment even in the tedious, time-consuming, and often frustrating detailed work, without active participation even in boring meetings, nothing can be done. I mention our Congress President Karl August Bushe as a positive example with regard to work within the World Federation of Neurosurgical Societies. Karl-August Bushe stepped into the breach for German neurosurgery when the world congress had to be organized in Europe.

In general, we German neurosurgeons are in the midst of a change of generations with regard to our figureheads. The trailblazers within the European Association of Neurosurgical Societies (EANS) are either dead, like Hans-Werner Pia, retired, like Kurt Schürmann, or (like Rolf Wüllenweber and myself) about to retire. You know that, as a member of the commission which started the preparation of a constitution in the mid-1960s, I am one of the founding fathers of the EANS, and that I have been an ever-present member of its Administrative Council. It is absolutely necessary that the work of the "middle generation" (of whom Mario Brock, Rudolf Fahlbusch, and Hans-Dietrich Herrmann, in alphabetical order, are examples) be taken up and continued by the younger generation.

The idea that only the occupants of university chairs can be "figureheads" of German neurosurgery must also be abandoned. I would mention Madjib Samii as an example who is at the same time provocative in several respects. Even if he is not German born, he has become a figurehead of German neurosurgery and therefore should have an appropriate position in the hierarchy of our society.

The term "phase of defence" which I used before points in another direction. The struggle for their daily bread among practicing physicians, a consequence of the irresponsible extension of training capacities which was forced on the medical faculties, has not yet impinged upon the established heads of larger departments, but will in one way or another affect the younger members of this audience who represent the future of our discipline. I would like to present some examples:

The surgical treatment of lumbar and cervical disk prolapses, of spinal root compression syndromes of other etiology, and of tumors of the spinal cord and vertebral columns has mainly been developed and improved by neurosurgeons. In this context, I should like to mention Henk Verbiest as an example for many others.

My colleague Caspar and myself have also contributed in this field. However, those neurosurgeons who stare spellbound at a vertebral disk like a rabbit at a snake and for whom narrow lateral recesses and instability are unknown concepts, who equate vertebragenic pain with disk prolapse and root compression and do not consider the many causes of pseudoradicular pain, promote the prevailing tendency to subsume the whole of spinal surgery within the domain of the orthopedic surgeons under the heading "neuro-orthopedics." It is important that we neurosurgeons have mastery of all forms of surgical treatment and that we can offer them to patients. Only then is an optimal choice of method of treatment ensured – a choice appropriate to the special features of the individual case.

The situation is similar for the "failed back syndrome." As a rule, one can learn more from failures than from cases with positive progress. If we leave it to the orthopedic surgeons to carry out all fusion operations necessary because of postoperative instability or to the anesthetist to use other methods of pain relief, we need not be surprised when they conclude that they can also carry out the original operation better than we can. We can only hold our ground if we are better. Every day fresh efforts are necessary to become better and to remain better. This, of course, applies not only to "neuro-orthopedics" but also to all neighboring disciplines: competition for peripheral neurosurgery with the traumatologists, for surgery of the base of the skull with the otosurgeon, for tumors of the orbit with the maxillofacial surgeon and ophthalmologist, for carotid surgery with vascular surgeons, and for vascular malformations with neuroradiologists. Nobody (not even a head of a department) should consider himself to be so good that he does not need to learn anything new or to consult someone who is better in a subspecialty or refer patients to him. We must also learn to appreciate that for the optimal solution of some special problems it may be reasonable to look for coöperation with neighboring disciplines, and, for example, to operate jointly. The very interest of our specialty and the well-being of the patients entrusted to us come together in this fundamental attitude.

For any scientific progress, teamwork has become an indispensible prerequisite. Experience teaches that science is fertilized by contacts with neighboring disciplines. This applies both to diagnostic problems and to scientific methods. We can also learn from Tönnis in this respect. From the beginning, he aimed to relate clinical problems and sophisticated scientific methods. He was able to achieve this by the link between the Max Planck Institute and the hospital. Depending on the local situation, other solutions may be possible, but they all have in common the integration of methodological knowledge and expertise (e.g., of neuropathologists, neurophysiologists, and neurochemists as well as the experience of neurologists, neuroradiologists, and radiotherapists) in the solution of clinically important questions. Real research can no longer be carried out at half steam, after finishing the daily routine and on one's own.

The term "phase of defence" should not be misunderstood as meaning that we must hold on to everything which has been initiated by neurosurgeons as a neurosurgical "property asset" under all circumstances. As illustrated by the example of neuroradiology, which was initially developed and practiced by neurosurgeons, when close contact is maintained the development of independent disciplines may be appropriate and make possible further developments which could not otherwise have been achieved. Tönnis also provided examples of this.

A lot more could be said about appropriate strategies in the "phase of defence" which has now started. For reasons of time, I will confine myself to two aspects:

1. It is important to participate in our professional organizations, e.g., the Chamber of Physicians (Ärztekammer) and the Association of Medicare Physicians (Kassenärztliche Vereinigung). Many crucial decisions take place there, extending up to the distribution and evaluation of medical fees (Gebührenordnung). Our professional association (Berufsverband), which has now become independent, will concern itself with these material questions in future.

Until yesterday, the Deutsche Gesellschaft für Neurochirurgie was the mouthpiece for *all* the concerns of neurosurgery in our country, from science to the representation of economic interests. Our section for professional questions (Abteilung für Berufsfragen), which I inaugurated at the beginning of the 1960s because of the first amendment of the scale of fees (Gebührenordnung), which was impending at that time, was our professional association, but simultaneously remained an integral constituent of our Society. In this way, we could always act in unison. The organizational separation adopted yesterday reduces the German Society of Neurosurgery to a specialist scientific society – before, it was more comprehensive and unequivocally more than it is now – and creates an independent professional association despite personal intercalation. The danger of an inherent trend to develop in different directions, which is intrinsic to all corporations, is great. This may lead to mutual competition and thus weaken the position of neurosurgery. It is of crucial importance that such a development be avoided, but this has become more difficult since the separation effected yesterday.

2. It is also important that as many of us as possible take part in refresher courses for general practitioners and speak there in terms intelligible to non-neurosurgeons. It was never beneath Tönnis's dignity to speak even to a county medical association, quite apart from specialist congresses of neighboring disciplines. I have myself invested a great deal of time and effort and remember a year in which I presented more than 70 papers at medical training meetings. In future, such a burden should be distributed over many shoulders.

However, let us return to Wilhelm Tönnis, in whose memory this lecture has been presented. I have reported on the successful neurosurgeon Tönnis. Yet what sort of a man was he in personal terms? Some personal experiences may shed some light on various facets of his personality.

One side was that of the absolute monarch who set exact dates for scientific papers which could only be met through long nights of work despite clinical work which more than filled the day, and who angrily told me to pack my bags because some slides had not been prepared to an adequate quality prior to a medical meeting. Besides her many other merits, it was the calming protective hand of Mrs. Tönnis which abrogated this disastrous verdict.

He was a boss one could not contradict. For example in an internal staff discussion the neurosurgical "infant" Loew, inexperienced in dealing with big bosses, argued ingenuously against the thoughts of Tönnis regarding a scientific project until Tönnis abruptly ended the discussion with an order as to how we were to proceed. A little later, during a hospital party and after consuming some alcohol, he drew me into a corner and said with a slurred tongue, and now I am quoting literally: "You are a silly person. You cannot contradict me in the presence of other people." The lesson was learnt. On the other hand, he was as sensitive as a seismograph, sensing implicit unarticulated dissent and taking it into consideration. It also happened that he responded to a new idea presented to him only with indifference or even rejection, yet a few weeks later asked reproachfully, "Why didn't you do it like *that*, we always do it like that," referring to the idea, which had been inwardly accepted in the meantime.

Under excessive stress, as well as under particular surgical tension, he tended towards outbursts of violent temper: "You are holding the spatula in the brainstem" or "You are killing my patient." The resigned complaint "Nobody is helping me" also resounded in the ears of patients who were only under local anesthesia. However, what he had said during the operation was no longer valid afterwards.

A quite different side of this man was his often shyly hidden humanity: during and after the war, he protected and shielded a communist woman who had been sought by the authorities; he took in nurses and doctors of his war team in the Knappschaftskrankenhaus in Langendreer after the collapse of the Nazi regime; he helped a colleague who had concealed the fact that he had not finished his studies and wrongly claimed an academic degree to complete his professional qualification discreetly and without loss of his post; and I shall never forget how he told me during the initial period of starvation after the war: "As long as I have something to eat, you will not starve."

Wilhelm Tönnis was like a king in times gone by. He combined both the claim of Louis XIV "L'état c'est moi" ("I am the state") and that of the great King of Prussia "I am the first servant of my state."

He radiated natural authority and distance, but identified with his staff and gave them considerate protection. He was more than a great neurosurgeon. He was a great, strong personality – a great person.

Decompression and Stabilization of the Cervical Spine

Indications and Complications of Cervical Spine Stabilization

A. Weidner, S.T. Chioe, and H.W. Schumacher[1]

Introduction

Fusion of the cervical spine is indicated for instability. If instability of the cervical spine is not treated, kyphosis with neurological deficits may develop. One complication of spinal fusion is pseudarthrosis. This develops when micromovements prevent vascular budding, which is necessary for incorporating the bone grafts in the fusion [2]. In a fusion involving one intersegmental level the shearing forces are moderate. However, in a fusion involving several segments these movements are more extensive, so that immobilization is indicated. External immobilization, even with a halo, has only a limited effect in reducing micromovements [6]. By performing an internal fixation with plates and screws, the shear forces upon the boundary surfaces to be fused are effectively reduced.

This study reports the complications in 251 internal fusions within the cervical spine (Table 1). Also, long-term follow-up results of 116 patients with ventral AO plate fusions of the middle and lower cervical spine are reported.

Table 1. Internal fixation: number of cases

Craniocervical junction	51
Middle/lower cervical spine	
Dorsal	11
Ventral	189
	251

[1] Paracelsus Klinik, Neurochirurgische Abteilung, Am Natruper Holz 69, D–4500 Osnabrück

Advances in Neurosurgery, Vol. 18
K.-A. Bushe, M. Brock, and M. Klinger (Eds.)

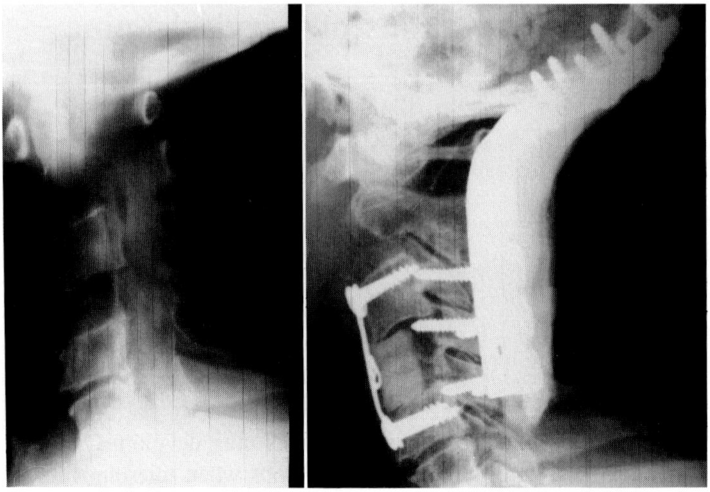

Fig. 1. *Left:* Metastasis of carcinoma of the prostate at C2 and C4 with therapy-resistant neck pain. *Right:* Removal of metastasis at C4; PMNA block; securing with a ventral AO plate. In the same operation additional dorsal stabilization with two Roy-Camille occipital plates [4] was performed

Craniocervical Junction

Internal fixation of the craniocervical junction was performed in 51 patients. In 28 patients the cause of instability was atlantodental laxity with spinal cord compression documented by MRI. In 14 patients trauma was the etiology for instability and in four patients an anomaly of the odontoid bone was the cause. Rarely were metastases (in three cases) or postoperative instabilities (in two cases) stabilized with metal plate and screws. Fusion of the occiput to the upper cervical spine was indicated in 13 cases. In rheumatoid arthritis, occipital-cervical fusion was performed to manage instability only when considerable destruction of the atlanto-occipital joint was present, producing impending pseudobasilar impression. Wire cerclage was employed four times and a Roy-Camille plate was used in nine cases (Fig. 1). In seven patients stabilization of C2-3 instability was performed with dorsal plates, which were fixated with transpedicular screws.

Most frequently, C1 was fused with C2 (33 cases). On three occasions, the fusion was reinforced with a Roosen-Trauschel clamp. In 20 patients wire cerclage was performed around C1 and C2 bilaterally. This procedure has been described by Brooks and Jenkins [1]. During the past year, we have exclusively employed the modified technique of Magerl and Seemann [3] to treat C1-2 instability. From a dorsal approach the joints of C1 and C2 are screwed together (Fig. 2). Double-threaded screws are utilized which exert compression on the articular surfaces.

Additionally, a wire cerclage in the midline is necessary. In metastases we use bone cement for reinforcement. In all other fusions autologous iliac crest bone graft is used.

Intraoperative complications did not occur. Pseudarthrosis developed in one patient with a Roosen-Trauschel clamp and in three patients with Brook's fusion. To avoid these complications we now exclusively perform fusions in accordance with the technique of Magerl and Seemann (Fig. 2).

Median/Lower Cervical Spine

Compared to the upper cervical spine, we rarely stabilize the middle and lower cervical spine from a dorsal approach (Table 1). We recognize indications for the dorsal approach in the management of pseudarthrosis of vertebral fusions, in dorsally positioned tumors, and in combined ventral and dorsal instability. The stabilization technique we employed is the technique of Roy-Camille with transpedicular fixation of the plates [4]. With this technique, operative complications did not occur. Loosening of plates or screws has not been observed in our small number of cases.

We have reinforced 187 ventral spondylodeses with AO plates. The operative technique was identical in all fusions [5]. One advantage of the ventral approach is easier patient positioning. This is particularly important for polytraumatized patients. Medial sequesters and bone fragments can be removed with less risk by employing the ventral approach.

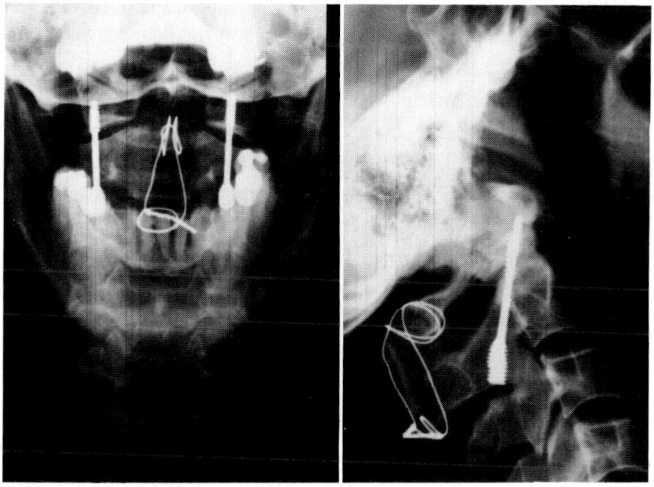

Fig. 2. Dorsal stabilization of C1-2: Transarticular screw fixation with double-threaded screws and additional wire cerclage in the midline

Table 2. Reasons for removal of instrumentation

Pseudarthrosis	2
Loosened screws	2
MRI	2
Infection	1
Young age	1
Bad position of the plate	1
Surgery one level above	1
	10

Between 1983 and 1986, 116 patients were stabilized with ventral AO plates. The average follow-up examination was over 37 months (24–72 months). In our series 57 patients received an AO plate to treat instability following spondylectomy. The indications included progressive kyphosis or stabilization after spinal cord decompression for cervical myelopathy. Thirty-three patients with postoperative instability, 17 with trauma, 6 with tumor, and 3 with rheumatoid arthritis required stabilization.

Removal of the implants was necessary in eight cases within the first 2 years and in two patients after 2 years (Table 2). In this group two patients developed pseudarthrosis. In one patient the plate was not positioned in the midline. We removed the original plate and replaced it with a new plate and graft with good results. In another patient with marked osteoporosis fusion could not be obtained. To relieve the patient's pain we removed the plate and performed a dorsal fusion. A total of 12 patients died. No deaths were from complications secondary to the cervical spine operation. Also, complications associated with the metal plate did not occur.

In only four patients was removal of the screws or plate indicated due to loosening of the screws. This corresponds to 3.4% of the total number of cases. Three of these four patients were females and one of them had marked osteoporosis.

The position of the screws was measured in every case. The sagittal diameter of the vertebral body was divided into thirds. In 80% of the cases, the tips of the screws were positioned within the dorsal third of the vertebral body; in 9% the dorsal vertebral body cortex was perforated. Only three screws could not be positioned into or past the middle third. Our results demonstrate that the posterior wall of the vertebra does not have to be perforated by the screws. However, the dorsal third of the vertebral body must be traversed by them.

The angle of the plate relative to the midline of the cervical spine is also an important factor. The average angle for all our cases was 5°. In two patients with screw loosening, the angle was 8° and 25° respectively. So in these cases the screw was exposed to increased stress which resulted in screw loosening.

Screw loosening always occurred within the first 6 months after surgery. Radiological follow-up studies were performed after 6 days, 6 weeks, and 6 months.

Four patients developed dysphagia after plating, which did not last longer than 3 months. In the patients in whom a screw had loosened, dysphagia was not noted.

One patient developed a deep infection within the first postoperative week necessitating removal of the plate and application of a halo. This patient's course was otherwise uneventful.

Indication for Internal Fixation

Marked osteoporosis is a contraindication to internal fixation. Instability of the craniocervical junction should be stabilized by the dorsal approach. In the middle and lower cervical spine, the stabilization is performed dorsally if decompression of the spinal canal can only be performed dorsally or if the instability is exclusively localized dorsally. All other fusions can be performed by a ventral approach with an AO plate.

A ventral diskectomy at one level without further instability does not require additional osteosynthesis.

The advantages of internal fixation are that it is a safe method and it produces sufficient stability that all patients can be mobilized without external fixation.

References

1. Brooks AL, Jenkins EB (1978) Atlanto-axial arthrodesis by the wedge compression method. J Bone Joint Surg 60A:279–284
2. Kaufman HH, Jones E (1989) The principles of bony fusion. Neurosurgery 24:264–270
3. Magerl F, Seemann PS (1987) Stable posterior fusion of the atlas and axis by transarticular screw fixation. In: Kehr P, Weidner A (eds) Cervical spine I. Springer, Wien New York, pp 322–327
4. Roy-Camille R, Mazel C, Saillant G (1987) Treatment of cervical spine injuries by a posterior osteosynthesis with plates and screws. In: Kehr P, Weidner A (eds) Cervical spine I. Springer, Wien New York, pp 163–174
5. Weidner A (1989) Internal fixation with metal plates and screws. In: Sherk HH et al. (eds) The cervical spine, 2nd edn. Lippincott, Philadelphia, pp 404–421
6. Whitehill R, Richman JA, Glaser JA (1986) Failure of immobilization of the cervical spine by the halo vest. J Bone Joint Surg 68A:326–332

Osteosynthesis of the Upper Cervical Spine and the Craniocervical Junction

P. Knöringer[1]

Introduction

The aim of surgical treatment of instabilities of the upper cervical spine and the craniocervical junction is the reconstruction of the physiological anatomy. Repositioning and stability are required to relieve the spinal cord, nerve roots, and vertebral arteries and to protect them from continuous microtrauma. One concern with regard to spinal motility is that osteosynthesis should produce no or only the least possible loss of function. This is of paramount importance, because this region, with the rotation and the nod joints of the head, has the greatest motility of the cervical spine.

Fractures of the Dens Axis

Osteosynthetic repair of odontoid fractures with a pair of screws introduced from the anteriolateral aspect has the advantage over the dorsal methods [4, 8] that the atlantoaxial joint is maintained or reconstituted [2, 3]. Compared with the transoral procedure [6, 7], there is less intra- and postoperative burden on the patient, and the risk of infection is reduced. Whereas conservative treatment necessitates long periods of constraint in a halo-fixateur or a Minerva cast, screw fixation of the axis, without placing undue strain on the patient, leads to quick and stable healing results after about 6 weeks of wearing a removable neck brace [11]. In the hands of a practiced surgeon, therefore, screw fixation of acute odontoid fractures presents a true alternative to conservative treatment and is to be seen as the method of choice among the surgical procedures. The use of double-threaded screws has resolved some of the disadvantages inherent in the original procedure.

The heads of the generally employed small fragment spongy bone screws almost invariably irritate the segment C2/3, leading to painful inhibition of ventral flexion combined with posterior headaches and osteophytic reactions at C2/3 [13, 14]. The

[1] Neurochirurgische Abteilung der Universität Ulm im Bezirkskrankenhaus Günzburg, Ludwig-Heilmeyer-Str. 2, D-8870 Günzburg

Advances in Neurosurgery, Vol. 18
K.-A. Bushe, M. Brock, and M. Klinger (Eds.)

double-threaded screws (Fig. 1), which are almost completely embedded in the axis, do not produce mechanical irritation of the segment C2/3. With regard to the diameters and length of the threaded portions, the dimensions of the double-threaded screws are exactly adapted to the anatomical conditions of the fracture types encountered at the axis. The cannulated prebore for the back thread increases the diameter of the bore hole from 1.8 to 3.7 mm, so that it can be easily found when the screw is introduced. This, together with the fact that it is no longer necessary to prethread the bore hole, serves to simplify the procedure.

Double-threaded screws can also be used to treat injuries with fracture lines running from the upper back to the lower front of the axis, if the upper vertebral joints of C2 can be repositioned and the fracture does not reach too deeply into the anterior, ventral part of the axis body. In these fractures, the posterior thread prohibits a ventrocaudal dislocation of the odontoid. An antiglide plate, as used in Böhler's procedure, is not required.

Hangman's Fractures

Hangman's fractures are generally treated conservatively by repositioning and a 6- to 12-week course of immobilization in a halo-fixateur. An indication for primary intercorporeal fusion of C2/3 with instrumentation is usually assumed when there is considerable damage to the intervertebral disk C2/3 with corresponding dislocation (fusion according to Cloward and Smith-Robinson with plate osteosynthesis). This is also the case for fractures which have achieved bony stability after conservative therapy, but with residual discoligamental instability of C2/3. These cases only require intercorporeal fusion without metal implants.

If the instability is primarily of a bony nature, as is the case in most injuries, transpedicular screw fixation [9, 16] can provide stability without loss of function, so that this method can be considered an alternative to conservative therapy.

Transpedicular screwing is accompanied by risk of injury to the vertebral arteries, the dura, and the spinal cord. This risk can be decisively reduced if the following points are observed. Besides plain X-ray imaging, computed tomography of C1-3 should be performed with bone window adjustment and 2-mm layers, the sections being parallel to the line of the upper arch of C2. This not only provides more precise visualization of fractures but also reveals anomalous vertebral arteries, recognized by alterations in the transverse foramina. As a rule, preoperative angiography is not needed. The exact subperiosteal view of the medial borders of the arches and the localization of the vertebral artery by intraoperative Doppler ultrasonography [15] permit relatively safe preboring and screw implantation between these two landmarks.

The risk of injury is further reduced by using double-threaded screws (Fig. 2) instead of the cortical and small fragment spongy bone screws [19] used to date.

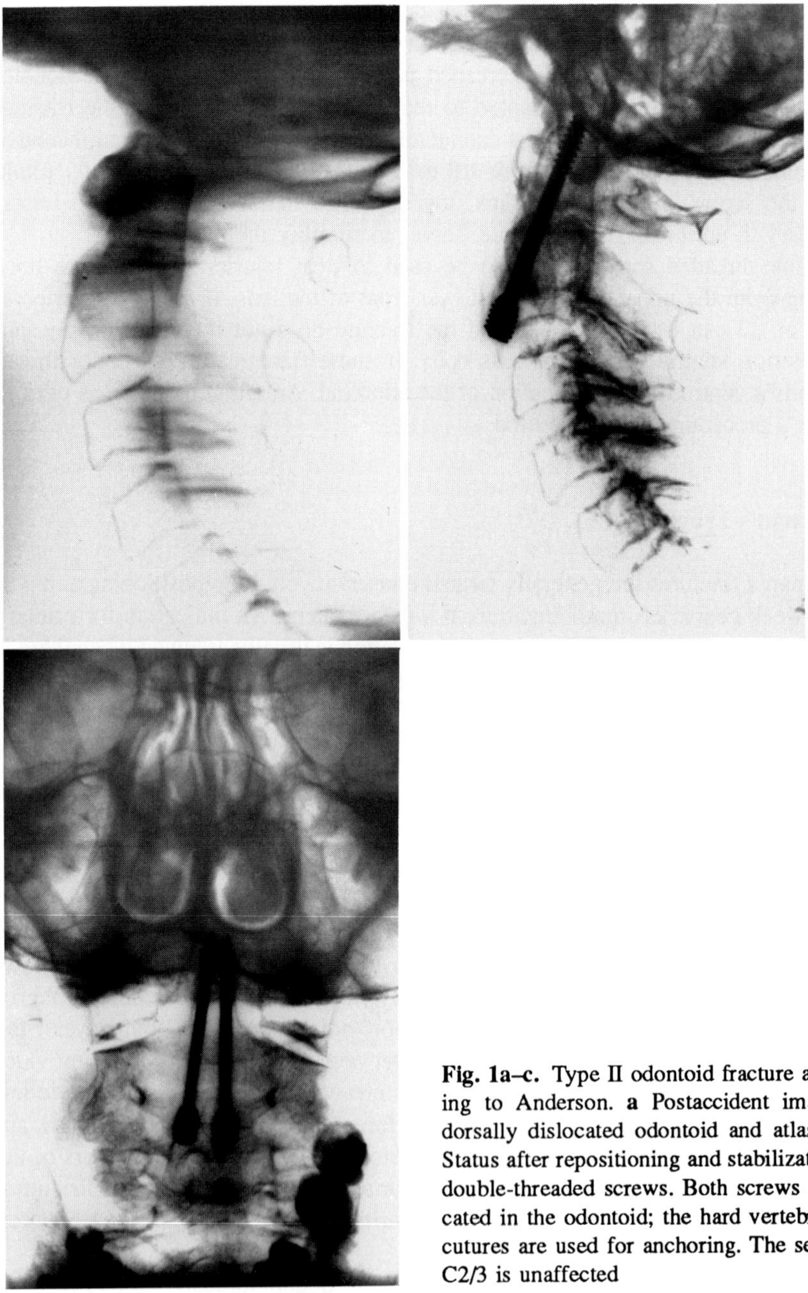

a b c

Fig. 1a–c. Type II odontoid fracture according to Anderson. **a** Postaccident image of dorsally dislocated odontoid and atlas. **b, c** Status after repositioning and stabilization by double-threaded screws. Both screws are located in the odontoid; the hard vertebral strcutures are used for anchoring. The segment C2/3 is unaffected

The preboring for the implantation of the double-threaded screws is done with a 1.8-mm drill wire with a chisel tip, whereas the other two screw types require a 2.0- or 2.5-mm spiral drill. While the thread does not need to be precut for osteosynthesis with double-threaded screws, this is necessary for the other two screw types. The risk of injury is thus reduced not only by the smaller diameter of the drill wire and its smooth surface when compared to a spiral drill, but also by the fact that a threadcutter, which probably presents the greatest intraoperative risk, can be dispensed with. The double-threaded screw also has a distinct spatial advantage after implantation. The site with the greatest injury potential is in the vertebral arch between the vertebral artery and the dura mater. At this site, the diameter of the implanted double-threaded screw is 2.0 mm, that of the cortical screw 3.5 mm, and that of the small fragment spongy bone screw 2.4 mm (in the unthreaded portion) or 4.0 mm (in the threaded portion). Since the small fragment spongy bone screw has too long a threaded portion for this indication, one can assume that at least part of this will come to lie in the danger zone, so that this screw type has the least suitable dimensions. Even if the transverse foramen is opened, the chisel tip and the smooth drill wire are likely only to displace the vertebral artery, whereas the spiral drill and particularly the threadcutter are more apt to cause injury to the vessel.

After the screws are tightened, the vertebral arteries should be examined for patency by means of Doppler ultrasonography. In the case of stenosis or occlusion by compression, immediate correction of the screw position can reinstate the patency of an undamaged vessel. This can prevent serious circulatory problems.

Atlantoaxial Instabilities

Atlantoaxial instabilities due to trauma or to rheumatic or congenital disorders (Table 1) are an indication for transarticular screw fixation and interarcual fusion of C1/2 [16, 17]. The advantages of osteosynthesis with double-threaded screws are the same as described in transpedicular C2 screwing (Fig. 3). Compared with

Table 1. Indications for transarticular screw fixation of C1/2

Jefferson fracture
Combined odontoid and Jefferson fracture
Odontoid pseudarthrosis
Rotatory subluxation of C1
Posttraumatic incongruence of C1/2
Rheumatic arthritis
Atlas assimilation and instability

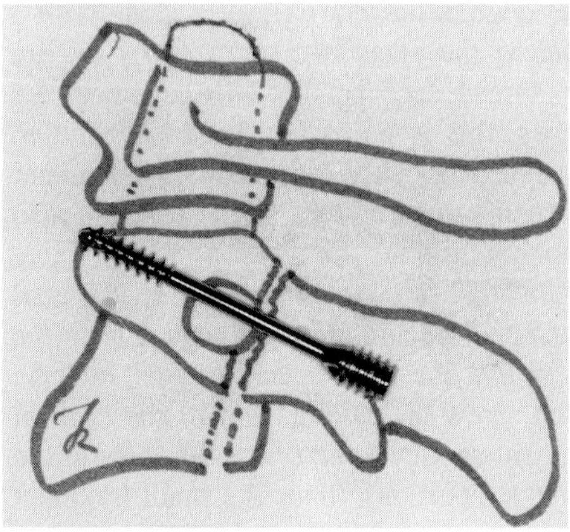

Fig. 2

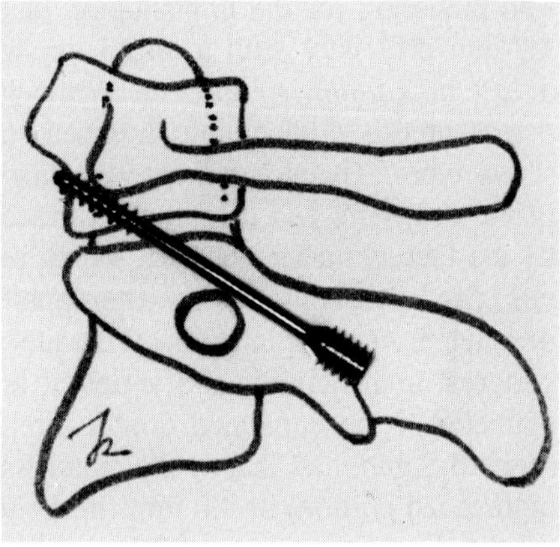

Fig. 3

Fig. 2. Schematic presentation of osteosynthesis of a hangman's fracture by means of a transpedicular C2 double-threaded screw. The screw is located medial to the vertebral artery. At this point it has its smallest diameter of 2 mm. The screw does not affect the segment C1/2 and its threaded parts are anchored in the anterior and posterior bone of the axis and the lamina

Fig. 3. Schematic presentation of transarticular double-threaded screw fixation of C1/2. The screw direction is sagittal and more inclined than for transpedicular C2 screw fixation. The screw is behind the verteberal artery and is anchored in the cortical bone of the lamina of C2 and the massa lateralis atlantis. The combination of interarcual fusion with bone graft and wiring gives a three-point fixation which ensures against loss of repositioning

lateral atlantoaxial screw arthrodesis [5], which needs to be performed bilaterally, this procedure has the advantage of a single access, which is furthermore generally known and technically uncomplicated. In contrast to the other methods, like fusion according to Gallie [8] or Brooks and Jenkins [4] and compression clamp spondylodesis [18], this procedure ensures that once repositioning has been performed, it will be maintained. Good repositioning and its maintenance are of the utmost importance, because only then is it guaranteed that the spinal canal will recover its physiological width and that the spinal cord will be completely free of pressure. All described fusion procedures concerning C1/2 lead to the loss of the rotation joint, but preserve the nod joint of the head.

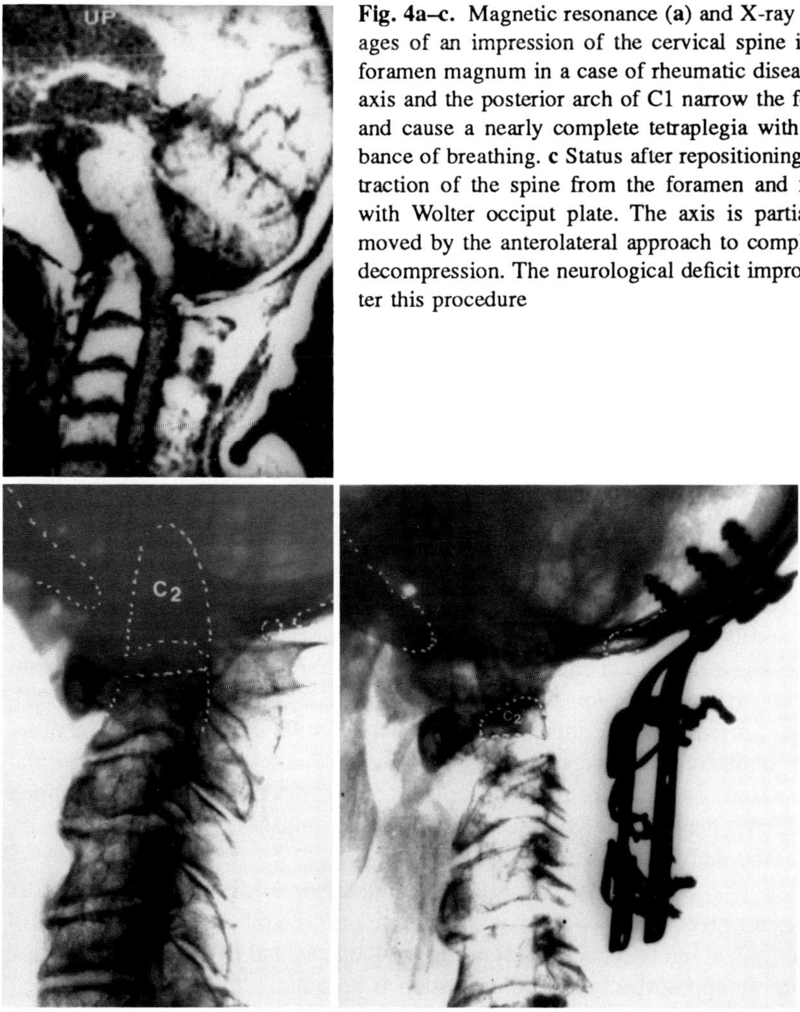

a

b

c

Fig. 4a–c. Magnetic resonance (**a**) and X-ray (**b**) images of an impression of the cervical spine into the foramen magnum in a case of rheumatic disease. The axis and the posterior arch of C1 narrow the foramen and cause a nearly complete tetraplegia with disturbance of breathing. **c** Status after repositioning by extraction of the spine from the foramen and fixation with Wolter occiput plate. The axis is partially removed by the anterolateral approach to complete the decompression. The neurological deficit improved after this procedure

Atlanto-occipital Instabilities

Occipitocervical fusion with instrumentation is necessary in cases of malignant extradural tumors of the craniocervical region, when there is rheumatic or traumatic impression of the cervical vertebral column into the foramen magnum (Fig. 4), and in the few patients who survive with traumatic axial instability of C0/1. We prefer the dorsal approach and use the Wolter occiput plate screwed onto the occiput and wired with the arches C1, 2, and 3. Permanent fusion is achieved by interposing spongy bone chips between the occiput and the arches up to the caudal end of the plate. This procedure gives great stability but causes loss of the rotation and nod joints of the head.

Discussion

Among the methods of surgical treatment of recent type II and III *odontoid fractures* as classified by Anderson and D'Alonzo [1], osteosynthesis with double-threaded screws constitutes the procedure of choice. Odontoid screw fixation preserves or reestablishes proper head motion. Since double-threaded screws are almost completely embedded into the axis, the irritation of C2/3 movement regularly induced by the heads of small fragment spongy bone screws or cortical bone screws is avoided. Double-threaded screws are exactly adapted to the anatomy of the axis and to the fracture types encountered at this site. The operative procedure is simplified by the instruments, which include two X-ray translucent retractors [12].

Hangman's fractures with dominant discoligamentous instability require intercorporeal fusion of C2/3, which causes an insignificant loss of cervical motility. Dorsal fusion of C1–3 [10] is unacceptable because of the loss of the rotation joint. Like odontoid screw fixation, transpedicular C2 screwing of hangman's fractures is a procedure associated with no functional loss of vertebral motility, so that this method fulfills the demands for an optimal osteosynthesis.

Due to the technique of three-point fixation, transarticular C1/2 screwing combined with interarcual fusion in the treatment of *atlantoaxial instabilities* of various origin provides the advantage of osteosynthesis with stability in all directions, in which maintenance of repositioning is ensured. The advantage of the double-threaded screws in the latter two surgical procedures lies not only in their simpler implantation when compared with the screws used until now, but also in the reduction of operative risk.

In cases of *craniocervical instability*, occipitocervical fusion using the Wolter occiput plate gives good stability even in cases of ventral defects. This kind of spondylodesis severely reduces the motility of the head and is therefore to be used only when an appropriately severe indication is present.

References

1. Anderson LD, D'Alonzo RT (1974) Fractures of the odontoid process of the axis. J Bone Joint Surg 56A:1663–1674
2. Böhler J (1981) Schraubenosteosynthese von Frakturen des Dens axis. Unfallheilkunde 84:221–223
3. Böhler J (1982) Anterior stabilization for acute fractures and non-unions of the dens. J Bone Joint Surg 64A:18–27
4. Brooks AL, Jenkins FB (1978) Atlanto-axial arthodesis by the wedge compression method. J Bone Joint Surg 60A:279–284

5. Du Toit G (1976) Lateral atlanto-axial arthrodesis, a screw fixation technique. S Afr J Surg 14:9–12
6. Estridge MN, Smith RA (1967) Transoral fusion of odontoid fracture. J Neurosurg 27:462–465
7. Fang HSY, Ong GB (1962) Direct anterior approach to the upper cervical spine. J Bone Joint Surg 44A:1588–1604
8. Gallie WE (1939) Fractures and dislocations of the cervical spine. Am J Surg 46:495–499
9. Judet R (1969) Colleque sur les fractures et luxations recentes du rachis cervical. Rev Chir Orthop 55:71
10. Karimi-Nejad A (1983) Operative Behandlung der degenerativen und der traumatischen Halswirbelsäulenveränderungen. Aktuel Neurol 10:45–51
11. Knöringer P (1984) Zur Behandlung frischer Frakturen des Dens axis durch Kompressionsschraubenosteosynthese. Neurochirurgia 27:68–72
12. Knöringer P (1986) X-ray translucent retractors in spine surgery. J Neurosurg 64:520–521
13. Knöringer P (1986) Osteosynthese der Densfraktur durch Zweigewindekompressionsschrauben. Hefte zur Unfallheilkunde. Springer, Berlin 181:93–97
14. Knöringer P (1988) Operative Versorgung der Densfraktur von ventral durch Doppelgewindeschrauben. In: Hohmann D, Kügelgen B, Liebig K (eds) Neuroorthopädie 4. Springer, Berlin Heidelberg New York, pp 199–206
15. Knöringer P (1988) Intraoperative Doppler-sonography during surgery of the cervical spine. In: Walter W, Brandt M, Brock M, Klinger M (eds) Advances in neurosurgery, vol 16. Springer, Berlin Heidelberg New York, pp73–79
16. Knöringer P (1989) Osteosynthesis of injuries and rheumatic or congenital instabilities of the upper cervical spine using double-threaded screws. Neurosurg Rev
17. Margerl F, Seemann PS (1987) Stable posterior fusion of the atlas and axis by transarticular screw fixation. In: Kehr P, Weidner A (eds) Cervical spine I. Springer, Wien New York, pp 322–327
18. Roosen K, Trauschel A, Grote W (1982) Posterior atlanto-axial fusion: a new clamp for laminar osteosynthesis. Arch Orthop Trauma Surg 100:27–31
19. Tscherne H, Muhr G, Op den Winkel R (1980) Frakturen und Luxationen der HWS – operative Behandlung und Ergebnisse. In: Burri C, Rüter A (eds) Verletzungen der Wirbelsäule. Hefte zur Unfallheilkunde. Springer, Berlin 149:89–94

Surgical Treatments to Stabilize the Upper Cervical Spine

R. Preger, H.-E. Clar, and M. Schmutzler[1]

Introduction

The occipitoatlantoaxial complex serves as a transitional zone between the standard vertebral joint structures and the radically different skull. The three columns of lines of forces from the lower spine change to two columns at the level of the atlas [7, 10]. Nearly horizontal joints, loss of disks, and a special ligamentous fixation allow extreme head motions [3, 7, 10] and give the spinal cord and adjacent structures optimal protection. Instability of this region may occur as a result of accidents, degenerative diseases, and tumors affecting the atlas, the axis, and their joints alone or in any combination. The extent of destruction and the clinical symptoms make necessary differentiated radiological diagnosis followed by conservative and surgical treatment. We will report our experiences in the treatment of 24 patients between April 1985 and April 1989.

Clinical Findings

Nineteen patients were admitted to our hospital with traumatic instabilities of the upper cervical spine. One patient suffered a combined severe head injury and died of its consequences 2 days after admittance. Another young patient suffered from paralysis of both arms because of bilateral cervical nerve root extractions. In a 73-year-old patient a fracture of the odontoid process was hidden by Bekhterev's disease. Neurological deterioration with hemiparesis caused radiological diagnosis and transfer to our hospital (Fig. 1). Only one patient suffered from an incomplete quadriplegia and respiratory impairment that made tracheotomy necessary. Resuscitation was performed successfully at the scene of the accident. A 27-year-old patient developed a hemiparesis due to traumatic carotid artery occlusion. In all other cases only pain and loss of functions were present as main symptoms.

We observed no cases of occipitoatlantic disturbance. There was one Jefferson's fracture and three cases of combined injuries with atlas fracture or atlantoaxial joint

[1] Neurochirurgische Klinik, Klinikum Ingolstadt, Krumenauer Str. 25, D–8070 Ingolstadt

Advances in Neurosurgery, Vol. 18
K.-A. Bushe, M. Brock, and M. Klinger (Eds.)

instability and odontoid process fractures of type II [1]. An isolated fracture of the odontoid process was treated in 14 patients. One female patient has developed an odontoid process pseudarthrosis with severe dislocation 5 years after her accident. One patient was admitted with an odontoid fracture of type III [1].

A pathological fracture occurred in one female patient with cystic tumor of the axis. Histology remained unknown. A transient spastic tetraparesis led to radiological examination. In the follow-up over a 3-year period the disease healed without any complications after the complete destruction of the axis.

We treated four patients with odontoid process destruction and dislocation caused by rheumatoid arthritis. The average age was 70 years. Upon admittance no neurological deficits could be found, but there were some complications (see below).

Diagnostic Procedures

In spite of exact neurological assessment, radiography is the most important diagnostic method. Virtually no problems are encountered in handling the conscious patient while taking plain films. However, great care must be taken in positioning the unconscious patient to prevent additional neurological complications. Most fractures in the upper cervical spine can be detected by anteroposterior, lateral, and oblique radiographic views. Tomography may be helpful in evaluation of fractures without displacement or for careful functional diagnosis. Computed tomography has replaced myelography. The patient can be examined in an upside down position without dangerous maneuvers. The pictures give detailed information about bone destruction, dislocation, and the width of the spinal canal. For further information the reader is referred to the literature [6, 13]. We have no experience with NMR findings.

Approaches and Techniques

Anterior Approaches

Twice we used the transoral approach for stabilization. The patient with the Jefferson's fracture was treated by interposition of a bone graft. In another patient with fracture of the odontoid process and displacement of the atlantoaxial joint we fixed the atlas and axis by means of a metal plate. In both cases additional posterior fixation was performed as described later. The transoral approach is limited to circumscribed surgery of the atlas and axis. For more extended surgery the transcervical approach is to be preferred [9].

In 1982 Böhler [2] described the technique of screwing odontoid fractures from a transcervical approach. With this method it is possible to stabilize these lesions

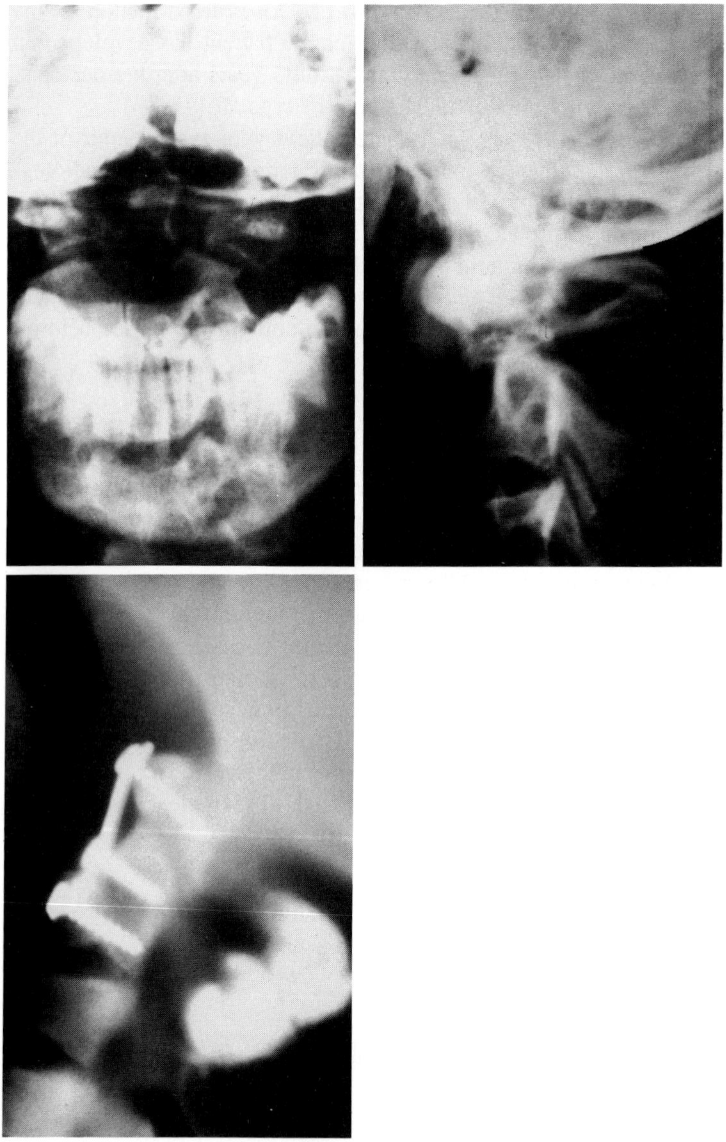

Fig. 1. 73-year-old patient with Bekhterev's disease and a hidden odontoid fracture of type II. Dorsal stabilization with compression clamps. The odontoid process is in an ideal position

without any loss of function. Knöringer [5] modified the screws and the instrumentation. We used this instrumentation in 13 of 14 patients. In 12 patients it was the only surgical treatment and was used in combination with external fixation by means of a cervical brace.

Posterior Approach (Fig. 2)

Posterior fixation of upper cervical spine instability is a well-proven method. Wires, bone grafts, and alloplastic materials have been used. The disadvantage, especially in fractures of the odontoid process, has been the loss of function of this important anatomical structure. For dorsal fixation we used special compression clamps designed by Roosen et al. [12] with interposed bone grafts. In the patient with the axis tumor we placed the clamps into the occipital squama and the arch of C3. We used the same technique to stabilize the odontoid fracture in the patient with the accompanying Bekhterev's disease, and in the four patients with rheumatoid arthritis. One patient with odontoid fracture of type II was treated in the same way in 1985, before we used anterior screw fixation. As mentioned previously, we used this additonal fixation in two patients with ligamentous and atlantoaxial joint instability. In the patient with the Jefferson's fracture we performed fixation with a metal sheet on the arch of the atlas.

Screw fixations of the axis to the lateral mass of the atlas are described in the literature [14]. We have never yet used this technique.

Halo Brace

Only one of our 24 patients was treated without surgery by the halo-fixateur externe (Fig. 3). This patient suffered an odontoid fracture of type III which healed in good position within 8 weeks. The halo device is very comfortable in the early treatment of instabilities of the upper cervical spine. Traction, repositioning, and fixation are well controlled by radiography [4]. For this reason we often use the halo brace as an initial means of stabilization. The halo is always applied under local anesthesia. With the halo orthosis, patients can be mobilized immediately. It is possible to prepare patients for intubation and surgery without any danger and to perform stabilization from each approach.

In some cases of anterior approach for surgery we only used a Gardener head clamp extension for repositioning of the dislocated fracture and fixation.

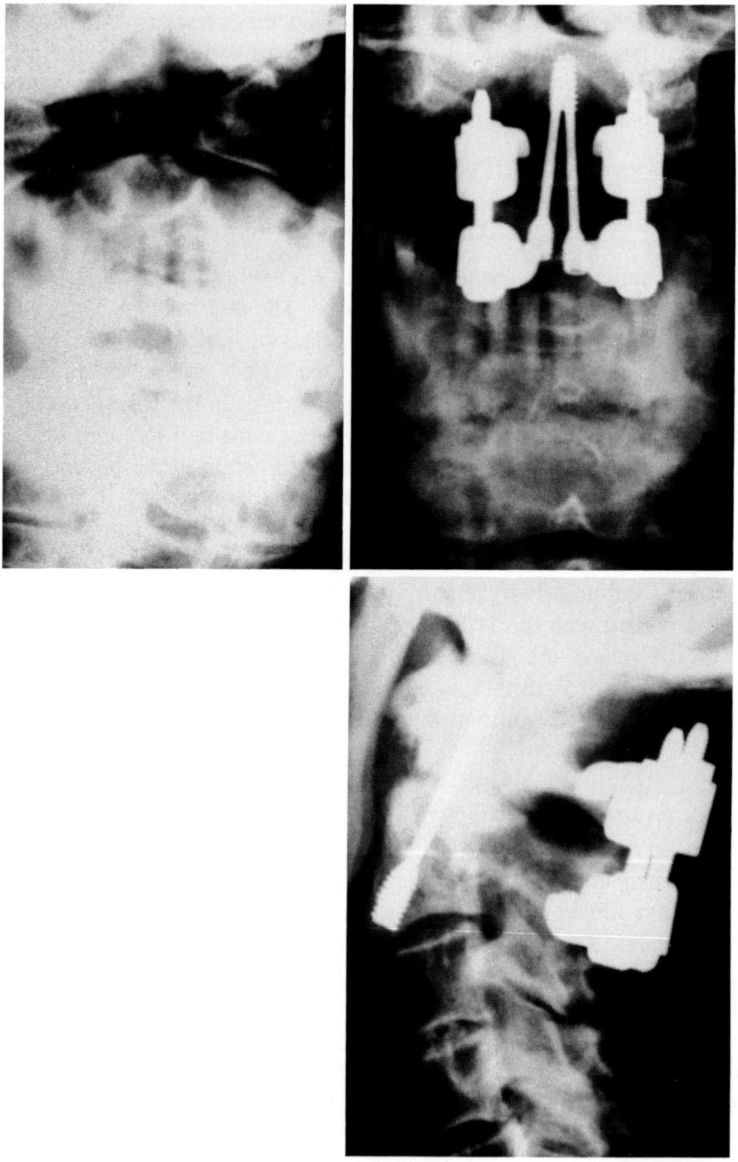

Fig. 2a–c. Odontoid fracture of type II (**a**) combined with atlantoaxial displacement (**b**) in a 17-year-old patient. Treatment with transoral and posterior fixation (**c**)

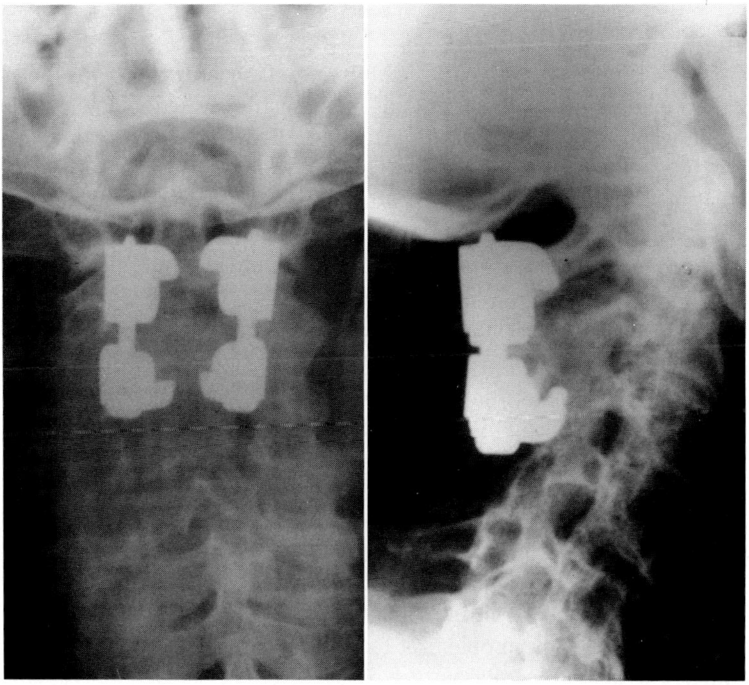

Fig. 3a, b. Transverse dislocation of an odontoid fracture of type II combined with a torn atlantoaxial joint in a 76-year-old patient (**a**). The odontoid process and atlantoaxial joint were replaced by a halo brace and fixed with dens screws in the normal position; dorsal fixation gave additional stabilization (**b**)

Complications

We have never observed any complication using anterior approaches and techniques. In some cases the radiological findings after odontoid screwing showed small gaps that healed within 6 months. Only the female patient with pseudarthrosis of the odontoid fracture retained her gap, and some slight dislocation of no clinical importance can be seen 3 years after operation.

When the posterior approach was employed, we observed three complications. The patient with the axis tumor showed secondary wound healing. In two of the four patients with odontoid destruction due to rheumatoid arthritis a second operation was necessary because of dislocation of the clamp. Unfortunately one patient died of an intraoperative brain stem infarction during the second operation. Compression clamps dislocated because bone grafts as well as the vertebral arches were weakened by long-term cortisol treatment [8].

Conclusion

Although lesions in the upper cervical spine can be very dangerous, as demonstrated by the case of tetraparesis caused by odontoid fracture and the intraoperative complication of brain stem infarction, most of our patients had no neurological deficit. If instability of the upper cervical spine is diagnosed, two questions have to be answered: (a) Is it possible to stabilize the fracture conservatively or surgically without loss of function, especially in young patients? (b) Is it necessary to immobilize the whole region because of fracture combined with joint and ligamentous lesion?

There are two possible ways of immobilizing patients before surgical treatment is performed. In young patients it is possible to stabilize the fracture temporarily by means of a head clamp extension. Elderly patients should not be immobilized. The halo brace is a useful device in these patients; we demonstrated its advantages especially during anesthesia and surgery. After surgical stabilization of the spine, patients received plastic braces to avoid early mobilization of this region.

In our opinion type II fractures of the odontoid process should only be treated by axis/odontoid screws. In our hospital we use Knöringer's double-threaded compression screws. The decision regarding surgical technique for tumors of the upper cervical spine must be reached on an individual basis. Our case report is only one example of treatment. The same problem is faced in treatment of combined atlas and axis fractures or ligamentous and joint lesions. It has to be decided individually whether surgical stabilization, for example by odontoid screwing or the halo brace, is sufficient or whether anterior and posterior stabilization is necessary. In circumscribed lesions the short way to C1 and C2, via the transoral approach, should be used if necessary.

If complete immobilization of the upper cervical spine is required, best results can be achieved by dorsal stabilization techniques performed with Roosen-Trauschel compression clamps, especially in patients with rheumatoid arthritis. As mentioned above, combination with anterior stabilization techniques can be necessary.

An additional nonsurgical form of treatment is available in the halo-fixateur externe, which in our case report was used to stabilize a type III fracture of the odontoid process.

References

1. Anderson LD (1983) Fractures of the odontoid process of the axis. In: Bailey RW et al. (eds) The cervical spine. Lippincott, Philadelphia
2. Böhler J (1982) Anterior stabilization for acute fractures and non-unions of the dens. J Bone Joint Surg 64A:18–28

3. Jofe MH, White AA, Panjabi M (1983) Kinematics. In: Bailey RW et al. (eds) The cervical spine. Lippincott, Philadelphia
4. Kleinfeld F (1981) Zur Behandlung von Frakturen der Halswirbelsäule mit dem Halo-Fixateur-externe. Unfallheilkunde 84:161–167
5. Knöringer P (1984) Zur Behandlung frischer Frakturen des Dens axis durch Kompressionsschraubenosteosynthese. Neurochirurgia 27:68–72
6. Knöringer P (1985) Diagnostischer Wert der Computertomographie der spinalen Verletzungen. Unfallchirurg 88:63–74
7. Lang J (1981) Klinische Anatomie des Kopfes. Springer, Berlin Heidelberg New York
8. Lesoin F, Duquesnoy B, Destee A et al. (1985) Cervical neurological complications of rheumatoid arthritis. Acta Neurochir 78:91–97
9. Lesoin F, Jomin M, Pellerin P, Pruvo JP, Carini S, Servato R, Rousseaux M (1986) Transclival transcervical approach to the upper cervical spine and clivus. Acta Neurochir 80:100–104
10. Louis R (1985) Die Chirurgie der Wirbelsäule. Springer, Berlin Heidelberg New York
11. Pierce DS, Barr JS Jr (1983) Fractures and dislocations at the base of the skull and upper cervical spine. In: Bailey RW et al. (eds) The cervical spine. Lippincott, Philadelphia
12. Roosen K, Trauschel A, Grote W (1982) Posterior atlanto-axial fusion: a new compression clamp for laminar osteo-synthesis. Arch Orthop Trauma Surg 100:27–31
13. Wackenheim A (1985) Schädel-Hals-Übergang (RX, CT). Springer, Berlin Heidelberg New York
14. Whitesides TE Jr (1983) Lateral retropharyngeal approach to the upper cervical spine. In: Bailey RW et al. (eds) The cervical spine. Lippincott, Philadelphia

The Caspar Plate as a Means of Stabilizing the Upper, Middle, and Lower Cervical Spine – Results in 45 Cases

A. v. Stockert, R. Schultheiß, and D.-K. Böker[1]

Ventral stabilization of the cervical spine by means of the Caspar plate, complemented by interbody fusion of the vertebrae after Cloward, Smith, and Robinson, is an established and proven method for the treatment of cervical instabilities of heterogeneous origin. From June 1984 through September 1988, 45 patients were operated on using this method at the Neurosurgical Clinic of Bonn University (Neurochirurgische Universitätsklinik Bonn). The distribution between the sexes was fairly even, with 53% female and 47% male patients. The average age of the female patients was 53 years – 12 years older than the average of the male patients, at 41 years; the overall average lay at 46 years.

In 35 cases the indication for operation was based on a fracture/fracture-dislocation. In two cases it was performed because of degenerative instability, in two cases because of PCP causing osteolysis, in two cases after surgery on a cervical disk with subsequent dislocation of the seal, and in four cases because of osteolytic metastases in various cervical vertebrae.

The height of the fusion lay 6 times at C2/3, 6 times at C3/4, 7 times at C4/5, 20 times at C5/6, 8 times at C6/7, and once at C7/T1.

In 23 cases only two vertebrae were fused, while in 18 cases the fusion bridged more than two vertebrae; not in all cases were all the disks within the length of the fusion removed. In four cases a complete vertebral body had to be supplanted because of tumor metastases.

The material used for the interbody fusion was Sulfix in 31 cases, bone in 8 cases, and hydroxyapatite in four cases. In three cases no material was used.

The primary neurological diagnosis was a complete paraplegia in 8 cases, without a postoperative tendency to improve, and in 13 cases an incomplete paraplegia which improved to varying degrees after the operation in 70%. Radicular defects (in 12 cases motoric, in 16 cases sensory) also improved, in 67% and 81% of cases respectively. Subjective complaints of cervicobrachialgia and tension, reported by 50% of the patients, improved in well over 70%. Forty percent of our patients had concomitant injuries, and some were polytraumatic.

[1] Orthopädie I, Rehabilitationskrankenhaus Langensteinbach, D–7516 Karlsbad

Advances in Neurosurgery, Vol. 18
K.-A. Bushe, M. Brock, and M. Klinger (Eds.)

Roentgenograms obtained immediately after the operation showed good repositioning and stabilization in 98% of the cases. In one case the result was not satisfactory. Further radiological monitoring in the weeks and months following the operation showed a loosening of the implanted plate or renewed instability necessitating reoperation in 8 of the 45 cases (just under 18%).

Loosening of screws was seen in ten cases, i.e., 22%. Forty-four percent (8 of 18) of the multiple fusions bridging more than two vertebrae were thus affected, yet less than 9% (2 of 23) of the simple fusions of just two vertebrae.

In four cases we found neurological deterioration; a temporary respiratory insufficiency also affected four patients. Five patients died during the further postoperative course, yet in only one case did the cause of death remain doubtful, and in none of the cases was it linked with the operation itself. We found sinterings of the vertebral bodies in three cases, dislocation of the Sulfix seal in two, and an infection of the wound in two. A CSF fistula, an allergic reaction, and postoperative hemorrhage were complications in one case each.

In 14 cases additonal operative treatment was necessary: in five cases to remove material, in eight cases to achieve a re-fusion because of renewed instability, and in one case for a laminectomy. In just under 70% postoperative treatment was carried out in other hospitals, while in 26% it was performed in the Neurosurgical Clinic of Bonn University. The length of our catamnestic monitoring averaged 31 months.

Results

The postoperative treatment resulted in immediate and complete rehabilitation in 14 cases; further clinical rehabilitative treatment was given in eight cases, and outpatient physical therapy in seven cases. Seven patients were placed in other hospitals; in four patients no further treatment was given; and in four patients the further treatment is unknown.

A questionnaire reached nearly half of the patients or at least their general practitioners or orthopedists. A catamnestic check within the context of the aims of this study was made in 22%, and roentgenography of the cervical spine was performed in over 50% of the patients. Neurological deficits remained in 75% in the form of pareses and numbness, and in 81% reflexes were weakened or absent.

Definite stability of the cervical spine could be established in 70% of the cases. Eleven percent of the patients have died, and we have no knowledge about the further development in 14%. Five percent are still under treatment.

The subjective appraisal of the results by the patients is positive in 50%, negative in 20%, and indifferent or unknown in 30%.

Conclusions

1. The radiological and functional stability of the cervical spine has only a limited influence on the neurological outcome. Insofar as there is any influence at all, one may find a connection between spinal stability and pain syndromes.

2. Stabilization of the cervical spine by surgical means is also indicated after primary paraplegia as it allows sufficient stability for training and thus helps early mobilization. A precondition for this, however, is a reduction of the rate of complications. Some complicating factors have already been eliminated, as most of them stem from the pioneer days of this surgical technique. It is to be expected that the results will improve even further along with a better mastery of this surgical method.

3. Postoperative loosening of screws was found five times more often in bridgings of more than two vertebrae than in simple fusions of only two. The distance bridged by fusion ought therefore to be kept as short as possible; if the indication is not unequivocal, only a simple fusion should be undertaken. In any case of multiple fusion all the disks within the length of the fusing distance must be removed and supplanted by interbody fusion, as the otherwise possible micro-movements work on a longer leverage and thus heighten the risk of loosening of screws.

4. The slotted holes of the Caspar plate themselves further micro-movements and can contribute to the relatively frequent loosening of screws in multiple fusions. A modified form of the Caspar plate ought to be considered, in which the slotted holes are replaced by several interlinked holes.

5. On the whole, stabilization of instabilities of the cervical spine using the Caspar apparatus is a well-established method. The surgical technique, however, and the instruments used, can be improved upon.

References

Caspar W (1986) Anterior cervical fusion and interbody stabilisation with the trapezial osteosynthetic plate techniques. Aesculap Scientific Information Leaflet S-039, edn VII, p 36. Aesculap Instruments Corp., Burlingame CA

Smith GW, Robinson RA (1958) The treatment of cervical spine disorders by anterior removal of the intervertebral disc and interbody fusion. J Bone Joint Surg 40A:607–624

Anterior Reconstruction and Stabilization After Cervical Vertebral Body Resection

G. Meinig, W. Kahl, K. Leyendecker, P. Eysel, and M.J. Melzer[1]

Introduction

Prerequisites for the regeneration of compressed or traumatized structures of the spinal cord and nerve roots are rapid repositioning and the best possible decompression, i.e., the restoration of the anatomical structure of the cervical vertebral column [4]. In many cases generous ventral vertebral body resection with subsequent substitution of spinal vertebrae and anterior plate fixation is the logical and adequate treatment.

Patients and Methods

Between 1986 and 1988, 155 ventral fusions of the spine were performed at the BG Unfallklinik, of which 40 involved additional anterior plate fixation necessitated by trauma; in a further 37 patients anterior plate fixation followed vertebral body resection and substitution of spinal vertebrae. Of these 37 patients, 23 had spinal trauma, nine had tumors, and five had degenerative disorders with spinal stenosis. In the patients with tumors the substitution of spinal vertebrae was completed using Palacos; in all other cases it was achieved with corticospongiosa bone grafts from the region of the pelvic crest.

In none of the cases was additional dorsal stabilization undertaken.

Results and Discussion

In all of the cases of tumor with incomplete paraplegia the neurological symptoms improved. However, we found improvement in only one-third of traumatically inflicted conditions.

[1] Berufsgenossenschaftliche Unfallklinik Frankfurt am Main, Friedberger Landstr. 430, D–6000 Frankfurt am Main 60

Advances in Neurosurgery, Vol. 18
K.-A. Bushe, M. Brock, and M. Klinger (Eds.)

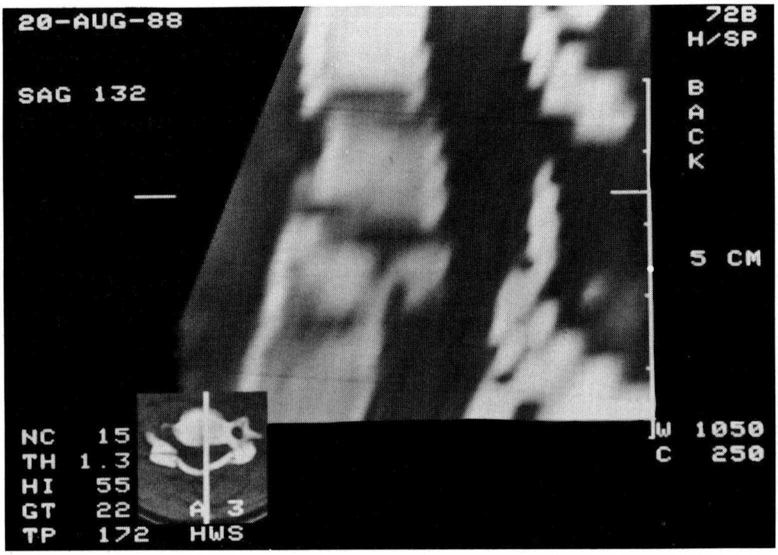

Fig. 1. Spinal stenosis caused by a C6/7 fracture

Figure 1 demonstrates a fracture of C6/7 with high-grade spinal stenosis. By reference to this case almost all aspects of the treatment of vertebral column injuries, including complications, may be illustrated.

Immediate surgery was performed, which involved resection of the body of C6 and partially also that of C7. Because of segment instability a diskectomy at C5/6 followed. At C5/6 a stabilizing lock was installed by means of a square dowel (according to Smith-Robinson). The large defect at C6/7 was bridged by a rectangular dowel which was extracted from the pelvic crest with the help of an oscillating saw. Subsequently extended anterior plate fixation was completed.

Three weeks after the operation a fistula developed in the region of the postoperative scar. A conservative approach involving excision of the necrotic area and the insertion of a Septopal chain failed. Finally a revision was performed under halo extension. The entire transplant was removed. Three weeks later, after the inflammatory symptoms had subsided, a spongiosa refill was performed.

The patient, who had initially suffered complete motor paralysis, was able to walk without the use of crutches 6 weeks after the accident. Figure 2 shows the stable situation 3 months after the accident. By this time the right leg had regained normal strength. Only the left leg still showed a minimal diminution in strength.

In the case of metastasis or spinal tumor the treatment is, as a rule, ventral vertebral body resection rather than laminectomy. Substitution of spinal vertebrae

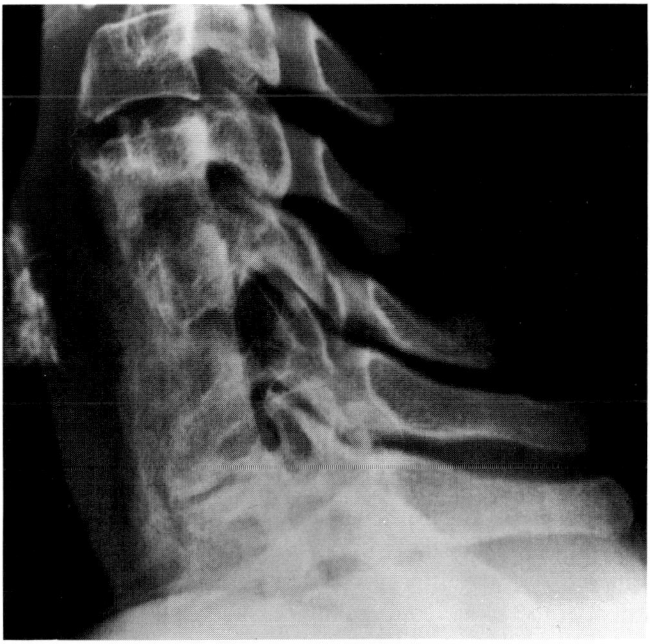

Fig. 2. Situation 3 months after vertebral body resection; substitution of spinal vertebra and ventral osteosynthesis, revision under halo extension because of infection, and restructuring by means of a spongiosa transplant (same patient as in Fig. 1)

involves the use of Palacos. Additional stabilization is achieved by anterior plate fixation.

Also in some instances of unsuccessful initial ventral fusion, as in cases of high-grade spinal stenosis because of degenerative conditions, vertebral body resection and stabilization by means of anterior plate fixation was successful. In this context it is to be kept in mind that kyphotic angulation after ventral fusion according to Busch [1] occurs in one-quarter of cases (average kyphotic angle: 15°). Since vertebral body resection poses the danger of secondary kyphosis, we regard additional anterior plate fixation as inevitably necessary. Even though Wörsdorfer [8] supports dorsal stabilization on the basis of the results of experimental biomechanical research, long-term clinical observation in the case of more than 100 ventral fusions with anterior plate fixation has shown that ventral treatment alone generally guarantees sufficient stabilization, so that in some cases an additional orthosis is dispensable. Nevertheless there are several preconditions which need to be respected when the operation is undertaken: The plate – installed at the utmost convexity – ought to be bent forward so that it is slightly kyphotic, and under certain circumstances even a bit farther than the physiological lordosis normally permits so that tensional and bending forces may be transformed into axial pressure forces. Among others, Weller [7] is to be mentioned in this respect. Finally the cortical screws definitely ought to engage the rear spinal corticalis.

Figures 3 and 4 show CT scans of a 46-year-old patient who fell when cycling and who was completely paraplegic at admission. Immediate repositioning under Crutchfield extension followed diagnosis by CT, and subsequently ventral fusion and anterior plate fixation were performed. Within a few days the paraplegia had resolved. Three weeks after the accident the patient was able to walk normally. There is no residual paresis.

Generalizations cannot be drawn from this one example, but it is in agreement with the experimental results of Tarlov [6] and Dolan et al. [2], who not only ascertained that the ability of paraplegia to resolve is dependent on the compression on the spinal cord or the intensity of the traumatic forces involved, but also established an interdependence between the resolution of paraplegia and the duration of the compression. Our conclusion in respect of clinical procedures is: the sooner decompression is accomplished, the better the chance of regeneration of the spinal cord, especially when paraplegia is incomplete. This means immediate repositioning and decompression, which in our opinion should be ventral, since a complete decompression of the spinal cord is only possible in this way and it also seems of major importance that the anterior spinal artery is decompressed. Certainly, for the resolution of neurological deficiencies, reperfusion and restoration of the microcirculation are decisive, and it seems as if the most important events happen in the postcapillary region. At this point reference must be made to the pathophysiology of spinal cord lesions and the contribution of Leyendecker and Schirmer [5], among others, in this area. Finally, precise research is necessary concerning the use of adjuvant medication to counteract the development of serious secondary reactive changes of the marrow.

Summary

Complete decompression of the spinal cord can be achieved by vertebral body resection in patients with fractures of the spine, tumors, serious degenerative stenosis, or an unsuccessful initial operation for complete spinal stenosis. In the case of incomplete paraplegia due to a tumor, we experienced significant resolution of the neurological symptoms. Only in one-third of the cases of vertebral trauma were we able to observe an improvement in the paraplegia accompanied by an improvement in the radicular compression symptomatology. Clinical findings and experimental research impressively confirm that both the intensity of the traumatic forces and the duration of the compression are of decisive importance for the recovery of the spinal cord. Obviously reperfusion, for instance via decompression of the anterior

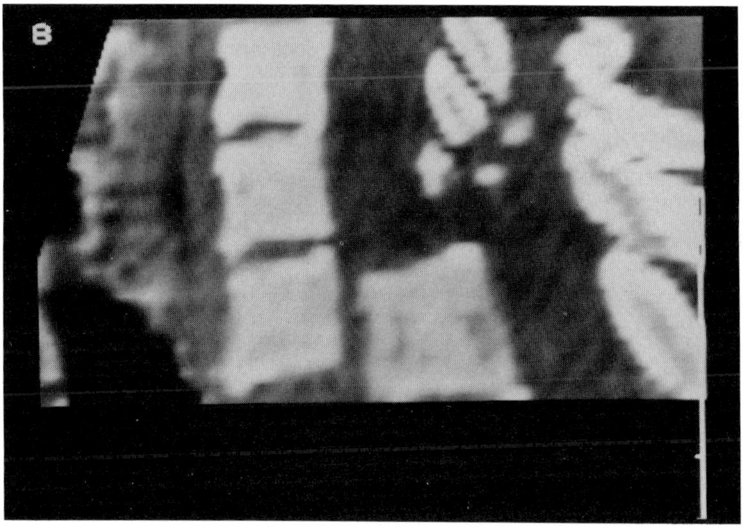

Fig. 3

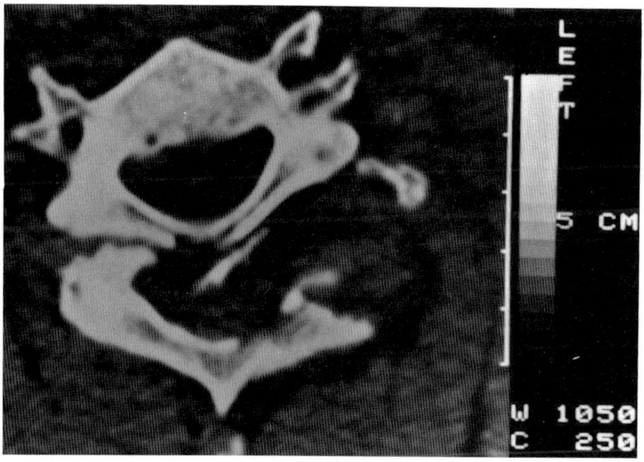

Fig. 4

Figs. 3, 4. Marked displacement after fall from bicycle with primary complete sensorimotor paraplegia, but full neurological recovery after immediate surgery and ventral fusion spinal artery, is significant. As a consequence of these considerations, in many cases we perform immediate surgery with complete decompression and stabilization, as long as there is no evidence which allows us to establish the prognosis immediately after the trauma.

Only in one case did a surgical complication develop, due to an infection which led to a revisional operation under halo extension. However, even in this case an optimal clinical result was achieved at the second attempt.

References

1. Busch G (1978) Anterior fusion for cervical spondylosis. J Neurol 219:117–126
2. Dolan J, Tator H, Endrenyi L (1980) The value of decompression for acute experimental spinal cord compression injury. J Neurosurg 53:749–755
3. Jäger J, Hundshagen W (1986) Heilung einer traumatischen Tetraplegie durch sofortige Operation. Beitr Orthop Traumatol 33:428–431
4. Knöringer P (1985) Zur operativen behandlung von Verletzungen der unteren Halswirbelsäule. Neurochirurgia 28:213–220
5. Leyendecker K, Schirmer M (1986) Traumatische Rückenmarksschädigungen. In: Schirmer M (ed) Querschnittlähmung. Springer, Berlin Heidelberg New York, pp 236–237
6. Tarlov JM (1957) Spinal cord compression. Thomas, Springfield, Ill
7. Weller S (1981) Biomechanische Prinzipien in der operativen Knochenbruchbehandlung. Akt Traumatol 195–220
8. Wörsdorfer O (1981) Operative Versorgung der thorako-lumbalen und lumbalen Wirbelsäule: Vergleichende biomechanische Untersuchungen zur Stabilität und Steifigkeit verschiedener dorsaler Fixationssysteme. Habilitationsschrift, Klinisch-medizinische Fakultät Ulm

Long-Term Results After Cervical Interbody Fusion with Polymethylmethacrylate

D.-K. Böker, E.M. Probst, and R. Schultheiß[1]

Introduction

Twenty years after its introduction, the technique of cervical interbody fusion using polymethylmethacrylate (PMMA) is still controversial. The main points of criticism are:

- Fear of insufficient stabilization of the cervical spine [6]
- Carcinogenic potential of PMMA [8, 9]
- Possible damage to neural or bony structures due to the exothermal reaction during the polymerization of PMMA

One result of this study may be anticipated: Damage to neural structures attributable to the exothermic polymerization reaction has not been registered. This is thought to be due to the excellent thermoinsulating ability of the gelfoam which is routinely used to cover the dura and the nerve roots prior to introduction (pouring) of PMMA into the intervertebral space [12].

Materials and Methods

Between 1967 and 1973 cervical interbody fusion with PMMA was performed in 409 patients. Of these, 100 could not be contacted at the time of this study; the remaining 309 patients were sent a questionnaire and asked to come for a follow-up examination. A total of 176 patients or their relatives answered our questions.

In the meantime 41 of the contacted patients had died; moreover, 44 patients did not agree to a reexamination, but of these 31 sent recent X-ray films of the cervical spine. Ninety-one patients could be reexamined in this clinic. Overall, plain X-ray films for evaluation were obtained from a total of 103 patients. Follow-up time was 16–21 years. For distribution of patients into different subgroups according to clinical symptoms, see Table 1.

[1] Neurochirurgische Klinik der Universität Bonn, Sigmund-Freud-Str. 25, D–5300 Bonn 1

Advances in Neurosurgery, Vol. 18
K.-A. Bushe, M. Brock, and M. Klinger (Eds.)

Table 1. The distribution of patients into different subgroups according to clinical symptoms

Diagnosis	n	Male	Female	Age (years)	Level
Local cervical pain	70	33	37	40–50	C4/5
Radiculopathy	186	113	73	40–60	C5/6
Myelopathy	109	65	44	40–70	C5/6
Trauma	44	32	12	20–30	C5/6

Table 2. Our results compared with the literature (Odom classification)

Diagnosis	Authors	Odom		
		1 + 2	3	4
Radiculopathy	Cloward (1963)	98.0		2.0
	Robinson et al. (1962)	72.7	21.8	5.5
	Dereymaeker et al. (1963)	64.5		35.5
	Hirsch et al. (1964)	83.0		17.0
	Our results (1989)	78.0		22.0
Local cervical pain	Roosen (1984), 26 authors	72.4	16.4	12.9
Radiculopathy Trauma	Our results (1989)	73.4		26.6
Myelopathy	Roosen (1984), 9 authors	55.7	17.6	26.7
	Our study (1989)	57.9	5.3	36.8

In accordance with Odom the clinical outcome was classified as: excellent (1), good (2), satisfactory (3), or poor (4) [7].

The X-ray films were evaluated for the following criteria:

1. Posture of the cervical spine
2. Height of fused vertebral bodies
3. Stability of fusion
4. Presence, absence, or development of spondylotic spurs
5. Position of PMMA implant
6. Phlogistic/resorptive alterations of vertebrae
7. Development of neoplastic disease.

Results

Using the Odom classification, 78% of our patients with radicular phenomena had an excellent or good outcome, whereas 22% showed a poor result. These results were better than the ones achieved by the patients with myelopathy. In this group 57.9% had an excellent or good outcome, 5.3% showed a satisfactory result, and 36.8% a poor result.

In 1984 Roosen reviewed the literature and reclassified the results from 26 authors using the classification of Odom [13]. He classified the patients with discogenic symptoms, radiculopathy, and spinal cord symptoms into one group. To be able to compare our results with the literature we did the same with our patients. We found that 73.4% had an excellent or good outcome, while 26.6% showed no difference or were worse than before the operation. For results, see Table 2. These results were comparable to the subjective judgment of the patients themselves.

One main criticism of cervical fusion using PMMA is the development of phlogistic/resorptive alterations of the vertebrae. In 32 cases plain films taken before and soon after surgery could be compared. In ten cases we had films showing the findings before surgery and at follow-up, and in another 25 cases we were able to evaluate films taken before and soon after surgery as well as at the time of follow-up. Thus a total of 139 vertebrae adjacent to PMMA implants in 67 patients could be examined. In 12 patients, and 16 vertebral bodies, we found a diminution in size immediately following surgery, which was due to operative conditions. In these cases more cuneiform shapes of the affected vertebrae were registered. At follow-up we found a diminution in size in a total of ten vertebrae, which had not been observed before. Specifically, this was characterized by a parallel approach of the cranial and caudal borders of the vertebral body.

Bony engraftment of the PMMA implant is the essential criterion for a stable fusion. X-rays suitable for evaluation were obtainable from 76 patients, 55 of whom had had vertebral interbody fusion in one segment, 20 in two segments, and one in three segments, resulting in a total of 98 fused segments. A maximal ossification was seen in 90% at the dorsal and in 65% at the ventral aspect of the implant. Figure 1 gives an example of an optimal postoperative result.

In those patients who had had vertebral fusion in more than one segment, radiological results were generally better than in those operated on only in one segment.

Complications

As permanent complications there were three cases of Horner's syndrome and one of tinnitus. One patient showed a fistula 3 years after surgery while another patient had aseptic putrid necrosis which we considered to represent a rejection of the implant. In both cases a cure was achieved after removal of the PMMA implant.

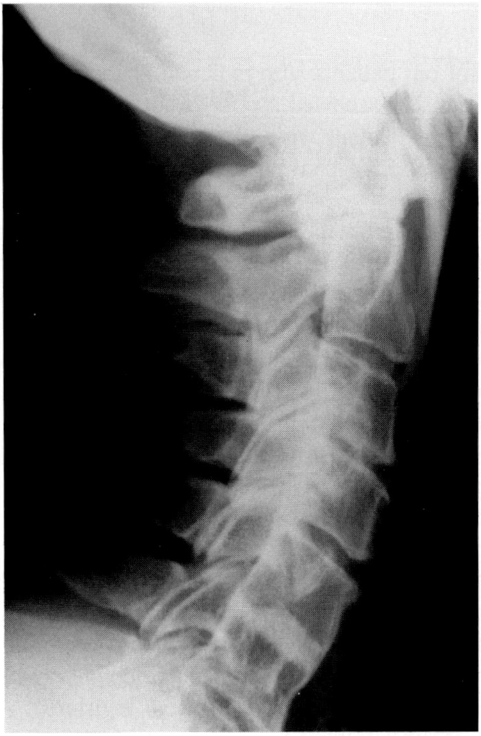

Fig. 1. Vertebral interbody fusion of C5/6. Optimal engraftment

Discussion

Our excellent and good long-term clinical results after cervical interbody fusion with PMMA are comparable with those found in the literature [2, 3, 5, 10]. However, with respect to total numbers we found a higher percentage of patients with a poor result. In our opinion this is due to the longer interval of observation, i.e., 16–21 years after surgery. During this time the natural degeneration process continues. This independently confirms our observation that symptoms reappear several years after surgery.

The main points of interest are the bony changes in the fused segments, stability of cervical fusion, and the development of neoplastic disease after implantation of PMMA. The latter has not been found in any case, but the follow-up period as well as the number of investigated cases may not be sufficient to make a final judgment on this point.

In 1979 Distelmaier et al. described radiological signs of resorption/necrosis in parts of the vertebral bodies adjacent to the PMMA implant [4]. They found these alterations in about 2% of their patients. Alterations of the size of the vertebral

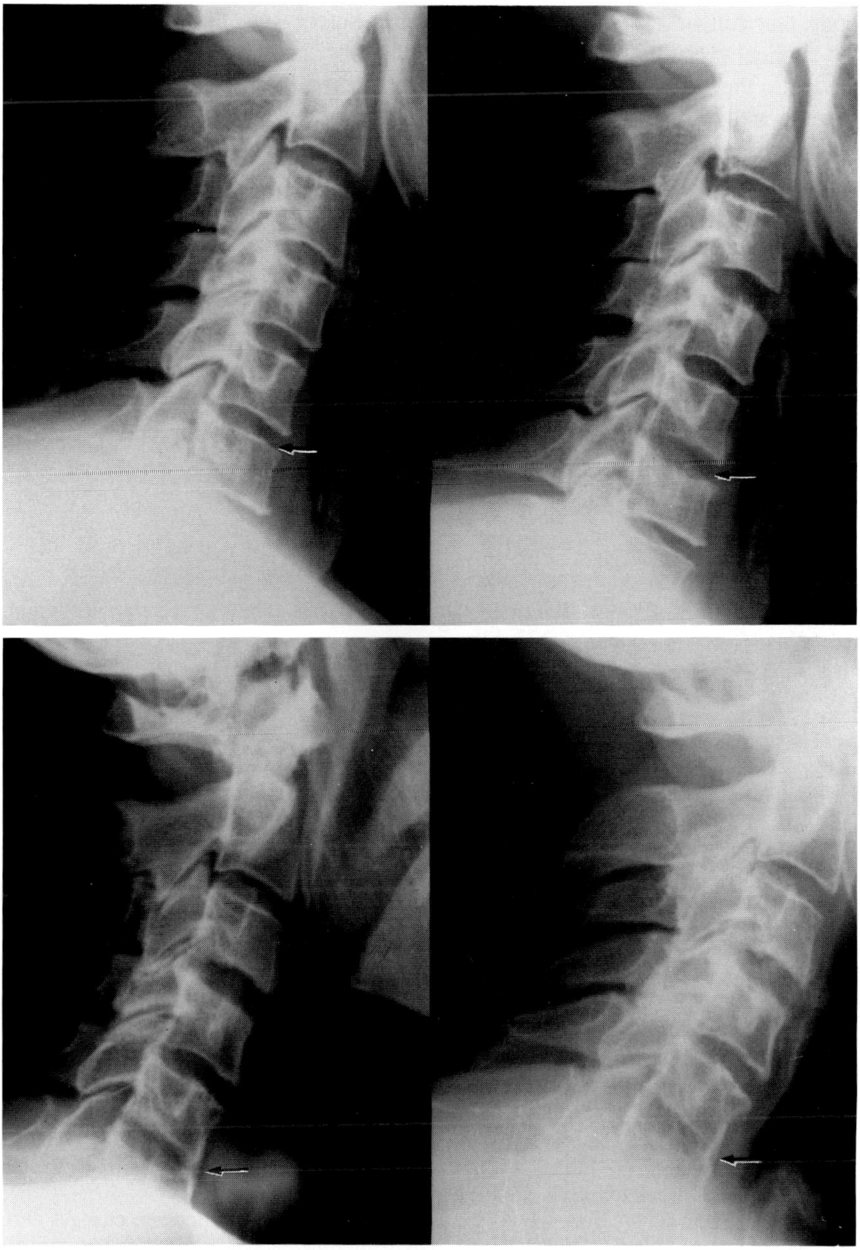

Fig. 2. Radiological signs of resorption/necrosis which heals and proceeds into a stable fusion. *Top left:* before fusion. *Top right:* 2 months after fusion. *Bottom left:* 3 years after fusion. *Bottom right:* 15 years after fusion

body that can be observed immediately after surgery are likely to be caused by the surgical procedure and technique. On the other hand, resorptions which might be due to the exothermic polymerization reaction of PMMA can be first registered some months after surgery and are characterized by a parallel approach of the cranial and caudal borders of the vertebral body, whose demarcation shows a foggy appearance. This process, however, does not continue but heals and proceeds into a stable fusion (Fig. 2).

Stable fusion with PMMA was achieved in 90% of the cases, which is comparable to the results obtained by fusion with autologous bone grafts or without any implant [1]. However, this technique avoids the complications related to removal of the bone graft from the iliac spine. Complications are mainly due to the anterior approach during the surgical procedure and are therefore not specific to the PMMA implantation technique [11].

Our results show that, among the techniques of vertebral interbody fusion, the implantation of PMMA is an alternative that deserves to be seriously considered.

References

1. Bertalauffy H, Eggert HR (1988) Clinical long-term results of anterior discectomy. Acta Neurochir (Wien) 90:127–135
2. Cloward RB (1963) Surgical treatment of traumatic cervical spine syndromes. Chir Rep 7:148–185
3. Dereymaeker A, Ghosez JP, Henkes R (1963) Le traitement chirurgical de la discopathie cervicale. Neurochirurgie 9:13–20
4. Distelmaier P, Vlajic I, Wappenschmidt (1979) Necrosis of vertebrae after Cloward's operation of the cervical spine using "Palacos" for fixation. Adv Neurosurg 7:160–163
5. Hirsch C, Wickbom J, Lindstrom A, Rosengren K (1964) Cervical disc resection. A follow-up of myelopathic and surgical procedure. J Bone Joint Surg 46A:1811–1821
6. Kazner E, Kollmannsberger A (1978) Differenzierte neurochirurgische Behandlung des zervikalen Bandscheibenvorfalls. In: Flügel KA (ed) Neurologische und psychiatrische Therapie. Perimed, Erlangen, pp 102–107
7. Odom GL, Finney W, Woodhall B, Durham NC (1958) Cervical disc lesions. JAMA 166:23–28
8. Oppenheimer BS, Oppenheimer ET (1958) The latent period in cancerogenesis by plastics in rats and its relation to the presarcomatous stage. Cancer 11:204–214
9. Ott G, Vollmer J, Hieronymie G (1963) Krebsgefährdung nach Implantation von Kunststoffen. Langenbecks Arch Klin Chir 302:608–628
10. Robinson RA, Walker AE, Ferlic DC, Wiecking DK (1962) The results of anterior interbody fusion of the cervical spine. J Bone Joint Surg 44A:1559–1587
11. Roosen K, Grote W, Bettag W (1975) Komplikationen zervikaler, ventraler Fusionsoperationen. Neurochirurgie 18:1–11
12. Roosen K, Grote W, Liesegang J, Linke V (1978) Epidural temperature changes during anterior cervical interbody fusion with polymethyl-methacrylate. Adv Neurosurg 5:373–375
13. Roosen K (1984) Experimentelle, klinische und radiologische Langzeituntersuchungen zum Ersatz zervikaler Bandscheiben durch Knochenzement. Habilationsschrift, Essen

Cervical Spine Injuries in Childhood: Long-Term Results After Surgical and Conservative Therapy; Morphological and Functional Aspects

F. Rauhut, H. Wiedemayer, L. Gerhard, and W. Grote[1]

Introduction

During infancy and childhood, cervical spine injuries are rare; in our group of patients with cervical spine lesions they represent less than 5% of cases. Surgical treatment in childhood was assessed critically bearing in mind the potential iatrogenic disturbances of growth and posture.

Patients and Methods

Thirty-five children (10 girls, 25 boys) with cervical spine injuries were treated over a period of 17 years. Their ages ranged from 3 to 17 years, with a median of 10.3 years. Twenty-one patients received conservative treatment involving Crutchfield extension or external stabilization, while 14 underwent surgery. Different surgical methods were carried out depending on the type of spinal lesion (Tables 1, 2). The disks and surrounding ligaments were investigated histologically in 14 cases, and in two cases autopsies were carried out.

Table 1. Surgical methods. PMMA, polymethylmethacrylate

Treatment	n
Anterior interbody fusion (autologous bone graft)	3
Anterior interbody fusion (PMMA)	1
Anterior interbody fusion + osteosynthesis	4
Posterior fusion (compression clamp or wiring cerclage)	5
Combined anterior + posterior fusion	1

[1] Universitätsklinikum Essen, Neurochirurgische Klinik, Hufelandstr. 55, D–4300 Essen

Advances in Neurosurgery, Vol. 18
K.-A. Bushe, M. Brock, and M. Klinger (Eds.)

Table 2. Types of cervical spine injury

Injury	n	
With spinal fracture	9	(26%)
Without spinal fracture	17	(48%)
Pseudosubluxation	9	(26%)

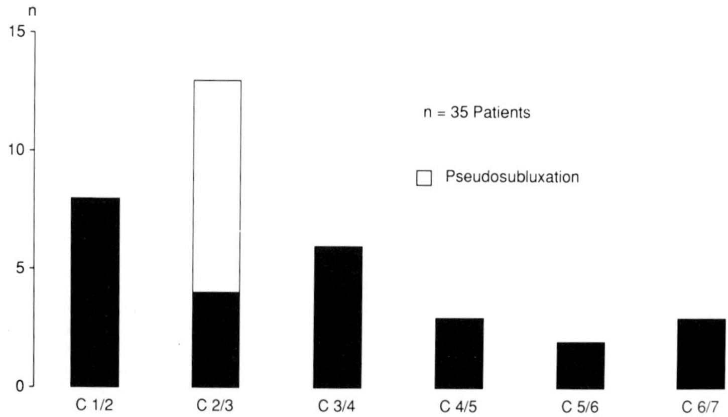

Fig. 1. Incidence and location of cervical spine injuries

Results

Overall 77% of the injured segments were located in the upper half of the cervical column (C1/2–C3/4). A total of eight cases involved the segments C4–C7 (Fig. 1). The types of injury are shown in Table 2. There were four radicular lesions, three incomplete transverse lesions with ability to walk, three incomplete transverse lesions with inability to walk, and four complete transverse lesions. Three patients died within 12–24 h postinjury, and one after 3 weeks. Two children (autopsies) suffered from cervical myelonecrosis without lesions of the bone and ligamentous structures. Two patients with initially severe transverse lesions were able to walk postoperatively, and in one case radiculopathy receded. In the case of a 4-year-old girl with a C1/2 fracture-dislocation, a severe transverse lesion with tetraparesis developed. Although the neurological deficit has remitted slowly, the child is still unable to walk. A 12-year-old boy suffering from a mechanical complication initially showed no pathological findings on X-ray examinations. A slight C2/3 dislocation had developed 4 weeks later. Despite external stabilization, dislocation progressed, and as a result an anterior interbody fusion of C2/3 was performed. A progressive C3/4 dislocation arose several weeks after surgery which required ventral (C3/4) and dorsal (C2/4) stabilization. In 12 cases no neurogenic or mechanical complications were observed. Figure 2 shows the follow-up times, with a median of 69 months.

In the conservatively treated group, the primary neurological deficit remained constant in 20 cases but was found to be aggravated in one case with secondary vertebral deformation.

Morphological examinations of the intervertebral disks showed that the disk structure in 3- to 4-year-olds is equivalent to that in young adults.

Discussion

The morphological characteristics are the cause of physiological hypermobility in childhood. The ligamentous and bone elasticity seems to be the main reason for the preponderance of discoligamentous lesions without spinal fractures. We believe that the incomplete maturation of collagen affects the elasticity. The intervertebral disks show fine ruptures in the annulus fibrosus in 3- to 4-year-olds and no vascularization, corresponding to the histological findings in young adults. Töndury reported that disk rupture develops between the 7th and 20th years [8, 9]. The hypermobility is maximal in the C2/3 segment, which leads to difficulties in diagnosis because there is no exact distinction between physiological and pathological mobility [2]. The physiological atlantodental distance is up to 5 mm in childhood [3, 4]. As in the

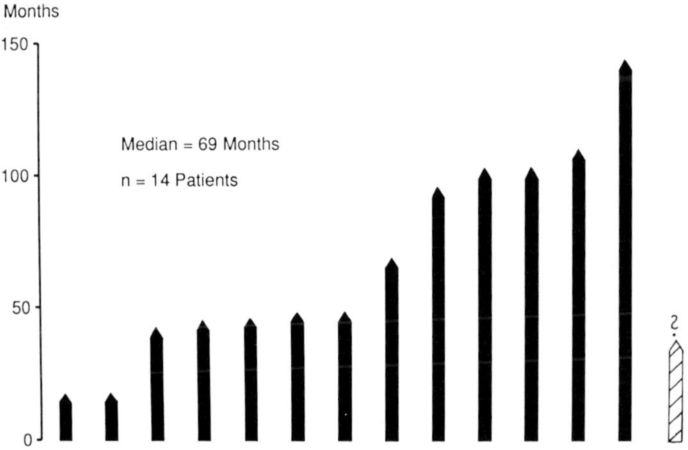

Fig. 2. Length of follow-up after surgery. One patient (?) could only be followed up for 3 years

case of congenital interbody fusion, surgical stabilization induced no deformation of the vertebral column and the interbody fusion did not lead to growth inhibition [5,6]. The age of the child is very important in the examination of the C1/2 region. The fusion between the dens axis and the dens body is completed after 7 years. Therefore, some authors have reported that epiphysiolysis is possible only up until this time [1,2]. In contrast to discoligamentous lesions in adults, the conservative treatment of such injuries in childhood, especially in the upper cervical spine, yields good results but requires immobilization (Crutchfield extension) over a period of 12–14 weeks. After primary C1/2 lesions, renewed dislocation occurs in up to 10% of cases [4]. The present study allows no comparison to be made between surgical and conservative treatment. In our experience surgery should be carried out in cases of (a) intraspinal hematomas or bone fragments, (b) unstable ligamentous lesions without spinal fractures after unsuccessful external fixation or extension (dislocation), (c) unstable fracture-dislocations with deformation, and (d) defective vertebral positions with secondary myelopathy. Compared with conservative treatment, surgery makes early rehabilitation possible, which is very important in cases of severe transverse lesions.

Summary

A total of 35 children with cervical spine injuries were treated over a period of 17 years. Overall 14 patients received different forms of surgical treatment depending on the type of spinal lesion. Long-term results after surgery and conservative treatment are presented and discussed; in addition morphological features have been investigated.

References

1. Bailey DK (1952) The normal cervical spine in infants and children. Radiology 59:712–719
2. Blockey N (1956) Fractures of the odontoid process of the axis. J Bone Joint Surg 38B:794–817
3. Hupfauer W, Konermann H (1980) Die sogenannte Pseudosubluxation der oberen Halswirbelsäule im Kindesalter. Orthop Praxis 2:106–109
4. Jonasch E (ed) Spezielle Frakturen und Luxationslehre, vol I/2. Thieme, Stuttgart, pp 1–27, 58–63
5. Rauhut F, Roosen K, Grote W (1984) Die operative Therapie von HWS-Verletzungen im Kindesalter unter Berücksichtigung der morphologischen und funktionellen Besonderheiten der kindlichen Halswirbelsäule. Z Kinderchir 39:155–159
6. Roosen K, Rauhut F, Maksoud M (1984) HWS-Verletzungen im Kindesalter – konservative oder chirurgische Therapie? Neurochirurgia 27:1–5
7. Roosen K, Trauschel A, Grote W (1982) Posterior atlantoaxial fusion: a new compression clamp for laminar osteosynthesis. Arch Orthop Trauma Surg 100:27–31

8. Töndury G (1958) Entwicklungsgeschichte der Wirbelsäule. In: Die Wirbelsäule in Forschung und Praxis, vol 7. Hippokrates, Stuttgart
9. Töndury G (1960) Entwicklung, Bau und Altersveränderungen der Zwischenwirbelscheiben unter spezieller Berücksichtigung der Halsregion. Therapiewoche 10:432–438

Bracing of the Dens Axis

D. Tertsch and M. Schneider[1]

Bracing of the dens axis was introduced in Halle at the Martin Luther University in 1984. The technique involves compression synthesis achieved by means of a plug screw which is pressed apart at its tip (Figs. 1, 2). M. Schneider was responsible for the construction as well as the technology of the device.

The screw is available with a total length of 4.5 or 5.5 cm. It is centrally punched and can be threaded onto a Kirschner's wire. This combination of screw and central wire can be subsequently introduced into the drill hole. The center carries a thread which presses apart the head of the screw. The head of the screw consists of a self-cutting thread. Distortion of the screw is prevented by a claw-like disk. After spreading out, the screw is drawn back by a hexagon nut. This is continued until the dens fragment located at the tip of the screw is pulled onto the vertebral body by the spreading out of the wire thread. Finally, a thread cap is screwed onto the end of the screw.

The advantage of this method is that one screw only is used. In addition, the operative procedure is performed via a Kirschner's wire in situ. Hence, bony consolidation is attempted, compressing the fragments and completely immobilizing them. The fracture is resistant to distortion and is stable under load.

The operative approach is performed according to Cloward. With the patient in a supine position, after repositioning, the cervical disk C2/3 is approached via a skin incision at the right side of the collum between the vessel-nerve bundle and the trachea/esophagus. Under X-ray guidance, the Kirschner's wire is drilled into the dens axis from the lower side of the body of C2. Drilling of the screw canal is performed using a crown drill, running along the Kirschner's wire. The dens screw is also inserted via the Kirschner's wire in position and led to the fragment. When the screw is correctly positioned, the Kirschner's wire is removed and the screw is secured against distortion by the claw-like disk. The expansion of the thread-bearing screw head by means of the center can be visually controlled on the X-ray screen. The pulling on of the fragments is monitored in the same way.

[1] Neurochirurgische Klinik der Martin-Luther-Universität, Leninallee 16, DDR–4020 Halle

Advances in Neurosurgery, Vol. 18
K.-A. Bushe, M. Brock, and M. Klinger (Eds.)

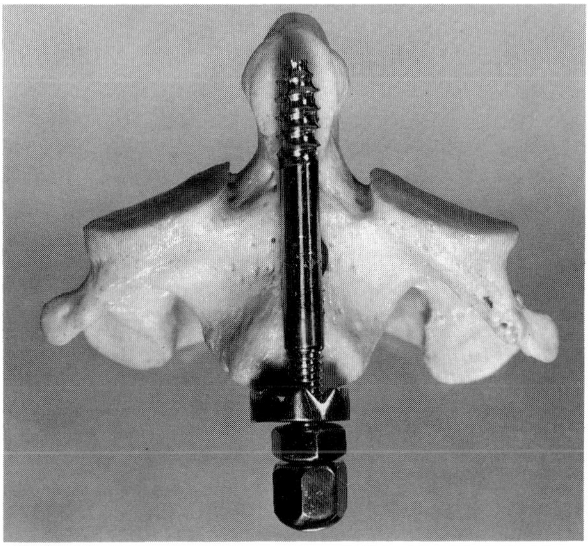

Fig. 1. Dens screw as a model

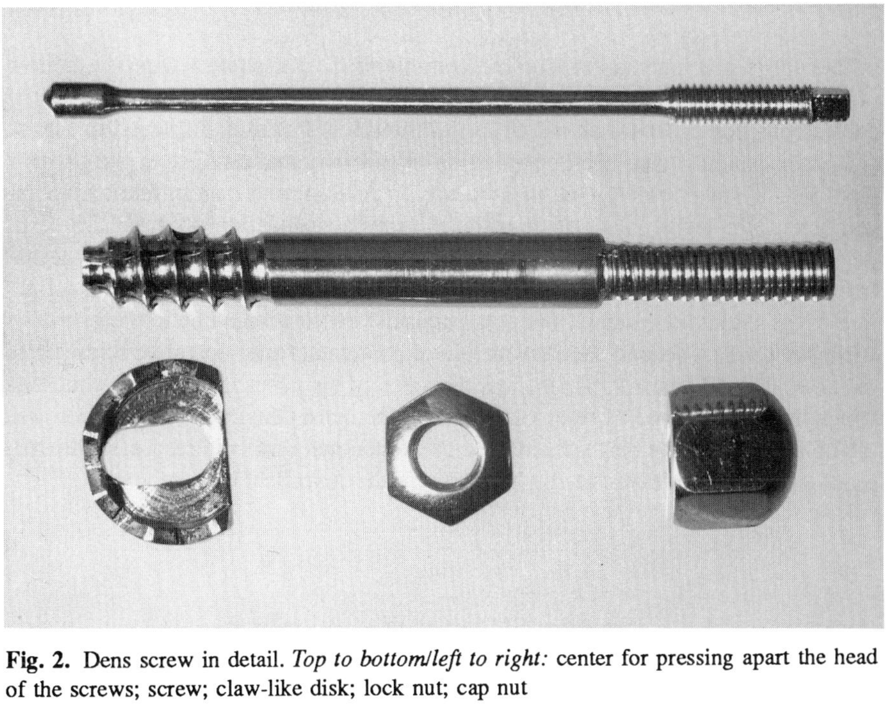

Fig. 2. Dens screw in detail. *Top to bottom/left to right:* center for pressing apart the head of the screws; screw; claw-like disk; lock nut; cap nut

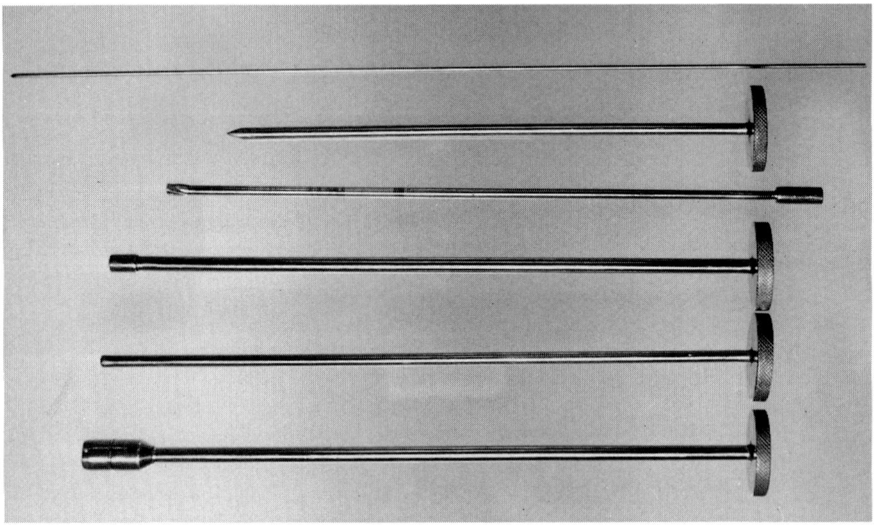

Fig. 3. Instruments. *Top to bottom:* Kirschner's wire; sharp hollow probe to guide Kirschner's wire; hollow crown drill; screwdriver for dens screw; screwdriver for center; screwdriver for lock nut

The entire instrumentation required consists of the Kirschner's wire, the hollow probe to guide the wire, the hollow crown drill, and the three screwdrivers for the dens screw, center, and lock nut (Fig. 3). The screw in situ is displayed in Fig. 4.

Sixteen patients have been operated on in this way to date. One of the patients died on the 9th day due to a stress ulcer. In addition, he had suffered from incomplete tetraplegia. In all other patients, bony consolidation occurred. Three of them showed pseudarthrosis. The indication for operation is a repositioned, sagittal fracture close to the base. However, one female patient was operated on for a fracture of the odontoid vertebra. She had persistent cervical pain due to the operative wiring of C1/2. A second patient was free of complaints after operative screwing of the dens. The 62-year-old patient was then free of his persistent cervical syndrome which had been caused by pseudoarthrosis of the dens. One local complication was the obliteration of the disk space C2/3. The functional results in respect of flexion, rotation, and lateroflexion of the head were either excellent or very good.

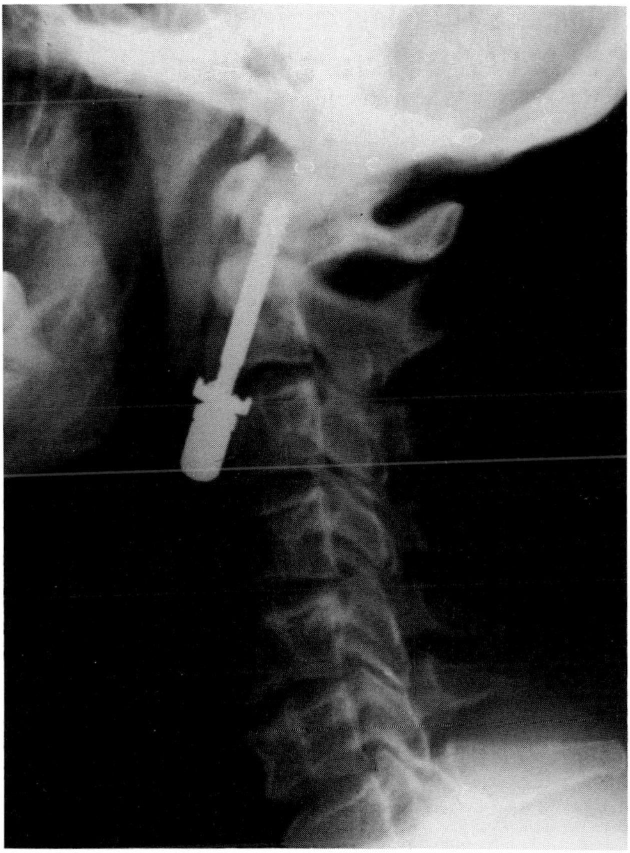

Fig. 4. Screw in situ

The Surgical Treatment of Atlas, Axis, and Combined C1 and C2 Fractures

R. Verheggen, J. Jansen, and E. Markakis[1]

Introduction

Between October 1984 and April 1989 23 patients suffering from fractures of the atlas and/or the odontoid vertebra underwent surgical stabilization in our department. Of these patients, 69% sustained injuries in automobile or motorbike/cycle accidents. The number of accidents with frontal collison exceeded those with lateral or rear impact or overturning of the vehicle. With regard to our patients, it is striking that a high percentage (43%) were passengers. With two exceptions, all injuries happened despite suitable precautions such as seatbelts, head rests, or crash helmets. In 17% of the cases a fall preceded the injuries. In 8% the cause of the trauma could not be reconstructed.

This study discusses our preferred osteosynthetic techniques and describes the functional restitution of our patients 4 months after surgery.

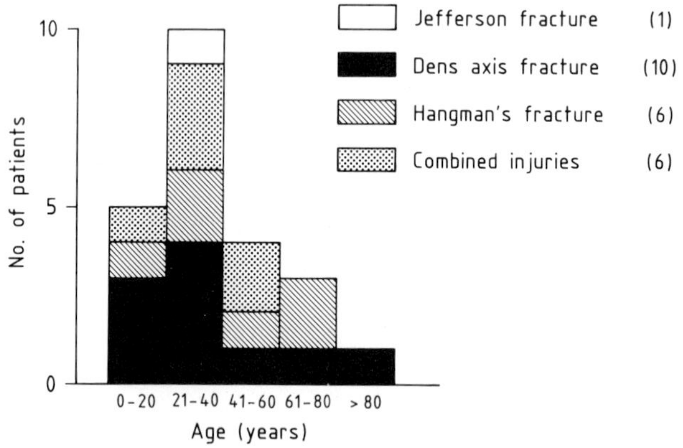

Fig. 1. Surgically treated C1 and C2 fractures in relation to age

[1] Neurochirurgische Universitätsklinik Göttingen, Robert-Koch-Str. 40, D-3400 Göttingen

Advances in Neurosurgery, Vol. 18
K.-A. Bushe, M. Brock, and M. Klinger (Eds.)

Patients

Twenty-three patients aged 15–85, suffering from fractures of the atlas and/or odontoid vertebra, were treated operatively. In the age group up to 20 years, in which there was a predominance of dens fractures, road accidents were the sole cause of the trauma, whereas in the age group above 60 years the upper spine injuries – mostly dens and hangman's fractures – were the result of falling (Fig. 1).

On hospital admittance, the neurological condition of 17 patients was unremarkable. They complained only of neck and shoulder pain with an occipital trigger point. In one patient a C6 symptomatology was observed in addition to the neck pain. One patient showed signs of incomplete quadriplegia while four suffered from complete quadriplegia.

Results

As assessed according to the classification of Anderson and d'Alonzo [1], eight cases of a type II fracture at the junction of the dens and the body of the axis and two cases of a type III fracture extending into the body of the axis with dislocation were secured by compression screw fixation as described by Böhler [2]. In order to reposition the dens, the head was clamped into the Mayfield fixture. In hyperextension an anteromedial approach was adopted with a half collar incision at the right side just above the cricoid cartilage to expose the anterior and upper borders of the odontoid vertebra. Using two image intensifiers, fusion of the dens was achieved using two converging screws.

In eight cases postoperative radiological control revealed appropriate, axis-oriented osseous consolidation. One case was complicated by pseudarthrosis due to the poor compliance of the patient. In another patient we removed the screws after 3 1/2 months because of their suboptimal position, at which time there was radiological proof of bony fusion. The further postoperative treatment involved wearing a plastic collar according to Kamp continuously for an average of 10 weeks.

With reference to Judet et al. [4], a screw fixation was carried out in six patients with fractures of the arch of the axis. By means of a dorsal approach two compression screws were fastened lateromedially through the pedicle in order to connect the arch with the body of the axis (Figs. 2, 3).

In two cases, a frontal collision with a submental impact caused an undisplaced bilateral fracture of the arch with sole involvement of posterior elements or a hangman's fracture type I according to Pepin and Hawkins [6] (Fig. 4).

In the four patients with a hangman's fracture type II, the demolition of the arch was combined with anterior displacement of C2 on C3, and therefore as a concomitant injury the anterior ligaments were disrupted and there was involvement of the C2-C3 disk bond.

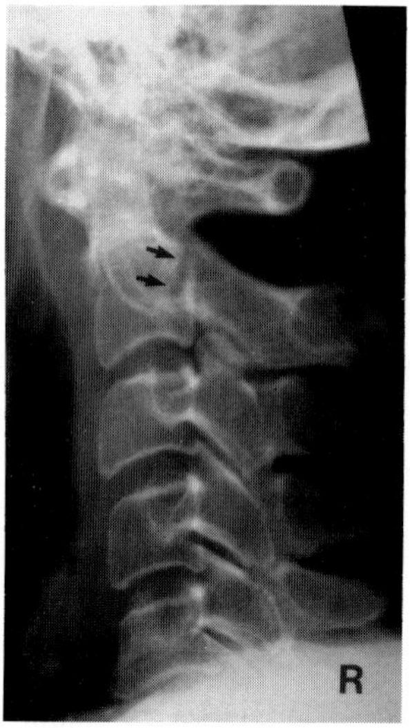

Fig. 2

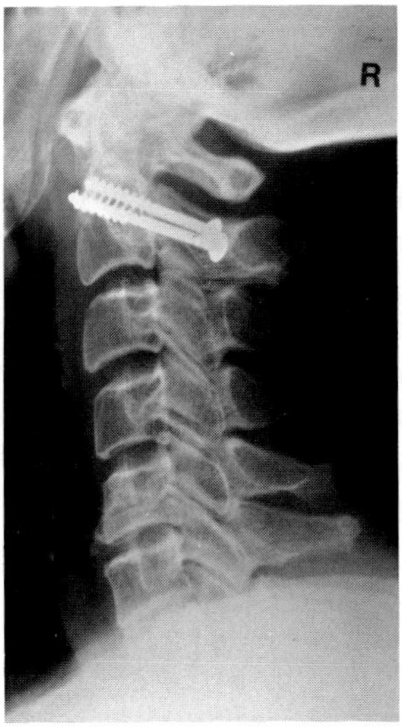

Fig. 3

Figs. 2, 3. A hangman's fracture type I in a 66-year-old woman before and after screw fixation

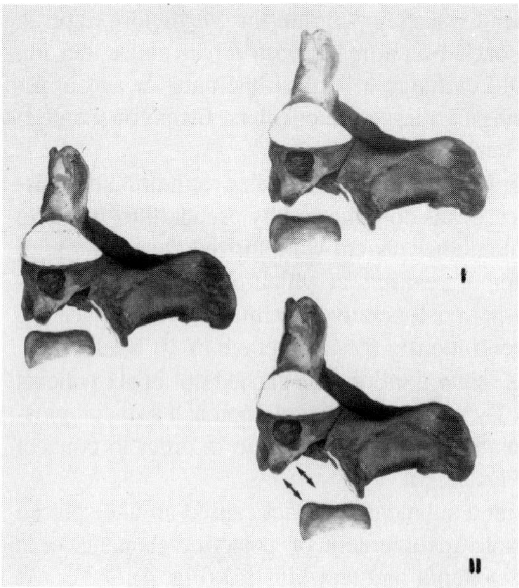

Fig. 4. Schematic diagram of the classification of hangman's fractures according to Pepin and Hawkins

As described, the further postoperative treatment included the wearing of a plastic collar. Control examinations 4 months later showed normalized mobility of the cervical spine, and the neurological status was unremarkable; only one patient complained of neck pain.

In six cases we preferred internal stabilization by the use of a Palacos wire-mesh fixed on the nodules of the occiput and extending to the sixth or seventh cervical vertebra, as described in detail by Jansen and Kunze [3]. There were five combined injuries: three atlas fractures associated with a hangman's fracture, one hangman's fracture type II with a dislocation of C2 on C3, and one burst fracture of the atlas involving both arches and the right lateral mass. In addition there was one case of a young woman requiring dialysis with osteomyelitis of C2 and osteolysis of the dens.

The mobility of the head was completely curtailed in the upper spine although anteflexion and retroflexion of about 65° and 40° respectively were made possible by the lower segments; the rotation of the head to the right and to the left was reduced to as little as 5°.

Four to six months after this operation the Palacos wire-mesh was removed. After that only the right and left rotation of the head was reduced.

Summary

Twenty-three patients suffering C1 and C2 fractures were treated surgically. Postoperative complications were pseudarthrosis in one case and a bony consolidated non-axis-orientated dens fracture in another. In the other 21 cases the osseous consolidation was correct, as proved by roentgenographic and computer tomographic control examinations.

A later examination of patients with a screw fixation according to Böhler [2] and Judet et al. [4] revealed normal ante- and retroflexion, whereas the right and left rotation was free up to 65°.

The patients with combined injuries stabilized with a Palacos wire-mesh showed, after removal of the mesh, insignificantly limited ante- and retroflexion although the rotation to the left and right was generally reduced by 35°–70°.

References

1. Anderson LD, d'Alonzo RT (1974) Fractures of the odontoid process of the axis. J Bone Joint Surg 56A:1663–1674
2. Böhler J (1982) Anterior stabilization for acute fractures and non-unions of the dens. J Bone Joint Surg 64A:18–27

3. Jansen J, Kunze E (1987) Die Verwendung von Knochenzement für die dorsale Stabilisation der Halswirbelsäule. In: Willert H-G, Buchhorn G (eds) Knochenzement. Hans Huber, Bern
4. Judet R, Roy-Camille R, Saillant G (1970) Fractures dur raches cervical. Actualités de chirurgie orthopédique de l'Hôpital Raymond-Pioncarré 8:174–195
5. Landells CD, van Peteghem PK (1988) Fractures of the atlas: classification, treatment and morbidity. Spine 13:450–452
6. Pepin JW, Hawkins RJ (1981) Traumatic spondylolisthesis of the axis: hangman's fracture. Clinical Orthopaedics and Related Research 157:133–138

Association of Injuries to the Head and Spine: 34 Consecutive Cases over a 3-Year Period*

D. Moskopp, D.-K. Böker, L. Solymosi, M. Kurthen, and E. Elatan[1]

Introduction

Because of the specialized emergency system in the Bonn area, an increasing number of patients with severe and multiple injuries are reaching the University Hospital with at least residual vital signs. Furthermore, the necessity to simultaneously manage traumata to the neurocranium and the spinal column is not at all rare. Nevertheless, the literature of the last decade contains no quantified information concerning the epidemiology with respect to neurosurgical purposes. The aim of this report is to provide data on this subject.

Patients and Method

Inclusion Criteria

For the period of February 1986 to January 1989 all inpatients with the diagnosis of head injury (including *at least* a short period of proven unconsciousness) were registered prospectively. For the analysis under discussion, *only* those with *additional* trauma to the suprasacral spine were selected (n = 46).

Exclusion Criteria

Of these 46 patients, a dozen were excluded for four reasons:

1 Poor data regarding the primary event (delayed transport from abroad: n = 5)
2. Insufficient documentation of the combined injury (no X-rays because of brain death on admission: n = 2)

[1] Neurochirurgische Klinik der Universität Bonn, Sigmund-Freud-Str. 25, D–5300 Bonn

* Dedicated to Professor Dr. med. Dr. phil. R. Wüllenweber, on the occasion of his 65th birthday

Advances in Neurosurgery, Vol. 18
K.-A. Bushe, M. Brock, and M. Klinger (Eds.)

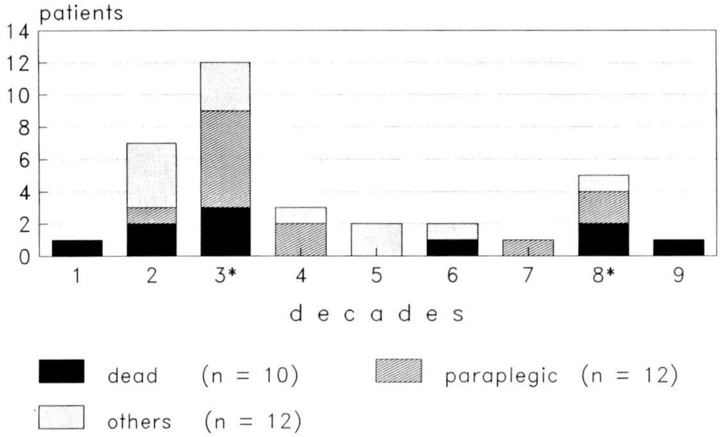

Fig. 1. Age and outcome (n = 34). *dead:* Corresponds to deaths as a direct result of the injury. * Two additional patients had died at the end of the observation period (4/89) due to KCN intoxication or pancreatic carcinoma. *paraplegic:* Summarizes both tetra- and paraparesis (n = 6, each). Most of the tetrapareses remained complete, whereas most of the parapareses were incomplete at the end of the observation period. *others:* Includes the outcome of at least four disabled patients at the end of the observation period: two hospitalized and two dependent at home

3. Transoral missile injury (n = 1)
4. Head injury and cervical root lesions *without* spinal fractures (n = 4)

Over the period in question an additional group of 56 patients suffering from a spinal cord injury *without* any clouding of consciousness were not reexamined for discrete neurological deficits as a result of the direct injury [6,9,16-19,29] or cervical vessel lesions [12,27].

Results

In general: Over the period of 3 years, 430 inpatients with head injuries were observed. Of these, 90% suffered from a primary and 10% from a secondary coma. On admission, the percentage ratio of coma/somnolence/mild clouding was 60/10/30. In a total of 8% (n = 34), the head trauma was accompanied by a spinal injury, regardless of whether the patient was still in coma on admission (n = 259). The cervical spine was affected in two-thirds of the cases (21/34). Information on the correlation between the clinical outcome (10 deaths, 12 paraplegics, 12 others) and some further traumatological parameters is given by the four figures. Figure 1

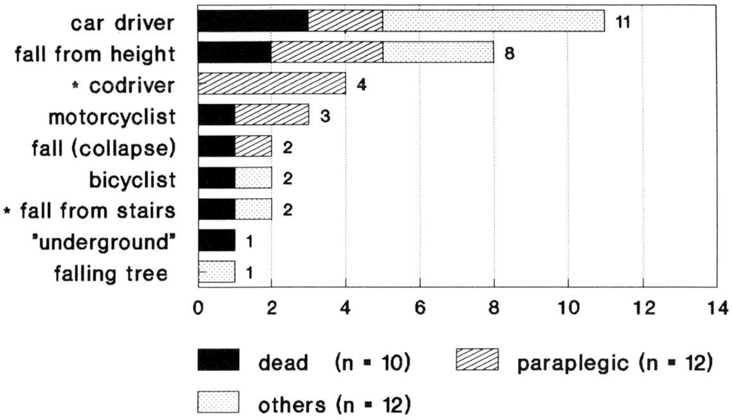

Fig. 2. Type of trauma and outcome (n = 34); cf. legend to Fig. 1. *"underground"*: Accidentally caught by the underground. Note that two-thirds of the accidents were traffic accidents, with motorcyclists not being the main group. In addition note the severe prognosis of codrivers, whose seating position within the car could not be evaluated after the event. Of the "falls from height," three were with a suicidal intention. At least two accidents were influenced by alcohol

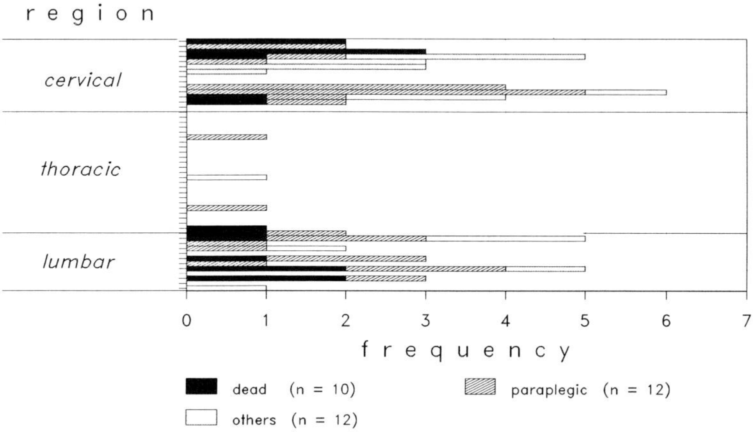

Fig. 3. Pathological vertebral segments (n = 63); cf. legend to Fig. 1. Note that the number of the indexed vertebral segments exceeds the number of human vertebrae because of numbering the vertebral (e.g., 5th cervical) and *inter*vertebral spaces (e.g., cervical 5/6)

shows the double-peaked age distribution (median: 27 years; males:females = 5:2). Figure 2 summarizes the circumstances of the accidents. Figure 3 depicts the topological pathology of the spinal injuries, and Fig. 4 shows the main diagnoses with regard to head and brain pathology. Basically, the main injury was either spinal (n = 15), cerebral (n = 10), or indistinguishable in this respect (n = 9). Neurosurgical operations were performed because of either urgent spinal (n = 15) or cerebral indications *alone* (n = 7). Twice a bilocular procedure proved necessary.

In detail: Because of the variability of these complex injuries, only half of the patients can be subdivided into fairly typical subgroups, each containing four to five patients:

1. Fatal injuries to the craniocervical junction, which were survived for only 1–7 posttraumatic hours, irrespective of the brain scan findings [7,13,15,20,25,30, 31]
2. Dislocated fractures of the fifth cervical vertebra with persistent tetraparesis in association with closed head injuries (coma between 15 min and 3 days)
3. Fractures of the thoracolumbal junction in association with multiple injuries
4. Fractures of the lumbar transverse process, as an epiphenomenon of the considerable force of the injury (survival time: 2 to 4 days).

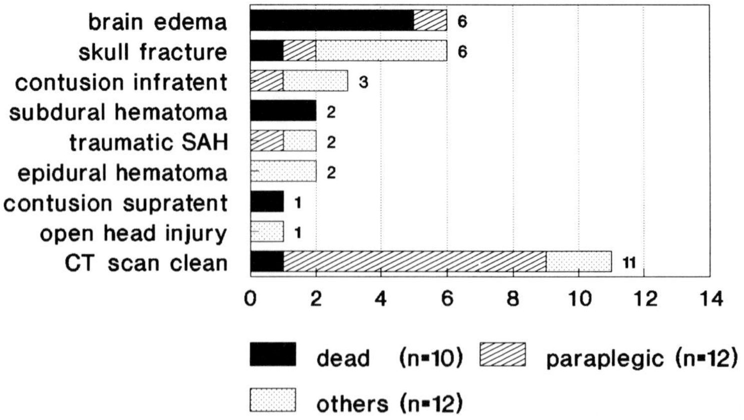

Fig. 4. Main diagnoses with regard to head and brain pathology (n = 34); cf. legend to Fig. 1. CT, cerebral computerized tomography; infratent, infratentorial; SAH, subarachnoid hemorrhage; supratent, supratentorial

Discussion

There were some early publications on the subject based on animal models [11,23,24]. As regards human traumatology, with few exceptions [e.g., 28] spinal and cerebral injuries have been reviewed separately. Insofar as this combined pathology is discussed at all, most authors avoid presenting data from their own series [e.g., 2, 5]. The terms, as well as the manner of classification of these injuries and their incidence, show considerable variation from one observer to another, seemingly depending on the author's specialty (neurosurgery, orthopedics, maxillofacial surgery, general surgery, or traumatology [1,3,4,8,10,14,21,22,26]). This survey merely aims to quantify the probability and topological distribution of these complex injuries, in the hope of providing helpful information for doctors on duty. In the state of unconsciousness, or in the presence of pharmacological residues from preceding resuscitation, clinical signs (e.g., motor response) may be doubtful or misleading. Therefore, an injury to the vertebral column is not unlikely to be overlooked prima facie.

The principal problem of overlapping clinical signs in coma patients can easily be demonstrated by the aforementioned head injuries associated with lesions of the cervical nerve roots: Half of these nerve lesions were detected secondarily, either due to an ipsilateral Colles' fracture (immobilization) or a contralateral subdural hematoma with central hemiparesis.

Conclusion

An associated vertebral fracture can be expected in about 10% of all head-injured patients (with at least a short period of unconsciousness). Two-thirds of these lesions occur in the cervical spine. In general, the risk of overlooking this diagnosis can be reduced by some simple methods: meticulous analysis of cerebral CT scans or conventional lateral X-ray of the skull *and* cervical spine, down to the seventh vertebra.

Acknowledgments. The authors are indebted to Mrs. Karin Mutlaq for providing the literature.

References

1. Arnold K (1985) Die Mehrfachverletzung und Mehrfachschwerverletzung (Bemerkungen zur Definition und Pathophysiologie). Z Ärztl Fortbild 79:963–966
2. Bone L, Bucholz R (1986) The management of fractures in the patient with multiple trauma. J Bone Joint Surg 68A:945–949

3. Börner M, Klemm K (1982) Eine Analyse motorisierter Zweirad-Unfälle – behandelt in der Berufsgenossenschaftlichen Unfallklinik Frankfurt am Main in den Jahren von 1977 bis 1980. Lebensversicherungsmedizin 34:194–197
4. Clark WK (1979) Trauma to the nervous system. In: Shires GT (ed) Care of the trauma patient, 2nd edn. McGraw-Hill, New York, pp 207–258
5. Cooper PR (1985) Resuscitation of the multiply injured patient. In: Wilkins RH, Rengachary SS (eds) Neurosurgery. McGraw-Hill, New York, pp 1587–1592
6. Davidoff G, Morris J, Roth E, Bleiberg J (1985) Closed head injury in spinal cord injured patients: retrospective study of loss of consciousness and post-traumatic amnesia. Arch Phys Med Rehabil 66:41–43
7. Dickman CA, Hadley MN, Browner C, Sonntag VK (1989) Neurosurgical management of acute atlas-axis combination fractures. J Neurosurg 70:45–49
8. Dittel KK, Weller S (1981) Zur Problematik des polytraumatisierten Patienten. Aktuel Traumatol 11:35–42
9. Erdmann H (1973) Schleuderverletzungen der Halswirbelsäule. In: Junghanns H (ed) Die Wirbelsäule in Forschung und Praxis, vol 56. Hippokrates, Stuttgart, pp 47–56
10. Garland DE, Rhoades ME (1978) Orthopedic management of brain-injured adults. Clin Orthop 131:111–122
11. Gosch HH, Gooding ER, Schneider RC (1973) An experimental approach to head and spine injuries. In: Schneider RC (ed) Head and neck injuries in football. Mechanisms, treatment, and prevention. Williams & Wilkins, Baltimore, pp 216–227
12. Gurdjian ES, Hardy WG, Lindner DW, Thomas LM (1963) Closed cervical cranial trauma associated with involvement of carotid and vertebral arteries. J Neurosurg 20:418–427
13. Jefferson G (1920) Fracture of the atlas vertebra. Report of four cases, and a review of those previously recorded. Br J Surg 7:407–422
14. Junghanns K (1970) Die kombinierte Verletzung des Schädels und der Halswirbelsäule. Monatsschr Unfallheilkd 73:97–101
15. Köhler T, Pauly T, Ludolph E (1987) Geschlossene atlantookzipitale Luxation (Dekapitation) nach Motorradunfall. Unfallchirurg 13:278–280
16. Krämer G, Hopf HC (1981) Zerebrale Störungen nach isolierten „HWS-Schleudertraumen" (zerviko-zephalen Beschleunigungstraumen). Aktuel Traumatol 11:114–119
17. Krämer G (1983) HWS-Schleudertraumen. Zur Pathogenese der zerebralen Beteiligung und persistierender posttraumatischer Störungen. Med Welt 34:1134–1140
18. Laubichler W (1987) Die Problematik einer Begutachtung von Verletzungen der Halswirbelsäule einschließlich cervico-cephalem Beschleunigungstrauma. Unfallchirurg 90: 339–346
19. Laubichler W, Spielmann A (1988) Zur Ätiologie des neurastheniformen „Psychosyndroms" nach Verletzungen der Halswirbelsäule, einschließlich cervicocephalem Beschleunigungstrauma. Beitr Gerichtl Med 46:439–44ß
20. Lee C, Rogers LF, Woodring JH, Goldstein SJ, Kim KS (1984) Fractures of the craniovertebral junction associated with other fractures of the spine: Overlooked entity? AJNR 5:775–781
21. Lewis VL Jr, Manson PN, Morgan RF, Cerullo LJ, Meyer PR Jr (1985) Facial injuries associated with cervical fractures: recognition, patterns, and management. J Trauma 25:90–93
22. McCabe JB, Angelos MG (1984) Injury to the head and face in patients with cervical spine injury. Am J Emerg Med 2:333–335
23. Ommaya AK, Faas F, Yarnell P (1968) Whiplash injury and brain damage. An experimental study. JAMA 204:285–289

24. Ommaya AK, Yarnell P (1969) Subdural haematoma after whiplash injury. Lancet II:237–239
25. Pang D, Wilberger JE Jr (1980) Traumatic atlanto-occipital dislocation with survival: case report and review. Neurosurgery 7:503–508
26. Rowe KW, Brooks CJ (1984) Head and neck injuries in Canadian Forces ejections. Aviat Space Environ Med 55:313–315
27. Schneider RC, Gosch HH, Norrell H, Jerva M, Combs LW, Smith RA (1970) Vascular insufficiency and differential distortion of brain and cord caused by cervicomedullary football injuries. J Neurosurg 33:363–375
28. Schneider RC (1973) (ed) Head and neck injuries in football. Mechanisms, treatment, and prevention. Williams & Wilkins, Baltimore, pp 2, 67–71, 77, 116–129
29. Torres F, Shapiro SK (1961) Electroencephalograms in whiplash injury. A comparison of electroencephalographic abnormalities with those present in closed head injuries. Arch Neurol 5:28–35
30. Traynelis VC, Marano GD, Dunker RO, Kaufman HH (1986) Traumatic atlanto-occipital dislocation. Case report. J Neurosurg 65:863–870
31. Wood-Jones F (1913) The ideal lesion produced by judicial hanging. Lancet I:53

Operative Treatment of Metastatic Cervical Spine Instability

W. Grote, R. Kalff, W. Kocks, K.-P. Schmit-Neuerburg, J. Pospiech, and M. Schax[1]

Introduction

Vertebral metastases occur in about 5% of patients during the course of cancer [1,2,11,12]. In almost all cases, local or radicular pain is the initial symptom [2,5,10]. If there is vertebral instability or compression of the spinal cord due to osteolytic destruction of the anterior or posterior elements of the cervical spine, progressive neurological deficit is unavoidable (Fig. 1). Therefore operative removal of the tumor in combination with internal stabilization is indicated [3,4,6,7,13,14,19,20]. Because of the impossibility of total surgical removal, local postoperative radiation therapy is necessary.

Patients and Methods

Thirty-three patients with metastatic cervical spine tumors were treated by operative stabilization because of cervical instability and impending neurological deficits (15% of all our cervical stabilization procedures). The mean age was 58 years, with a balanced sex distribution.

Breast cancer (12 cases) was the most frequent primary tumor. Only in two cases was the origin of the metastatic tumor unknown.

At the time of diagnosis of cervical spine metastasis, the mean duration of cancer was 4 years, with the mean duration of cervical symptoms being 5 months.

Clinical findings were classified into five groups (Table 1). Twenty-four patients had pain or radicular deficits. Five patients showed incomplete paraparesis, while another four were already unable to walk.

The distribution of metastatic localization showed a peak in C5 and C6 as well as in the axis. Monosegmental osteolysis was present in 20 patients. Two adjacent vertebrae were affected in six cases, while seven patients had a multisegmental involvement.

---

[1] Neurochirurgische Klinik, Universitätsklinikum der Gesamthochschule Essen, Hufelandstr. 55, D-4300 Essen 1

Advances in Neurosurgery, Vol. 18
K.-A. Bushe, M. Brock, and M. Klinger (Eds.)

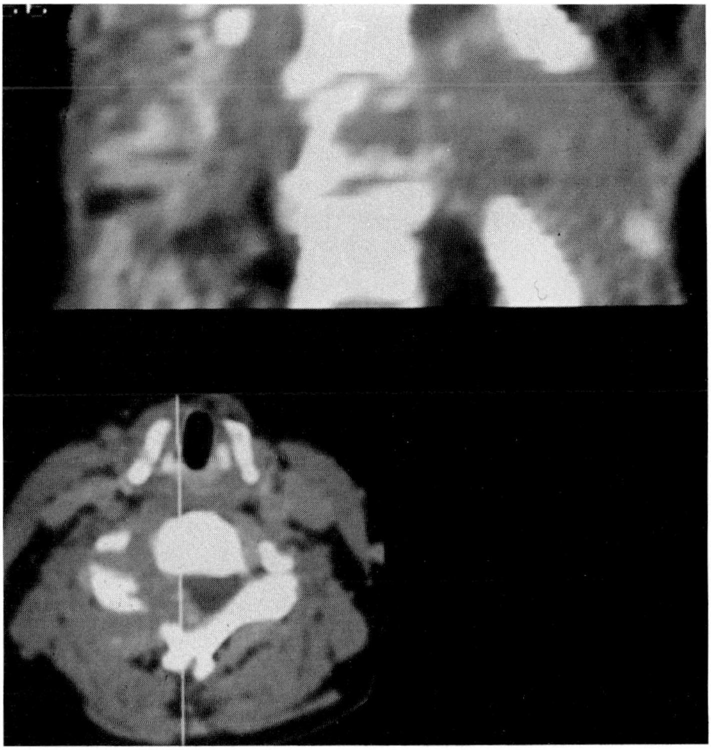

Fig. 1. CT scan showing tumoral infiltration of the vertebral body and into the spinal canal

Spondylectomy with subsequent anterior spondylodesis (Fig. 2) was performed in 15 cases, posterior spondylodesis only in five. A combination of ventral and dorsal stabilization was necessary in 13 patients because of global instability (Fig. 3).

Results

Two patients with radicular neurological deficits recovered completely. Another three patients with incomplete paraparesis were able to walk again after decompression and stabilization. Three patients died in the immediate postoperative course, one because of severe pneumonia and two because of acute pulmonary embolism. In one patient with normal neurological status preoperatively, radicular pain developed in the postoperative course.

The remaining 23 patients showed no change in their neurological status. Nineteen patients are still alive without any worsening of their neurological findings. In

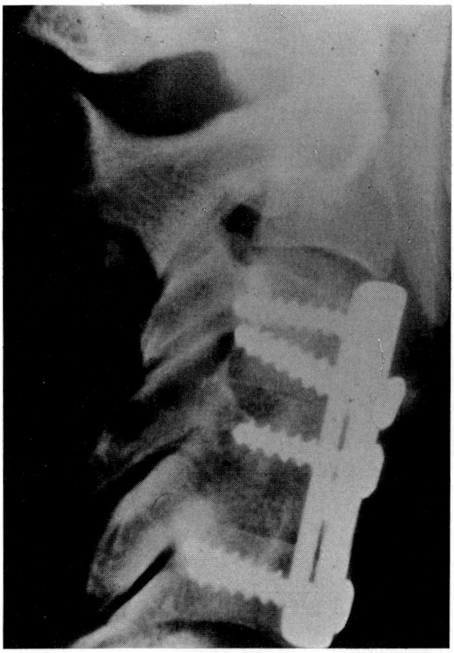

Fig. 2. Ventral plate stabilization of C3–C5 after removal of the body of C4 and reconstruction with a homologous bone graft

the meantime 14 patients have died; only in two of them did paraparesis occur in the final phase of their cancer (Table 1). The mean survival time was 10.5 months.

Discussion

In our opinion in cases of cervical spine metastasis, surgical stabilization is indicated at the onset of neurological deficits and/or in the event of instability of the spine. Life expectancy should be at least 6 months.

We regard as contraindications involvement of the cervical spine at multiple levels, rapid growth of the tumor under medical treatment, and complete paraplegia.

Nowadays ventral spondylodesis should be achieved with plates after removal of the vertebral body and reconstruction using bone cement or homologous bone graft. Ventral decompression and stabilization is often insufficient because of complementary dorsal instability caused by infiltration of the tumor into the pedicles and spinal canal. In these cases additional dorsal stabilization is necessary. For this

Table 1. Neurological deficit at time of admission and follow-up I, normal; II, radicular pain or deficit; III, incomplete paraparesis – able to walk; IV, incomplete paraparesis – unable to walk; V, complete paraparesis

Admission	Breakdown at follow-up	Classification at follow-up	Deaths
I = 11	I = 9	I = 12	⟶ 1
	II = 1		
	V = 1		
II = 13	I = 2	II = 12	⟶ 6
	II = 11		
III = 5	I = 1	III = 4	⟶ 2
	III = 2		
	V = 2		
IV = 4	III = 2	IV = 2	⟶ 2
	IV = 2		
V = 0		V = 3	⟶ 3 (postop.)

we prefer the compression clamps of Roosen and Trauschel [15–18], as mentioned in preliminary reports [8,9].

The aim of surgical treatment of cervical spine metastases is the avoidance of deterioration of neurological deficits. This can be achieved by rapid decompression of the spinal cord and nerve roots with subsequent internal spondylodesis. External fixation has to be avoided to allow rapid mobilization of the severely ill patients. The quality of life in the remaining short survival time can be preserved by saving the patients from confinement to bed. In our opinion progressive paraparesis in patients with cervical spine metastases does not need to be accepted as inevitable.

References

1. Barron KD, Hirano A, Araki S, Terry RD (1959) Experiences with metastatic neoplasms involving the spinal cord. Neurology 9:91–106
2. Black P (1985) Spinal epidural tumors. In: Wilkins RH, Rengachary SS (eds) Neurosurgery, vol 1. McGraw-Hill, New York, pp 1062–1069
3. Brenner A, Roosen K, Weiss H, Grote W (1984) Posterior stabilization of the vertebral column in cases of spinal metastases. Adv Neurosurg 12:51–57
4. Brenner A, Roosen K, Wissing HJ, Grote W, Schmit-Neuerburg KP (1986) Osteosynthesis with AO plates in the cervical and lumbar regions of the vertebral column in cases of spinal metastases. Adv Neurosurg 14:133–136

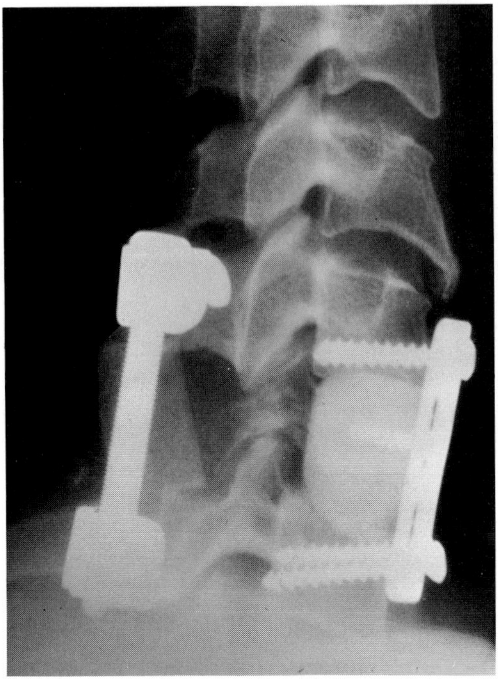

Fig. 3. Ventral plate stabilization and dorsal spondylodesis by means of compression clamps

5. Gilbert RW, Kim JH, Posner JB (1978) Epidural spinal cord compression from metastatic tumor: diagnosis and treatment. Ann Neurol 3:40–51
6. Heipertz W, Fohler N, Schmitt E (1974) Stabilisierung der Wirbelsäule durch Spondylodese bei Tumoren. Z Orthop 112:851–853
7. Herrmann HD (1976) Partieller Wirbelkörperersatz und Stabilisierungsmaßnahmen nach Exstirpation von Wirbelsäulentumoren. In: Schiefer W, Wieck HH (eds) Spinale raumfordernde Prozesse. Perimed, Erlangen, pp 249–256
8. Kalff R, Roosen K, Kocks W, Grote W, Schmit-Neuerburg KP (1988) Operative Stabilisierung bei Metastasen der Halswirbelsäule. Tumor Diagnostik und Therapie 9:142–148
9. Kalff R, Roosen K, Kocks W, Grote W (1988) Surgical treatment in cervical spine metastases. J Cancer Res Clin Oncol 114 (Suppl):130
10. Klinger M, Reitmeier H (1984) Should patients with extradural spinal tumors undergo operations? Adv Neurosurg 12:47–50
11. Kollmann G, Diemath HE, Strohecker J, Spatz H (1984) Spinal metastases as the first manifestation. Adv Neurosurg 12:44–46
12. Livingstone KE, Perrin RG (1978) The neurosurgical management of spinal metastases causing cord and cauda equina compression. J Neurosurg 49:839–843
13. Polster J, Hoeffert H (1974) Grundlagen für die Indikationsstellung zur vorderen und hinteren Spondylodese. Z Orthop 112:753–757

14. Röttgen P, Vlajic I, Schöter I (1976) Ersatz eines Wirbelkörpers durch Palacos. In: Schiefer W, Wieck HH (eds) Spinale raumfordernde Prozesse. Perimed, Erlangen, pp 263–267
15. Roosen K, Trauschel A, Grote W (1982) Posterior atlanto-axial fusion. A new compression clamp for laminar osteosynthesis. Arch Orthop Traumat Surg 100:27–31
16. Roosen K, Grote W, Trauschel A (1983) Dorsale Verklammerungsspondylodese der oberen Halswirbelsäule – ein neues technisches Konzept. Med Orthop Techn 103:18–21
17. Roosen K, Grote W, Trauschel A (1983) Modern treatment of the symptomatic os odontoideum. Neurosurg Rev 6:229–232
18. Roosen K, Grote W, Trauschel A, Kalff R, Castro W (1987) The role of the compression clamp for dynamic posterior fusion in rheumatoid atlanto-axial dislocation. In: Voth D, Glees P (eds) Diseases in the cranio-cervical junction. Walter de Gruyter, Berlin, pp 269–277
19. Schmitt O (1984) Indikation und Ergebnisse ventraler Wirbelsäuleneingriffe. Z Orthop 122:356–364
20. Umbach W, Oppel F, Janusch H (1976) Stabilisierung spinaler Gefügestörungen durch laminierte Endoprothesen. In: Schiefer W, Wieck HH (eds) Spinale raumfordernde Prozesse. Perimed, Erlangen, pp 257–261

On the Symptomatology and Neurosurgical Treatment of Senile Ankylosing Hyperostosis of the Spine (Forestier)

W.-D. Siedschlag, G. Nisch, and H.P. Molsen[1]

Spondylosis hyperostotica as an isolated ossification of the anterior longitudinal ligament of the vertebral column was first mentioned in 1824 by Wenzel and was described by Forestier and Rotes-Querol as an independent disease in 1950 [1]. Spondylosis hyperostotica occurs mainly in male patients during middle age. Ossification of the anterior longitudinal ligament in the cervical region is very rare, and for this reason we wish to describe some observations of our own:

1. A 56-year-old man had been suffering from dysphagia and dyspnea for 1 year. Twenty-six years previously he had had an accident involving the cervical vertebral column. Clinically we found somewhat reduced mobility in the cervical region. Neurologically we found a normal status. The indication for the operation was the severe dysphagia. The X-ray films of the cervical vertebral column showed an extended position with extreme ventral ossification from C2 to C4 and a diameter of the lumen of the trachea of 5 mm. Cervical myelography yielded no pathological findings. During surgery to remove the exostosis, we found marked degeneration of the disk C3/4. This disk was removed, and fusion after Cloward was performed in this region (Figs. 1–3).

2. A 59-year-old man had been suffering from a worsening cervicomedullary syndrome for 2 years. The X-ray films of the cervical vertebral column demonstrated marked ossification of the anterior ligament from C3 to the upper thoracic region. In the region of C5/6 there was a dorsal osteophyte with narrowing of the spinal canal. We performed a laminectomy from C5 to C7 and attained a remarkable improvement.

For the diagnosis of this disease, we favor the following steps: After an exact neurologic examination we take X-rays in two positions. In cases of extreme spondylophytes, X-ray examination of the esophagus with the aid of contrast medium should be performed. Myelography is no longer employed in our clinic. We now prefer CT examination, which makes possible very precise measurement of the transverse section of the vertebral canal.

[1] Neurochirurgische Klinik im Städtischen Klinikum Berlin-Buch, Karower Str. 11, DDR–1115 Berlin

Advances in Neurosurgery, Vol. 18
K.-A. Bushe, M. Brock, and M. Klinger (Eds.)

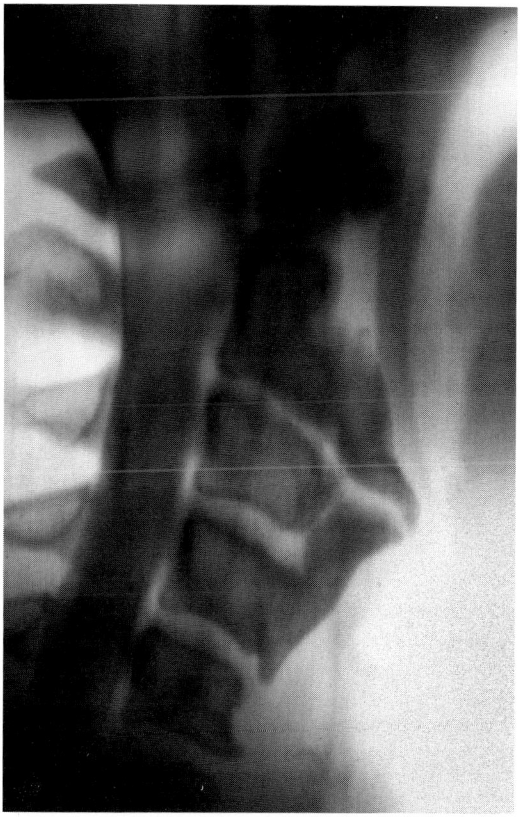

Fig. 1. Spondylosis hyperostotica in a 56-year-old man; myelography

Treatment of spondylosis hyperostotica in the cervical region has two aims:

1. To remove the spondylophytes if they compress the esophagus or trachea
2. To achieve stabilization of the vertebral column. This is possible by fusion after Cloward or, in cases of fracture, by osteosynthesis with metal pates.

Within the last 10 years we have seen four male patients with spondylosis hyperostotica. All were more than 50 years old; three of them suffered from diabetes mellitus and from hypertension. In two cases medullary symptoms predominated, twice respiratory embarrassment was the first symptom, and twice there was severe trauma of the cervical region in the anamnesis.

In all four cases symptomatology was reduced by our treatment.

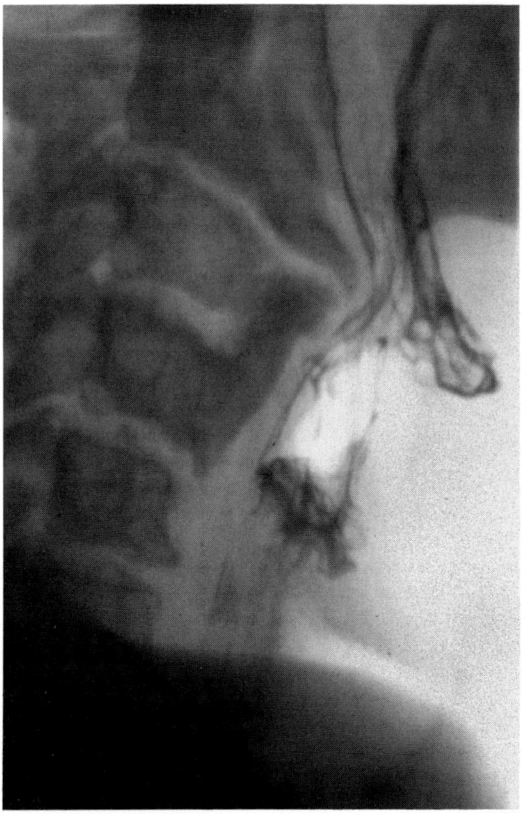

Fig. 2. Extreme stenosis of esophagus and trachea

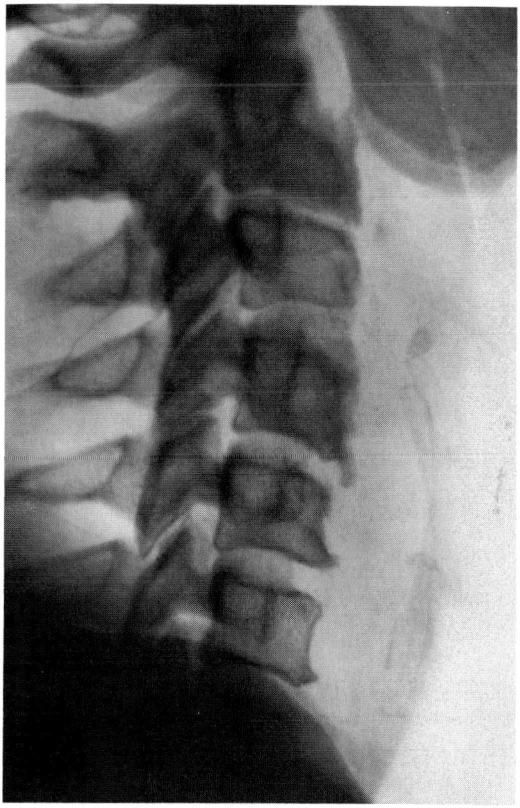

Fig. 3. Situation after removal of the exostosis and fusion in the region C3/4

References

1. Forestier J, Rotes-Querol J (1950) Senile ankylosing hyperostosis of the spine. Ann Rheum Dis 9:321–330
2. Grosch G, Schulz W (1961) Zur Kenntnis der senilen ankylosierenden Hyperostose der Wirbelsäule. Med Welt 48:2533–2535
3. Knauer S, Molsen HP, Meier U, Siedschlag W-D, Knopf W, Nisch G (1988) Halswirbelsäulenkomplikationen bei der Spondylosis hyperostotica (Morbus Forestier) aus neurochirurgischer Sicht. Zentralbl Neurochir 49:93–100

Cervical Spine Fractures in Ankylosing Spondylitis – An Imperative Indication for Surgical Spondylodesis

W. Kocks, R. Kalff, K.-P. Schmit-Neuerburg, J. Pospiech,
W. Grote, and M. Schax[1]

Introduction

Ankylosing spondylitis is a chronic and usually progressive inflammatory disease of joints and connective tissues [3]. The disease predominantly affects young men, with a sex ratio of 10:1. Hereditary factors play an important role in the development of ankylosing spondylitis, as indicated by the presence of HLA-B27 antigen in 95% of spondylitic patients [3, 4].

Histopathological changes such as cartilage destruction and later fibrosis and ossification affect the intervertebral disks, the paravertebral connective tissues, the diarthrodial joints of the spine, the sacroiliac joints, and in 30% of the patients, the hip, knee, and shoulder.

Ankylosis of the vertebral column changes its biomechanical properties and, alongside osteopenia and kyphosis, is one of the causes of the extreme susceptibility of these patients to vertebral injuries [2].

Patients and Methods

Particulars of cervical fractures and their therapy in spondylitic patients are deducible from the patients' data (Table 1). All patients had only minor trauma, but generally a long delay to diagnosis and treatment of the fracture (mean duration: 164 days). The lower cervical segments are predominantly affected. Fractures and dislocations are located typically at the level of an ossified disk (Fig. 1) and usually affect the whole vertebra (ventral and dorsal parts).

Two patients had minor neurological deficits, while four patients had incomplete para- (tetra-)paresis, one of whom was unable to walk. One patient had immediate tetraplegia.

Six of the seven patients underwent external traction therapy using the Crutchfield device, but in two cases repositioning had to be performed by facetectomy.

[1] Neurochirurgische Klinik, Universitätsklinikum der Gesamthochschule Essen, Hufelandstr. 55, D-4300 Essen 1

Advances in Neurosurgery, Vol. 18
K.-A. Bushe, M. Brock, and M. Klinger (Eds.)

Table 1. Patients' data

Age Sex	Duration of ank. spond.	Trauma	Duration Trauma – Op.	Localization	Ext. traction	Surgical procedures	Neurological status pre-op.	Neurological status outcome
43 y man	22 y	Fall while walking	13 m	C4	–	Ventral plate C3–5 Laminectomy C4 + 5	III	I
43 y man	12 y	Fall from bicycle	3 m	C6/7	7 d	Ventral plate C6–7 Facetectomy	III	I
30 y man	5 y	Fall down stairs	2 d	C4/5	1 d	Ventral plate C4–5	V	IV
64 y man	43 y	Mild traffic accident	21 m	C6/7	1 d	Ventral plate C6–T1 Laminectomy C6 Compression clamp	III	III
42 y man	10 y	Fall	15 d	C6/7	1 d	Ventral plate C6–T1 Compression clamp	II	I
64 y man	30 y	Fall	19 d	C6/7	1 d	Ventral plate C6–T1 Compression clamp	II	I
57 y man	10 y	Fall	1 d	C6/7	1 d	Ventral plate C6–T1 Facetectomy, Laminectomy C6 Compression clamp	IV	IV

Only in the case of complete tetraparesis was ventral spondylodesis in combination with external fixation sufficient. In all other cases combined ventral and dorsal operation was necessary, for decompressive laminectomy, facetectomy, or global spondylodesis (Fig. 2).

Results

That such extensive surgical procedures were indicated is supported by the good neurological outcomes (Table 2). Five patients showed improvement of their neurological deficits, while two patients remained clinically unchanged. All had immediate relief of pain. Two patients who preoperatively were unable to look in a horizontal plane benefited from correct anatomical repositioning.

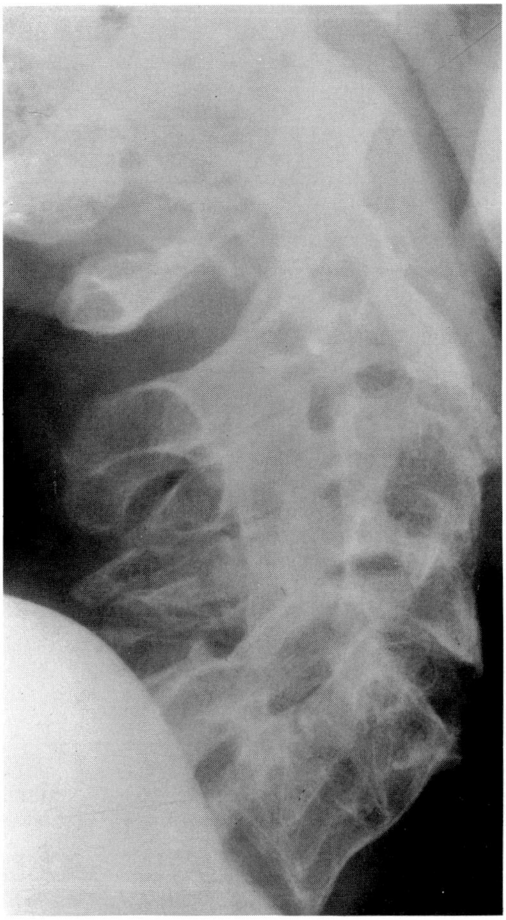

Fig. 1. Fracture-dislocation at the disk level C4/5. Note the typical "bamboo spine." The fracture affects the ventral and dorsal parts of the column

Discussion

Patients with ankylosing spondylitis need special attention in respect of vertebral fractures:

1. About 12% of all spondylitic patients incur a vertebral fracture at some time in their lives [7], 57% of which involve severe neurological deficits [5].

2. Spinal fractures in such patients are often diagnosed with remarkable delay [2,4,5,8]. The reasons for this delay are complex: Generally, the fractures are the result of a minor trauma, which normally does not cause any severe damage. The relatively mild force is amplified by the rigid lever arm of the ankylosed neigh-

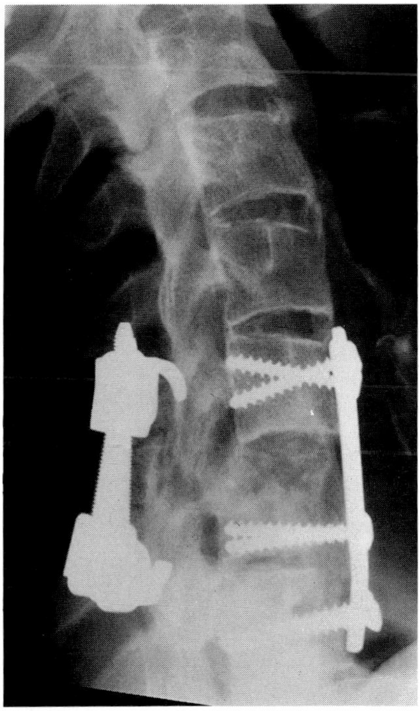

Fig. 2. Ventral and dorsal stabilization is necessary in the case of global instability

boring segments and is sufficient to cause a fracture in the osteopenic spine. The second reason is the predominant localization in the lower cervical segments [2, 5]; such fractures are often not well-visualized on radiographs or are misinterpreted because of distorted anatomy and osteopenia [2].

3. Healing of the injured segment in a dislocated position not only causes late neurological disturbances but also can reduce the quality of life by rendering the patient unable to look in a horizontal plane. Repositioning of the dislocated segments must be achieved, either by careful external traction or by facetectomy if necessary.

4. A high degree of instability is caused by the ankylosis of the adjacent segments (lever arm) and the fact that vertebral injury almost always involves the whole vertebra. Spondylodesis therefore has to be performed at the ventral as well as the dorsal parts of the injured segments [1, 6]. Facetectomies and decompressive laminectomies must also be performed if necessary. The surgical procedures are extensive but, as shown by our results, rewarding.

Table 2. Neurological course

Neurological grading		Preop.	Outcome
I	Normal		4[a]
II	Radicular deficits	2	
III	Incomplete paraparesis – able to walk	3	1[b]
IV	Incomplete paraparesis – unable to walk	1	2[c]
V	Complete paraparesis	1	

[a] Two preoperatively graded as II, two as III
[b] Preoperatively graded as III
[c] Preoperatively graded as IV and V

References

1. Böhler J (1977) Operative Behandlung instabiler Frakturen und Luxationsfrakturen der Halswirbelsäule. Unfallchirurgie 3:25–31
2. Broom MJ, Raycroft JF (1988) Complications of fractures of the cervical spine in ankylosing spondylitis. Spine 13:763–766
3. Gilliland BC, Mannik M (1977) Ankylosing spondylitis. In: Thorn GW (ed) Principles of internal medicine. McGraw-Hill, Kogakusha, Tokyo, pp 2059–2061
4. Hunter T, Dubo H (1978) Spinal fractures complicating ankylosing spondylitis. Ann Intern Med 88:546–549
5. Murray GC, Persellin PH (1981) Cervical fracture complicating ankylosing spondylitis. Am J Med 70:1033–1041
6. Roosen K, Wissing H, Grote W, Schmit-Neuerburg KP (1985) Operative Stabilisierung von Halswirbelsäulen-Frakturen bei Morbus Bechterew. Unfallchirurg 88:215–218
7. Weinstein PR, Karpmann RR, Gall EP, Pitt M (1982) Spinal cord injury, spinal fracture and spinal stenosis in ankylosing spondylitis. J Neurosurg 57:609–616
8. Young JS, Cheshire DJE, Pierce JA, Vivian JM (1977) Cervical ankylosis with acute spinal cord injury. Paraplegia 15:133–146

Primary Autologous Bone Transplantation in the Treatment of Cervical Spondylitis

D. Scale and E. Schmitt[1]

Although as recently as 1962 Max Lange warned against the operative removal of spondylitic foci of the cervical spine [2], this approach is today a routine procedure. The aims of the operation are the removal of diseased tissue resulting from the inflammatory processes, the procurement of suitable biopsy specimens with the advantage of being able to differentiate the spondylitis histologically, bacteriological analysis of smears, and the implant-free establishment of primary stability by an autologous bone transplant.

Materials and Methods

Twenty-six patients with cervical spondylitis were operated upon between 1971 and 1988. We chose left sided anteromedial access with ventral removal of foci and insertion of a corticospongiosa transplant taken from the pelvis. In five cases the operation was restricted to an exploratory excision since the bone substance of the affected segment in these patients proved to be firm.

A Minerva surgical plaster cast or plaster of Paris-strengthened Schanz bandage was applied postoperatively for 5–8 weeks in nonspecific spondylitis and for 8–12 weeks in specific spondylitis, in accordance with the intraoperative stability. Time of removal of the plaster cast and the duration for which the final head-neck orthotic device was carried depended on X-ray, clinical, and laboratory parameters. Antibiotic therapy, given for up to 12 weeks in patients with nonspecific spondylitis and for up to 18 months in patients with tuberculous spondylitis, was carried out according to principles of treatment of osteomyelitis. The operated patients, 14 female and 12 male, had an average age of 56.2 years. A graphic analysis of the age distribution at the time of operation (Fig. 1) shows that the risk of contracting the disease increases with increasing age. The frequency distribution of spondylitis in regard to the different segments of the cervical spine involved (Fig. 2) shows that there is a tendency for the lower vertebrae to be more readily affected. In the

[1] Orthopädische Universitätsklinik Friedrichsheim, Marienburger Str. 2,
D–6000 Frankfurt 71

Advances in Neurosurgery, Vol. 18
K.-A. Bushe, M. Brock, and M. Klinger (Eds.)

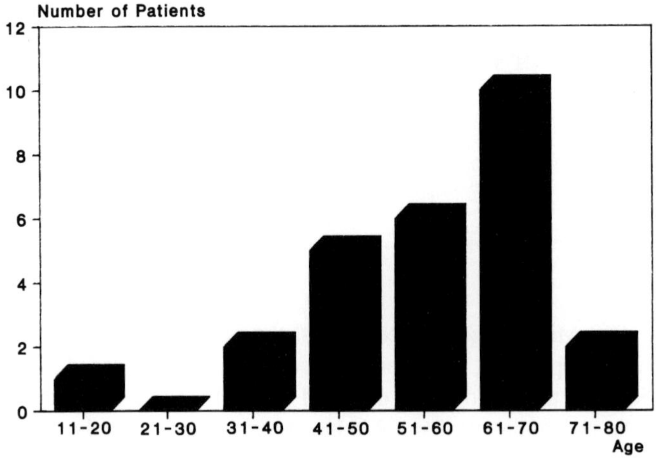

Fig. 1. Graphic analysis of age distribution at time of operation

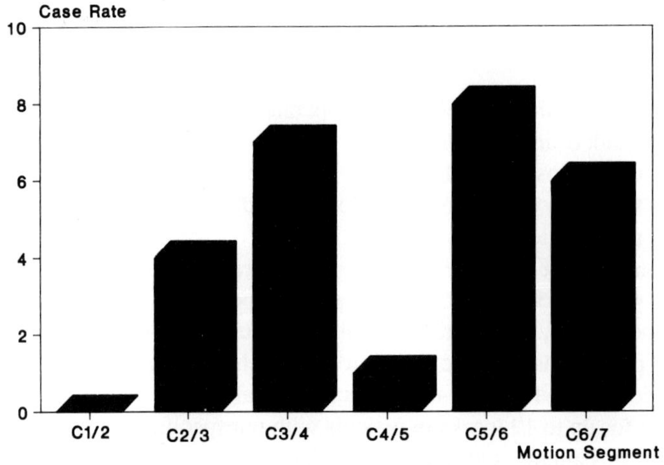

Fig. 2. Localization and incidence of cervical osteomyelitis

four cases with multisegmental spondylitic changes the classification was based on the segment involved in the primary disease.

Results

The most impressive results of the operation, occurring in 12 patients and seven patients respectively, were the disappearance within a few days of local pain, in most cases severe, and of nerve root stimulation. Furthermore, in five of six patients with root compression symptoms and loss of sensory and/or motor responses there was a full recovery, as was the case in a patient with preoperative paraplegia due to polysegmental involvement of the cervical spine in whom it was necessary to carry out a spondylodesis on C2 to T1. A no less important gain from the operation was being able to obtain representative tissue biopsy specimens. The histological differentiation of spondylitis carried out (Table 1) was important in the subsequent therapeutic strategy for two reasons:

1. Antibiotic therapy was confined to patients having suppurative subacute and chronic spondylitis; it was not given in nonpurulent, above all noninfected cases, where there was no evidence at the cellular level of inflammatory changes.

2. A change in treatment was carried out in four patients when it was found necessary to alter the initial, preoperative suspected diagnosis. Two patients with a suspected primary tumor had tuberculosis. A case of suspected nonspecific spondylitis with corresponding osteolytic changes proved to be a mixed infection tuberculosis. In a further patient a suspected diagnosis of primary tuberculosis based upon isolated narrowing of the disk space and a highly positive tine test was shown to be nonspecific spondylitis.

Bacteriological investigations were less useful. Evidence for the presence of pathogens was found in only 45.5% of the cases investigated in this way (*Staph. aureus* 3×, *Mycobacterium tuberculosis* 1×, *Pseudomonas aeruginosa* 1×), but these findings enabled targeted antibiotic therapy to be carried out.

Whereas a good postural correction could be attained when the autologous graft was inserted next to firm bone in the spondylitic segments, some loss of this correction occurred later. With an average preoperative kyphosis of 10° (min. 0°, max. 20°), an average intraoperative correction (Fig. 3) of 4° could be obtained (min. 0°, max. 15°). Following bone fusion and incorporation of the graft, however, the average kyphotic angle was 8° (min. 0°, max. 15°). This indicates that an average intraoperative improvement of 6° (min. 0°, max. 20°) was followed by an average postoperative correction loss of 4° (min. 0°, max. 12°). This correction loss, insofar as it can be estimated on the basis of our radiographic records, occurred within the first 3 postoperative weeks. Two factors were responsible for these observations:

Table 1. Results of histological investigation

Cervical osteomyelitis		n
Tuberculous		5
Bacterial		21
Subacute/chronic (purulent)	9	
Chronic	7	
Bland	5	

1. The cervical column in the plaster cast was not always in the fully erect position.
2. Complete immobilization, even using a Minerva plaster cast, could not be achieved [4].

Residual movement in the cast and faulty erection of the cervical spine result in increased mechanical loads on the graft and graft site, causing absorption processes and a loss of correction. In view of this fact it is important to achieve the best possible immobilization of the cervical spine and in the right position. According to present-day practices this is best guaranteed through use of a halo body splint.

The complication rate was low, with one case of recurrent paresis (a case involving a fracture graft) and one fistula requiring attention.

The operative treatment of five cases in which there was no clear histological evidence of spondylitis raises the question of present-day differential indications

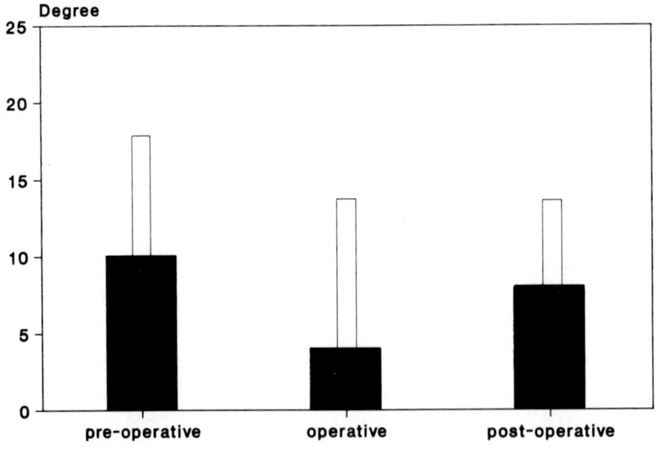

Fig. 3. Average degree of segmental kyphosis before, during, and after operation (minimum and maximum indicated)

in the treatment of spondylitis. The presence or worsening of neurological symptoms, X-ray or CT findings of instability due to excessive bone destruction, cases of abscess in which there is a need to prevent multisegmental spondylitis, problems associated with the differential diagnosis, and patients who fail to respond to conservative therapy represent absolute indications for operative treatment. We conservatively treat those patients with a high operative risk, e.g., patients with spondylitic syndromes of rheumatic origin and patients who show an improvement of the inflammation with radiographic evidence of supportive reactions following a short period of immobilization. Since inflammation parameters are unreliable for assessing local inflammatory activity [1], we tend, as a general rule, to set our indications for operative intervention relatively widely. There is also the potential advantage of achieving histological and bacteriological differentiation of the spondylitis and a reduction in the duration of treatment, especially in regard to tuberculous spondylitis, as we have been able to show from treatment of the thoracic and lumbar vertebral column [3].

Summary

The operative removal of spondylitis with subsequent primary stabilization through autologous bone grafts has a number of advantages: rapid improvement of the subjective and objective clinical symptoms, shortening of the healing process, at least in regard to tuberculous spondylitis, the possibility of histological and bacteriological differentiation of the spondylitis, and application of the findings in therapy. A postoperative loss, on average 4°, in the initial angle of correction of the vertebral locomotor segments must be taken into account.

References

1. Enderle A, Schmitt E, Zichner L (1981) Zur Diagnostik herdförmiger Wirbelsäulenerkrankungen – eine kritische Betrachtung. Z Orthop 119:193–226
2. Lange M (1962) Orthopädisch-Chirurgische Operationslehre. J.F. Bergmann, Munich, pp 197–198
3. Scale D, Starker M, Schmitt E (1988) Differentialindikation in der Spondylitisbehandlung. Vortrag 74. Tagung der DGOT in Saarbrücken
4. Schmitt E (1981) Ruhigstellung der Bogengelenke. Die Wirbelsäule in Forschung und Praxis 87:145–148

Treatment of Craniocervical Instabilities with the Halo-fixateur

R. Schultheiß, A. v. Stockert, and D.-K. Böker[1]

Introduction

The treatment of craniocerebral instabilities has become more and more an operative one. The newer osteosynthetic techniques have proved to be very effective, so conservative treatment seems to assume a minor role. On the other hand, there might be individual or local reasons for considering conservative treatment of a craniocervical lesion. The halo-fixateur externe, first described by Perry and Nickel [14], has proved to be the most effective device for ensuring maximum immobilization [11, 15], reducing flexion/extension and lateral bending to 4% and rotation to 1% of the normal motion between the occiput and first thoracic vertebra [10]. Restriction of rotational movement is the most important aim of external fixation considering the possible range of motion in the atlantoaxial joint [16]. The mean flexion/extension allowed between C1 and C2 is 3.4° [9] with a halo/body jacket. We reviewed our patients treated with a halo in order to evaluate its clinical value.

Material and Methods

Between 1980 and 1989 we treated 12 patients with a halo (body jacket type) at the Neurosurgical University Clinic, Bonn – four women (aged 18–54) and eight men (aged 24–90). Five patients needed external fixation after operative procedures performed on cervical fractures failed to achieve immediate stabilization. At that time, cervical fusions had to be performed without anterior osteosynthetic plate fixation. In three cases infectious complications after anterior fusions because of cervical disk disease led to a halo treatment. Three odontoid fractures and one case of atlantoaxial instability represented our group of craniocervical unstable lesions treated exclusively conservatively, all of them being neurologically without pathological findings.

The mean duration of halo application was 85 days (17 thereof in the hospital), the duration being shortest for the "adjuvant" therapy in operated dislocation-

[1] Neurochirurgische Universitätsklinik, Sigmund-Freud-Str. 25, D–5300 Bonn 1

Advances in Neurosurgery, Vol. 18
K.-A. Bushe, M. Brock, and M. Klinger (Eds.)

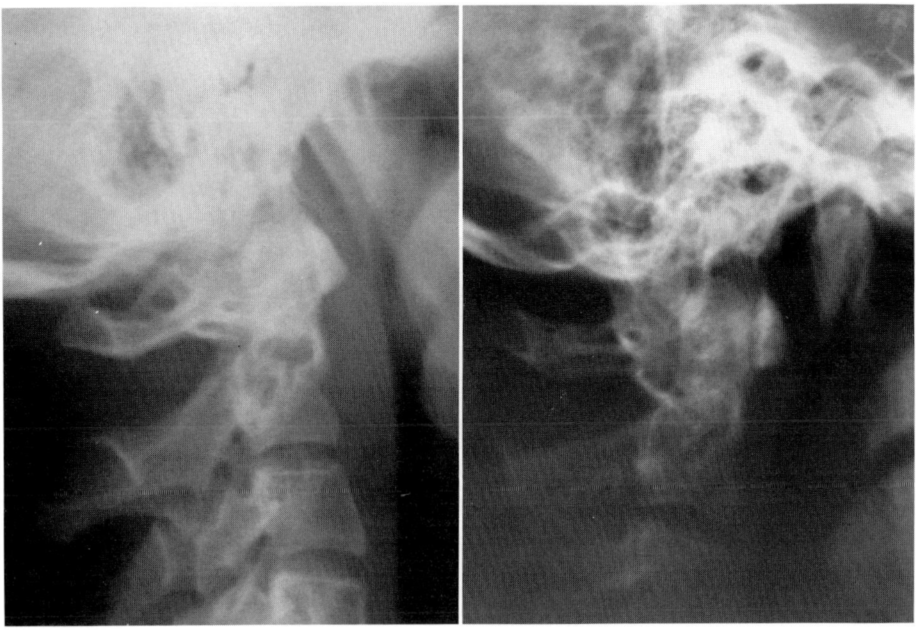

Fig. 1. Displaced odontoid fracture in a 24-year-old man before (*left*) and 2 months after (*right*) halo treatment for 170 days with good stability

fractures (61 ± 9 days) and longest in the odontoid fractures (141 ± 14 days, leaving out one death 2 weeks after the injury).

Results

Of the patients treated for an instability of the middle and lower cervical spine, a 78-year-old man suffering from a fracture of C5/6 died of pneumonia. Six others showed perfect stability after removal of the orthoses; only in one case did a luxation of C3/4 persist, requiring dorsal wiring.

In two patients two fixation pins had to be renewed because they became loose and were assumed to be starting to perforate the tabula interna of the skull. One patient missed her last regular checkup for control of the pins and reappeared after 5 weeks' treatment with a totally displaced halo ring which could be removed, as stability had already been achieved.

A 90-year-old man died of pneumonia and cardiac insufficiency 8 days after application of the halo, and 16 days after an injury leaving him with a type III [2] odontoid fracture.

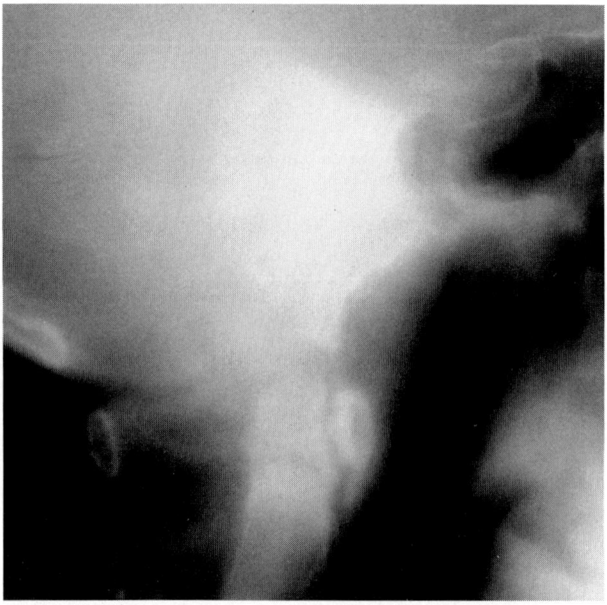

Fig. 2. Odontoid fracture in a 46-year-old man 1 year after halo treatment for 113 days with complete bony fusion in optimal position (cf. Fig. 3)

A 24-year-old man who suffered from a type II fracture after a car accident was treated for 170 days. The fracture became stable in a functionally good position (Fig. 1).

A 46-year-old man with a type II fracture after a bicycle accident remained in the halo jacket for 113 days. The fracture united in a normal position (Fig. 2). Initial displacement could easily be reduced by fine adjustment of the device (Fig. 3).

A 54-year-old female patient developed atlantoaxial instability reminiscent of atlantoaxial dislocation in rheumatoid arthritis [3, 6, 8, 12, 13]. She had undergone curettage and developed abdominal symptoms with temperature, pain, and leukocytosis after 5 days. As these complaints improved, a severe headache developed from the 9th day and persisted at an extreme level. X-rays of the craniovertebral junction showed atlantoaxial dislocation and anterior and posterior defects of the dens (Fig. 4). No serological evidence of rheumatoid arthritis could be detected at that time or later, and no other joints showed similar inflammatory signs. A halo was applied as a first measure and pain subsided immediately. The patient refused operative intervention; the halo was removed after 72 days. No clinical recurrence of the severe pain or radiological deterioration occurred during a follow-up period of 3.5 years to date.

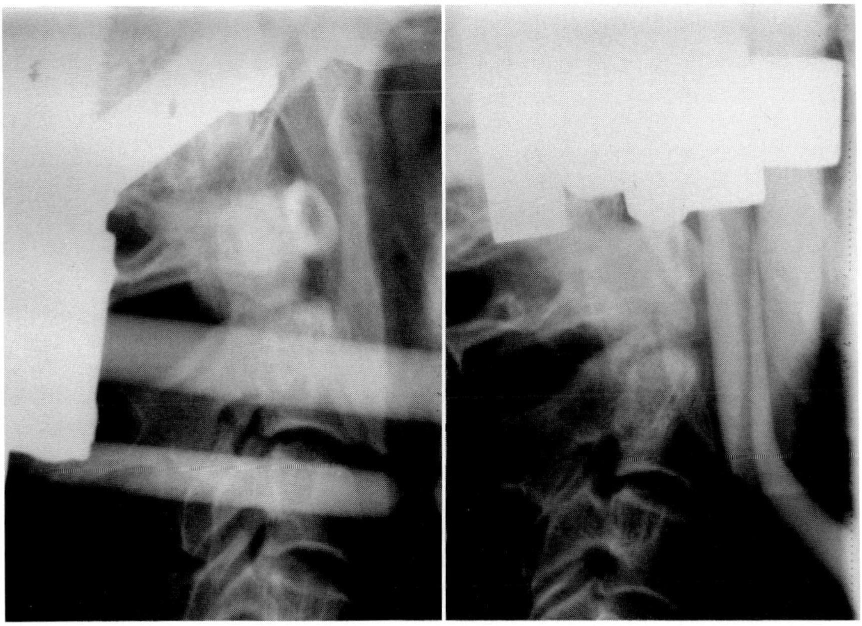

Fig. 3. Odontoid fracture fixed in a halo vest with displacement before (*left*) and correct position after (*right*) readjustment of the brace

Discussion

In a survey of the literature, Althoff and Bardholm [1] give a 31% incidence of nonunions in the conservative treatment of odontoid fractures, i.e., 168 of 535 cases. Anderson and D'Alonzo [2] analyzed their failures in halo treatment and reported good results for type I and III lesions, with only one nonunion out of 15 cases, whereas 8 of 22 type II fractures did not unite. The figures given by Ekong et al. [5] (7/17 nonunions), Ersmark and Kalen [7] (1/16), and Chan et al. [4] (1/15) seem to reflect improving results of conservative treatment. Two of our own cases document the easy applicability of the method with optimal results even in the type II lesions more apt to cause persisting instability. From our experience with these patients and the others treated with a halo we assume that with improved knowledge of the pitfalls of halo mounting, similar results may be achieved regularly.

Periodical control of all fixation pins with readjustment of the torsion force and a protocol of the number of turns and/or depth of pin penetration into the skull is important to secure optimal head fixation and to prevent or immediately detect perforation of the tabula interna.

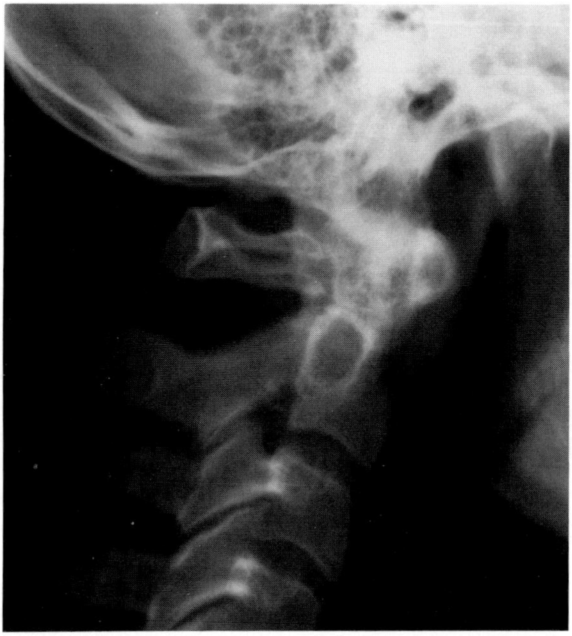

Fig. 4. Atlantoaxial dislocation and partial destruction of the dens in a 54-year-old woman with acute signs of an inflammatory process of the craniovertebral junction. No serological or clinical signs of rheumatoid arthritis were found. Clinical symptoms of severe headache subsided after 72 days of halo treatment. No clinical or radiological deterioration has been observed during the last 3.5 years

As the fur of the body jacket will give way during the first 2 weeks, regular controls and readjustments of the total halo geometry may be necessary to ensure precise positioning of the fractured bone.

Some of the low-profile halo types available neglect simple static principles: axial extension is the main therapeutic principle and is not guaranteed if short struts to the halo ring may bend where they should be secured to the transverse shoulder strut. Likewise, some systems lack a simple counter-screw fixation which allows maintenance of extension while readjusting the head position in an anterior-posterior direction: it may be surprising how many directions of motion are rendered free after the simple loosening of a single bolt. Thus, the use of a simple construction with only four axial struts fixing the ring to the jacket seems to solve many of the practical problems involved by reducing the "fiddling factor" to a minimum and at the same time freeing the neck from the troublesome transverse material that interferes with X-ray controls.

A comparison of conservative and operative treatment of craniocervical lesions today should be based on halo treatment exclusively as the optimal method; the older figures achieved with Minerva casts and other orthoses should be regarded as historical and disregarded as statistical proof. Optimal halo technology and application technique seem to promise a high rate of good results. Conservative treatment can still have a place in craniocervical lesions, even in unusual cases, as in the described atlantoaxial dislocation.

References

1. Althoff B, Bardholm P (1979) Fracture of the odontoid process. A clinical and radiographic study. Acta Orthop Scand [Suppl] 177:61–95
2. Anderson LD, D'Alonzo RT (1974) Fractures of the odontoid process of the axis. J Bone Joint Surg 56A:1663–1674
3. Bland JH (1974) Rheumatoid arthritis of the cervical spine. J Rheumatol 1:319–325
4. Chan RC, Schweigel JF, Thompson GB (1983) Halo-thoracic brace immobilization in 188 patients with acute cervical spine injuries. J Neurosurg 58:508–515
5. Ekong CE, Schwartz MD, Tator CH, Rowen DW, Edmonds VE (1981) Odontoid fracture: management with early mobilization using the halo device. Neurosurgery 9:631–637
6. El-Khoury GY, Wener MH, Menezes AH, Dolan KD, Kathol ME (1980) Cranial settling in rheumatoid arthritis. Radiology 137:637–642
7. Ersmark H, Kalen R (1982) The halo vest in the management of unstable cervical spine lesions. In: Vecsei V (ed) Traumatology of the spinal column. Informatica, Vienna, pp 242–244
8. Gschwend N, Scherer M, Munzinger U (1981) Entzündliche Veränderungen der Wirbelsäule bei der chronischen Polyarthritis. Orthopäde 10:155–168
9. Johnson RM, Hart DL, Simmons EF, Ramsby GR, Southwick WO (1977) Cervical orthoses. A study comparing their effectiveness in restricting cervical motion in normal subjects. J Bone Joint Surg 59A:332
10. Johnson RM, Owen JR, Hart DL, Callahan RA (1981) Cervical orthoses. A guide to their selection and use. Clin Orthop 154:34–45
11. Kock RA, Nickel VL (1978) The halo vest. An evaluation of motion and forces across the neck. Spine 3:103–107
12. Martel W (1961) The occipito-atlanto-axial joints in rheumatoid arthritis and ankylosing spondylitis. AJR 86:223–240
13. Mathews JA (1974) Atlanto-axial subluxation in rheumatoid arthritis. Ann Rheum Dis 33:526–531
14. Perry J, Nickel VL (1959) Total cervical-spine fusion for neck paralysis. J Bone Joint Surg 41A:37–60
15. Podolsky S, Baraff LJ, Simon RR, Hoffman JR, Larmon B, Ablon W (1983) Efficacy of cervical spine immobilization methods. J Trauma 23:461–465
16. White AA, Panjabi MM (1978) The clinical biomechanics of the occipitoatlantoaxial complex. Orthop Clin North Am 9:867–878

The Natural History of Unstable Traumatic Injuries of the Cervical Spine

D.J. Zeilstra, F.B. Niessen, A. ten Brinke, and L. Penning[1]

Survival and rehabilitation of patients after injury of the cervical spine have improved significantly since World War II due to increased interest in these lesions and to improved surgical management. Traumatic lesions of the cervical spine may be classified according to radiological or clinical criteria. The mechanism leading to the injury can also be taken into account [2]. A luxation is defined as loss of contact of articular facets, in the case of the cervical spine (C2–7) leading to uni- or bilateral interlocking of articular facets. Displacement of vertebral bodies is defined as retrolisthesis (backward slipping) or anterolisthesis (forward slipping). Another effect of injury may be local kyphotic angulation of the cervical spine. Instability implies an abnormal mobility in one or more positions [3], also defined as the inability of the cervical spine to prevent neurological damage under physiological conditions [6,7]. Instability is found to occur in 6%–12% of all conservatively treated cervical spine injuries. It is usually the result of a fracture and/or luxation, frequently of a rupture of posterior ligaments [4]. Unstable lesions become stable after a period of 10–12 weeks as a result of scar formation and bony fusion [1]. It is estimated that in 36%–66% of the patients a spontaneous fusion will occur. Indications for operation are:

- Possibility of open reduction
- Damage to posterior elements involving:
- Kyphotic angulation [5]
- Facet interlocking
- Fractures of the articular process
- Compression fractures
- Root symptoms [7]

Further arguments in favor of operative stabilization are: rupture of the intervertebral disk with failure to heal spontaneously and rapid immobilization with quick healing of a stable and painless neck. These arguments are mentioned frequently

[1] Department of Neurosurgery, Sophie Ziekenhuis Zwolle, Dr. van Heesweg 2, 8025 AB Zwolle, The Netherlands

Advances in Neurosurgery, Vol. 18
K.-A. Bushe, M. Brock, and M. Klinger (Eds.)

Table 1. Types of injury

Sprain	47
Sprain with compression	50
Craniocervical region	39
Other injuries	25

and are widely known as such. However, they lack clear proof and documentation [1]. There seems to have been a tendency in the last decade, with the introduction of new fixation methods, to operate at an early stage. We have investigated, in a retrospective study, the natural history of patients with unstable injuries of the cervical spine, treated conservatively.

Between 1976 and 1986 161 (128 male and 33 female) patients were admitted with a cervical spine injury. According to the criteria mentioned above, 94 patients were considered to have an unstable lesion. The types of lesion are given in Table 1. Of these patients, 48 were operated on, and 46 were treated conservatively. Patients with paraplegia were excluded and only injuries below the craniocervical region (C2–7) were considered. Twenty-two patients (18 male and 4 female) with a fully documented case history and follow-up were included in the study. The average age of these patients was 42 years. In 17 cases the cause was a traffic accident. Therapy consisted of traction (8 cases), a halo jacket (2 cases), or a cervical collar (12 cases). The patients were all seen on an outpatient basis during which the history was taken and a neurological examination was performed. All patients were examined radiologically, including flexion and extension films. On these radiographs the angles between two adjacent vertebrae were measured in both flexion and extension.

Before the accident four patients were no longer working and receiving compensation. Of the total number of 22 patients, 16 were not restricted in any way in their daily activities, including work. After the accident 40% of the patients complained of headache or pain in the neck, 33% of reduced mobility of the neck, and 27% of loss of sensation in the fingers. The average period of immobilization was 4.2 months (1–8 months). When asked how long it took the patient to reach a situation in which no further alteration of his or her condition occurred, this period varied from 0 to 48 months (average 11 months). The subjective result was considered good or very good by 64%, moderate by 9%, and poor or very poor by 27% (Table 2).

On neurological examination there was reduced mobility of the cervical spine (rotation less than 45° in either direction) in 36%. Nine percent had motor and 23% sensory deficit of the arms. Pain was still experienced by 36% of the patients. These figures may be compared with an operative series yielding similar results [1].

Table 2. Results

	Objective	Subjective
Very good (no complaints)	8	8
Good (minor complaints)	9	6
Moderate (pain, impaired in some activities)	3	2
Poor (much pain, unable to do most activities)	2	4
Very poor (completely disabled)	0	2

Table 3. Mobility of the cervical spine

	Injured segment	Adjacent segments
(Almost) None	8	6
Reduced (< 5°)	7	11
Normal (5°)	4	2
Increased (> 5°)	0	0

The radiological investigations showed loss of mobility in the injured segment in all cases. Furthermore there was a reduction of the mobility in adjacent segments (Table 3). There was no case of late instability.

Although the patients with injuries of the cervical spine were not randomized as to treatment in this study, it may be said that the final results are favorable, especially when considering the radiological criteria. There was no late instability, and almost two-thirds of the patients considered the result as "good" or "very good." The period of immobilization, on the other hand (average 4.2 months), is long in comparison to operative treatment with modern fixation methods. The same holds true for the period required to achieve a final situation (11 months). In our opinion operative treatment appears to lead to a quicker but not per se to a better result. In those situations in which conservative treatment for whichever reason is chosen, it seems to be an adequate therapy.

References

1. Aebi M, Mohler J, Zäch GA (1986) Indication, surgical technique, and results of 100 surgically-treated fractures and fracture-dislocations of the cervical spine. Clin Orthop 203:244–257

2. Braakman R, Penning L (1976) Injuries of the cervical spine. In: Vinken PJ, Bruyn GW (eds) Handbook of neurology, vol 25. North Holland, Amsterdam, pp 227–380
3. Cheshire DJE (1969) The stability of the cervical spine following the conservative treatment of fractures and fracture-dislocations. Paraplegia 7:193–203
4. Donovan WH, Kopaniky D, Stolzmann E, Carter ER (1987) The neurological and skeletal outcome in patients with closed cervical spinal cord injury. J Neurosurgery 66:690–694
5. Savini R, Parisini P, Cervellati S (1987) The surgical treatment of late instability of flexion-rotation injuries in the lower cervical spine. Spine 12/2:178–182
6. White AA, Southwick WO, Panjabi MM (1976) Clinical instability in the lower cervical spine. A review of past and current concepts. Spine 1:15–27
7. White AA, Panjabi MM (1984) The role of stabilization in the treatment of cervical spine injuries. Spine 9/5:512–522

Syringomyelia

Evolution of the Neurosurgical Management of Syringomyelia in the Last Two Decades

R. Van den Bergh[1]

The neurosurgical management of syringomyelia has undergone very important evolution in the last two decades, obviously in line with the development of our knowledge regarding its pathogenesis and with the introduction of new diagnostic procedures.

Evolution of Theories on Pathogenesis

In the early 1960s syringomyelia was considered mainly as a segmental congenital dysraphic anomaly of the spinal cord.

In 1965 Gardner [5] directed attention to the posterior fossa. He postulated that failure of normal perforation of the rhombic roof led to persistence of the central canal of the spinal cord in communication with the ventricular system. The arterial pulsations of the choroid plexus were supposed to give rise to dilatation of the central canal, eventually leading to hydromyelia and syringomyelia. Williams [15, 16] believed that the force which provokes the dilatation of the central canal originates from venous pressure waves.

In the 1970s, as a result of new neuroradiological techniques and growing neurosurgical experience, it became clear that Gardner's ideas did not match reality in many cases. Ball and Dayan [2] and later Aboulker [1] proposed an alternative explanation of the pathogenesis of syringomyelia. They believed that syringomyelic cavities are filled from the perimedullary subarachnoid space, either through small clefts and breaches along the dorsal roots or via the perivascular Virchow-Robin spaces.

In the meantime more and more cases of posttraumatic syringomyelia were published. This form of syringomyelia is almost exclusively a complication of thoracic or lumbar spinal cord injury and can only be explained by penetration of CSF into the injured cord substance below a block of the subarachnoid space.

[1] Department of Neurology and Neurosurgery, Catholic University of Leuven, Leuven, Belgium

Advances in Neurosurgery, Vol. 18
K.-A. Bushe, M. Brock, and M. Klinger (Eds.)

Evolution of Diagnostic Procedures

As long as syringomyelia was considered a segmental dysraphic anomaly, diagnostic exploration was limited to plain skull radiography, spine radiography, and positive contrast myelography.

As soon as the importance of CSF hydrodynamics was known, air myelography in different positions and iodoventriculography were performed in order to visualize collapsing cysts, to assess the posterior fossa relations, and to study the connection of the syrinx with the central canal and the ventricular system.

Computerized axial tomography (CAT), introduced in the 1970s, can visualize the syrinx, in communication with the central canal or not, mostly by direct penetration of contrast medium into the cord parenchyma.

Nuclear magnetic resonance (NMR) imaging, performed since 1983, has surpassed all other examination techniques. It permits direct and complete visualization of the syrinx and has demonstrated important facts: syrinxes can occur in patients without syringomyelia syndrome; there are several distinct spinal cord cavitations of different pathogenesis and with different prognosis; and the syrinx volume is not proportional to the severity of the symptoms and signs.

Evolution of Neurosurgical Procedures

The first neurosurgical treatment of syringomyelia was myelotomy, introduced by Elsberg [4] and Puusepp [12] and consisting in syringostomy, fistulization, and marsupialization.

After the publication by Gardner of a series of 74 cases, posterior fossa operations, which previously were applied mostly in arachnoiditis and in Arnold-Chiari malformation, were employed by Newton [9], Conway [3], and Hankinson [7]. These operations consisted basically in opening the fourth ventricle roof and in removing adherent membranes.

Because of the mortality and morbidity and the rather frequent recurrences after posterior fossa surgery, other types of operation were soon proposed. Gardner himself switched to terminal ventriculostomy [6], which is a variant of myelotomy with excision of the filum terminale and of the tip of the conus medullaris. Many indirect palliative procedures were introduced, e.g., the ventriculocardial [8], the ventriculoperitoneal, and the lumboperitoneal shunt. The former two make sense only in real communicating syringomyelia and preferably in the presence of hydrocephalus; the latter is indicated if there is a block of the spinal CSF compartment with pressure peaks caused by coughing, straining, etc.

Since 1980 the direct approach to the syrinx has come into vogue again. Taking advantage of new inert materials, some authors have returned to the use of one or other variation of syringostomy, for example the syringosubarachnoid shunt [13].

The syringoperitoneal shunt [10, 11] and more recently the syringopleural shunt have been added to the range of possibilities. The results of syringoperitoneal shunt are excellent, both in general and in our department. In our series this technique has a success rate of 80% without any mortality. The success rate of our posterior fossa operations was less than 50%, with a low mortality and morbidity. The choice of the cord incision for catheter insertion can be a problem: the incision may be on the midline between the dorsal funiculi, or in the dorsal root entry zone, on the left or right depending upon the exact position of the syrinx.

Today the syringoperitoneal shunt is considered the method of choice. We want to stress that, over and above the choice of operative procedure, the essential decision is whether to operate at all. In our series more than one-half of the patients were not operated upon and were treated conservatively. In our opinion formal indications for operation are: obvious progression of symptoms and signs, intractable pain, and signs of craniospinal pressure dissociation by paroxysmal tonsillar herniation. In this last indication a posterior fossa operation remains the procedure of choice [14], possibly followed later by a syringoperitoneal shunt. It has to be stressed that the syrinx visualization as such, by CAT scan or by NMR, is of secondary importance in the decision making.

Conclusion

Both the pathogenesis and the treatment of syringomyelia remain matters of doubt and controversy. In the first place it has to be stressed that in a large group of patients the illness does not show unlimited progression, and that spontaneous stabilization and even improvement can occur.

The syringoperitoneal shunt seems to be the treatment of choice at present. It has to be preceded, however, by posterior fossa decompression in cases of craniospinal pressure dissociation. It can be associated with a ventriculocardial or ventriculoperitoneal shunt if there is accompanying hydrocephalus.

References

1. Aboulker J (1979) La syringomyélie et les liquides intrarachidiens. Neurochirurgie (Suppl) 25:1–44
2. Ball MJ, Dayan AD (1972) Pathogenesis of syringomyelia. Lancet II:799–801
3. Conway LW (1967) Hydrodynamic studies in synringomyelia. J Neurosurg 27:501–514
4. Elsberg GA (1916) Diagnosis and treatment of surgical diseases of the spinal cord and its membranes. Saunders, London
5. Gardner W (1965) Hydrodynamic mechanism of syringomyelia; its relationship to myelocele. J Neurol Neurosurg Psychiatr 28:247–259
6. Gardner WJ, Bell H, Poolos P, Dohn D, Steinberg M (1977) Terminal ventriculostomy for syringomyelia. J Neurosurg 46:609–617

7. Hankinson J (1970) Syringomyelia and the surgeon. Mod Trends Neurol 5:127–148
8. Krayenbühl H, Benini A (1971) A new surgical approach in the treatment of hydromyelia and syringomyelia. The embryologic basis and the first results. J R Coll Surg Edinb 16:147–161
9. Newton EJ (1979) Syringomyelia as a manifestation of defective fourth ventricular drainage. Ann R Coll Surg Engl 44:194–213
10. Pecker J, Javalet A, Boutelis A (1983) La dérivation syringo-péritonéale. Neurochirurgie 29:171–173
11. Phillips TW, Kindt GW (1981) Syringoperitoneal shunt for syringomyelia: a preliminary report. Surg Neurol 16:462–466
12. Puusepp M (1926) Traitement opératoire dans deux cas de syringomyélie avec amélioration notable. Rev Neurol (Paris) 1:1171–1179
13. Tator C, Meguro K, Rowed D (1982) Favorable results with syringosubarachnoid shunts for treatment of syringomyelia. J Neurosurg 56:517–523
14. Van den Bergh R (1987) Indications for posterior fossa surgery in syringomyelia. In: Samii M (ed) Surgery in and around the brain stem. Springer, Berlin Heidelberg New York, pp 496–500
15. Williams B (1969) The distending force in the production of "communicating syringomyelia". Lancet II:189–193
16. Williams B (1970) The distending force in the production of "communicating syringomyelia". Lancet II:41–42

Syringomyelia: Aspects of Therapeutic Decisions

M. Westphal, N. Freckmann, D. Winkler, L. Cristante, and H.-D. Herrmann[1]

Introduction

Many more patients present with syringomyelia than will eventually turn out to be good candidates for surgery. Many different surgical procedures have been advocated but no treatment has become standard [2, 3, 5–9, 11–15]. Preoperative diagnostic procedures, however, have greatly improved with the introduction of magnetic resonance tomography (MRT) and gadolinium [1, 10]. Thus the selection of appropriate surgical treatment from the various advocated methods has been facilitated. It has been recognized that patients in whom intramedullary, intracavitary pressure is present will benefit most from draining/shunting procedures. In addition there are patients with progressive disease and intramedullary dissection but without evidence of pressure. Progression of neurological deficits may even occur long after intramedullary pressure has been successfully treated by shunting.

In this paper we report our experiences in a series of 45 patients with intramedullary cystic lesions but no evidence of tumor at the initial time of diagnosis. In these patients syringomyelia as a consequence of spinal trauma was present in 8 cases, nontraumatic, noncommunicating syringomyelia was present in 26, and a syrinx communicating with the fourth ventricle (hydromyelia) was present in 11. These patients were seen and treated between 1984 and 1989, i.e., after the introduction of MRT.

Diagnostic Criteria

For the reasons stated above, in selecting a patient for treatment there must be some evidence of intramedullary pressure. This is most conclusively indicated by obliteration of the subarachnoid space as visualized on MRT. Widening of the spinal canal on plain X-ray studies is seen only rarely and is an unreliable, often false-negative indicator. Direct puncture of the lesion and measurement of pressure

[1] Neurochirurgische Klinik, Universitätskrankenhaus Eppendorf, Martinistr. 52, D–2000 Hamburg 20

Advances in Neurosurgery, Vol. 18
K.-A. Bushe, M. Brock, and M. Klinger (Eds.)

[5] is also unreliable, is associated with morbidity, and is no longer indicated in the era of MRT. Only myelography in combination with CT scanning may allow some assessment of the subarachnoid space and may be used if MRT is completely unavailable.

To select the appropriate surgical procedure for the patient, it is mandatory to have some indication as to whether the cyst has communication with the fourth ventricle. In some patients this question can be unequivocally answered by MRT. Communication has to be suspected in those patients in whom an Arnold-Chiari malformation is incidentally present [3,4,8,12].

Finally, before the introduction of gadolinium as a contrast agent for MRT, very small angioblastomas, which may be the cause of large noncommunicating cysts, were missed and only found upon later secondary intervention and reinvestigation for progressive symptoms. Two such cases are part of our series. It is therefore mandatory nowadays that every nontraumatic, noncommunicating syrinx is thoroughly investigated by gadolinium-enhanced MRT.

Therapeutic Approaches and Results

Two alternative methods have been used in this series for drainage of the cysts: (a) drainage into the subarachnoid space (syringostomy, syringosubarachnoid shunt, SSS) and (b) drainage into the pleural cavity (syringopleural shunt, SPS). In those cases in which there is clearly a communication of the syrinx with the fourth ventricle, be it with or without accompanying Arnold-Chiari malformation, we now advocate primary plugging of the aperture at the obex with muscle and fibrin glue (Gardner's procedure) (Table 1).

While performing drainage procedures we have routinely included intraoperative filling of the cavity with contrast medium and X-ray examination. Using this approach it is possible to assess septum formation within the cystic lesion, which results in only partial drainage and failure surgical treatment. Most importantly, however, it is possible to assess upward communication, which in some cases of

Table 1. Therapeutic approaches in 45 patients with syringomyelia

SPS:	21 patients
SSS:	16 patients
Gardner's procedure:	8 patients
Revisions:	2, revisions of SPS for dysfunction
	2, revision of SPS to SSS – pleural effusions – bilateral hygroma
	2, eventually removal of angioblastoma

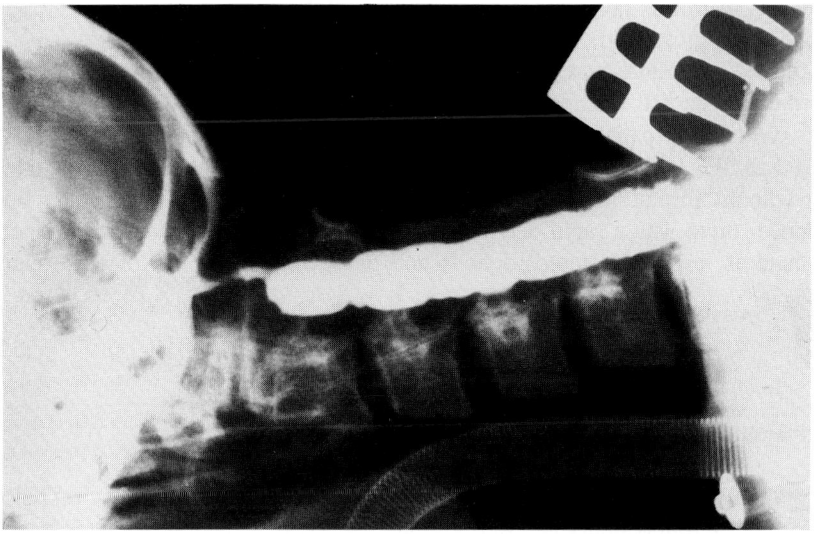

Fig. 1. Intraoperative X-ray examination with filling of the intramedullary cavity with contrast medium. Upward communication is visible and the patient was selected for SSS

cervical or cervicothoracic lesions cannot be excluded by preoperative neuroradiological examination (Fig. 1). It has been found that this is a safe proccdure if the latest generation of nonionizing contrast media such as isovist (iotrolan) is used. Before the advent of these contrast media, occasional seizure-like motor discharges could be observed postoperatively. They were usually easily managed with intravenous diazepam but should now be avoided by the use of only safe contrast media.

In those cases in which a noncommunicating syrinx can be clearly demonstrated intraoperatively or even by preoperative MRT, an SPS is advocated. In our series 21 cases have been treated in this manner with mostly favorable results (Table 2). Only in one case did the SPS have to be revised to SSS because of massive recurrent pleural effusions. In two more cases the shunt had to be revised because of dysfunction. In two cases the shunt became unnecessary because the patients eventually were shown to be suffering from intramedullary angioblastoma and had their tumors removed. One patient with a communicating cyst was erroneously drained by an SPS and as a result developed ventricular collapse associated with bilateral subdural effusions. In this patient, the shunt was eventually revised into an SSS.

An SSS was placed in 16 cases in which either there was communicating syrinx or the SPS or Gardner procedure was not the first choice of treatment (Table 2).

The results were less favorable. In two cases an SSS was later revised into an SPS because of progressive disease and failure of the syrinx to collapse.

Gardner's procedure was performed in eight patients. In seven patients it was performed as the first therapeutic measure without combination with any drainage of the syrinx. One patient underwent a Gardner's procedure 2 years after having received an SPS with no postoperative improvement and further deterioration later, albeit without formation of hygroma. In all cases in which this procedure was performed, there was a rapid decrease in cyst size on MRT. Rapid improvement of symptoms, especially motor deficits and pain, occurred in six of these eight patients.

Conclusions and Implications

Preoperative diagnostic procedures should be aimed at demonstrating the presence or absence of intramedullary pressure, communication with the fourth ventricle, and the presence of tumor to which the cyst may be secondary. This is best achieved by way of carefully performed MRT with gadolinium enhancement.

Upon demonstration of a tumor, removal alone is sufficient and no further drainage procedures are required. In the case of a noncommunicating cyst, we advocate an SPS. This procedure has the advantage of using the pressure gradients between two different compartments, thus allowing the intraspinal pressure to drive the fluid continuously out of the cyst. In the case of communicating syringomyelia we advocate either an SSS or, preferably, the procedure according to Gardner. If necessary, this may be performed at a later stage in combination with an SPS. It can be observed, however, that after closure of the central canal there is rapid and longlasting reduction of cyst size.

The operative treatment of syringomyelia is aimed primarily at the reduction of intramedullary pressure. Progressive neurological deficit arising from such pressure as one of the disease components will be reversed or arrested in most cases.

Table 2. Results of surgical treatment of syringomyelia in 45 patients

	Improved	Unchanged	Worse
SPS	14	5	2
SSS	7	9	0
Gardner's procedure	6	2	0
	27	16	2

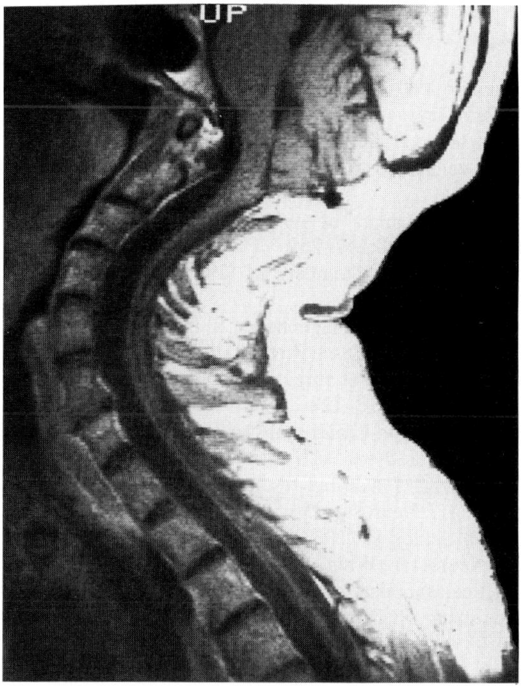

Fig. 2. MRT study of a patient who is not included in the series presented in this report. In this patient, Gardner's procedure was performed 8 years prior to MRT study. The patient experienced improvement of motor function postoperatively but after about 6 years again noticed a very gradual deterioration. It can be clearly seen that the syrinx has been reduced to a thin residual cavity and that there is no intramedullary pressure. This case illustrates the possibility of later disease progression despite alleviation of pressure

However, cavity formation/dissection may progress without cyst formation, causing further neurological deficit. This is illustrated in a case presented in Fig. 2.

References

1. Aichner F, Gerstenbrand F, Huk N, Dallun A (1984) NMR-Tomographie in der Diagnostik der Syringohydromyelia. Nervenarzt 55:324–327
2. Alvisi C, Cerisoli M (1984) Long-term results of the surgical treatment of syringohydromyelia. Acta Neurochir (Wien) 71:133–140
3. Barnett HJM, Foster JB, Hudgson P (1973) Syringomyelia. In: Waltan JN (ed) Major problems in neurology, vol 1. W.B. Saunders, Philadelphia
4. Batzendorf U (1988) Chiari I malformation with syringomyelia. J Neurosurg 68:726–730

5. Ditelmann JL, Babin E, Wackenheim A, Bonneville JF, Maitrot D (1982) Percutaneous puncture of spinal cysts in the diagnosis and therapy of syringomyelia and cystic tumors. Neuroradiology 24:59–63
6. Faulhauer K, Loew K (1978) The surgical treatment of syringomyelia. Long-term results. Acta Neurochir (Wien) 44:215–222
7. Filizzolo F, Versari P, D'Aliberti G, Arena O, Scotti G, Mariani C (1988) Foramen magnum decompression versus terminal ventriculostomy for the treatment of syringomyelia. Acta Neurochir (Wien) 93:96–99
8. Gardner WJ (1965) Hydrodynamic mechanism of syringomyelia: its relationship to myelocele. J Neurol Neurosurg Psychiatry 28:247–259
9. Gardner WJ, Bell HS, Pooles PN, Dohn DF, Steinberg M (1977) Terminal ventriculostomy for syringomyelia. J Neurosurg 46:609–617
10. Jacobi C, Stichnoth F-A, Knaak G, Neitzel L, Felgenhauer K (1987) Kernspintomographie: Indikationen bei ZNS-Erkrankungen. Dtsch Ärzteblatt 84:B65–B68
11. Krayenbühl H (1975) Evaluation of the different surgical approaches in the treatment of syringomyelia. Clin Neurol Neurosurg 2:110–128
12. Rhoton AL Jr (1978) Microsurgery of Arnold Chiari malformation and hydromyelia in adults. In: Rand RW (ed) Microsurgery, 2nd edn. C.V. Mosby, St. Louis
13. Tator CH, Meguro K, Rowed DW (1982) Favorable results with syringosubarachnoid shunts for treatment of syringomyelia. J Neurosurg 56:517–523
14. Vanquero J, Martinez R, Salazar J, Santos H (1987) Syringosubarachnoid shunt for treatment of syringomyelia. Acta Neurochir (Wien) 84:105–109
15. Williams B, Fahy G (1983) A critical appraisal of "terminal ventriculostomy" for the treatment of syringomyelia. J Neurosurg 58:188–197

Pathogenesis and Operative Treatment of Syringomyelia

H. Wiedemayer, H.E. Nau, F. Rauhut, L. Gerhard, V. Reinhard, and W. Grote[1]

Introduction

The term "syringomyelia" was first introduced by Charles P. Ollivier d'Angers in 1827. Pathological and clinical descriptions of this condition were given in earlier times by Estienne in 1546, Brunner in 1688, Morgani in 1740, and Portal in 1804 [3]. There has been extensive documentation of this entity in the literature. The clinical picture, diagnosis, and natural course of syringomyelia are well known. The diagnosis of syringomyelia has been significantly facilitated by the introduction of magnetic resonance imaging. In contrast, the pathogenesis and treatment of syringomyelia still present unsolved problems.

Pathogenesis of Syringomyelia

As regards the pathogenesis of syringomyelia, three conditions have to be clearly distinguished: syringomyelia due to intramedullary tumors, syringomyelia following spinal cord trauma, and so-called idiopathic or communicating syringomyelia.

In the first condition syringomyelia is simply an accompanying phenomenon of the fluid producing the tumor itself. Treatment and prognosis do not differ from those of other noncystic tumorous lesions.

Syringomyelia following spinal cord trauma was first studied extensively by Barnett et al. [2]. Delayed myelopathy occurring months or years after the initial trauma is caused by progressive spinal cord cavitation which most often develops upwards from the level of injury. The first important factor seems to be traumatic myelomalacia or hematomyelia. Local vascular components, spinal arachnoiditis with tethering of the cord causing chronic mechanical irritation, and glial structures surrounding the cavity presumably producing fluid are considered to be secondary factors contributing to the progressive extension of the cord cavitation [2, 3, 5].

[1] Neurochirurgische Universitätsklinik am Klinikum der Gesamthochschule Essen, Hufelandstr. 55, D-4300 Essen 1

Advances in Neurosurgery, Vol. 18
K.-A. Bushe, M. Brock, and M. Klinger (Eds.)

Table 1. Pathogenesis of syringomyelia

Primary factors	Abnormalities at the cervicomedullary junction:
	– developmental (e.g., cerebellar ectopia)
	– acquired (e.g., adhesive arachnoiditis)
	Hydrodynamic theory
Idiopathic syringomyelia	
Secondary factors	Proliferative reactive gliosis
	Local vascular problems
Posttraumatic syringomyelia	
Primary factors	Hematomyelia
	Myelomalacia
	Spinal arachnoiditis
	Tethering of the cord
Cystic intramedullary tumors	
(angioblastoma, ependymoma, glioma)	

A most interesting approach to the pathogenesis and operative treatment of idiopathic or communicating syringomyelia was introduced by Gardner [6] and Williams [13]. In contrast to earlier theories which considered syringomyelia to be a developmental anomaly of the spinal cord or a primary degenerative process with a proliferative component, they considered this condition to be due to a disturbance in the normal circulation of the cerebrospinal fluid. This "hydrodynamic theory" of syringomyelia is based on operative, radiological, and pathological-anatomical findings of abnormalities in the cervicomedullary junction. These may be developmental malformations, like ectopia of the cerebellar tonsils or ineffective perforation of the roof of the fourth ventricle, or acquired abnormalities, e.g., local adhesive arachnoiditis or tumors located in this region. As the CSF enters the fourth ventricle, whose outlets are obstructed, it is forced at the obex via a communication along the central canal into the medulla. Substantiated pressure differences are further necessary to cause progressive dissection of the medullary substance, forming the syrinx. These pressure differences may be due to the arterial pulsation of the CSF [6] or to venous craniospinal pressure dissociation [13]. Particularly in later stages of the disease secondary factors like reactive proliferative gliosis and local vascular problems may contribute to further progression of clinical symptoms. A schematic overview of the pathogenetic factors is given in Table 1.

Table 2. Clinical findings in 24 patients

Age: 7–67 years; mean: 44.8 years	
Sex: 14 male, 10 female	
Presenting symptoms:	
Paresthesia and sensory loss	13
Disturbance of gait	13
Radicular pain	8
Neurological examination:	
Radicular lesion only	3
– motoric	2
– sensory	1
Involvement of long spinal pathways	21
– motoric	1
– sensory	7
– both	13
Localization of the cord cavity:	
Cervical	7
Cervicothoracic	14
Thoracic	1
Thoracolumbar	2

Patients, Operative Treatment, and Results

Twenty-four patients with spinal cord cavitations and progressive neurological deterioration were treated operatively in our clinic. A summary of the clinical findings is given in Table 2. One patient was considered to be suffering from posttraumatic syringomyelia: 21 years after fracture of the third lumbar vertebra he developed symptoms of progressive spinal cord cavitation. Another patient had operatively verified spinal arachnoiditis following meningitis and developed progressive syringomyelia 10 years later. All other patients had idiopathic syringomyelia.

Laminectomy was in most cases restricted to one or two levels. Using the operating microscope an interlaminar approach was usually sufficient. In most instances a midline myelotomy in a relatively avascular area was performed and a Pudenz ventricular catheter was inserted several centimeters rostrally into the cavity. An overview of operative treatment, complications, and results is given in Tables 3–5. Eighteen patients had follow-up periods of more than one year. Four of these 18 patients improved. In one patient sensory disturbances and in three patients disturbances of gait regressed. Full improvement of neurological symptoms was observed in no case. Eight patients showed arrest of the preoperatively progressive neurological deterioration. One patient died several weeks after operation with unchanged neurological status, having already been transferred to a neighboring hospital for

Table 3. Operative procedures in the 24 patients

Syringostomy	1
Shunt to cisterna magna	4
Syringoperitoneal shunt	3
Syringosubarachnoid shunt	7
Syringopleural shunt	9

Table 4. Operative complications

Neurological deterioration	
Temporary	4
Permanent	1
CSF fistula	3
Thoracic drainage	2
Shunt displacement	3

Table 5. Results of operative treatment in 24 patients

	Length of follow-up	
	> 1 year[a]	< 1 year
Improved	4	1
Stable/unchanged	8	4
Worse	5	1
Dead	1	0
	18	6

[a] Follow-up in this group of 18 patients was for 1–14 years; mean = 3.1 years

physiotherapy. The cause of death was presumed to be pulmonary embolism. Five patients continued to worsen. In three of these shunt obstruction was detected. In two cases a second operation was performed because of shunt malfunction, but the course was further deterioration. In two patients with progressive deterioration postoperatively, magnetic resonance imaging demonstrated a collapse of the cord cavity, indicating sufficient drainage of the syrinx.

Discussion

As discussed above, several primary and secondary factors can influence the course of syringomyelia. The value of these factors may differ between different patients as well as in the same patient during the course of the disease. Particularly secondary factors, e.g., reactive spinal gliosis and local vascular problems, may influence the progression of neurological deficits in later stages of syringomelia. We feel that these factors contribute to some cases of failure of operative treatment. This is confirmed by our observation that operative treatment is less effective in cases with severe neurological deficits.

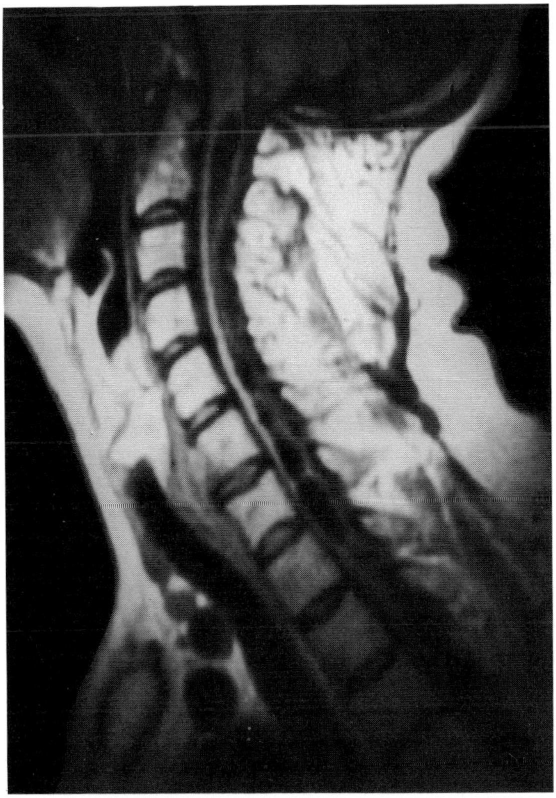

Fig. 1. Magnetic resonance imaging of syringomyelia demonstrating multiple "septa" within the cord cavitation. Communication throughout the whole syrinx was demonstrated intraoperatively when filling the cavity with contrast medium

Following the hydrodynamic theory, surgery of the posterior fossa (suboccipital decompression, drainage of the fourth ventricle, obstruction of the central canal) was suggested [6]. The reported results seemed to be excellent. Later on, in 1977, Gardner suggested another operative treatment called "terminal ventriculostomy" [7]. It is not well understood – as stated by Williams [13] – why he changed his operative strategy. A simple ventriculoperitoneal shunt was proposed by Benini and Krayenbühl [4]. Reports of long-term results are missing. Recently, most reports have dealt with direct exposure of the spinal cord cavity using various shunting techniques.

In our patients syringopleural and syringosubarachnoid shunts were used most commonly. As demonstrated by postoperative magnetic resonance imaging, collapse of the cord cavity is attainable using all the different modalities of shunting. But keeping a shunt system patent and sufficient over a long time constitutes a problem. We feel that drainage to a low-pressure compartment over a relatively short distance, as in syringopleural shunt, may offer some advantages, but further follow-up of these patients is necessary to confirm this.

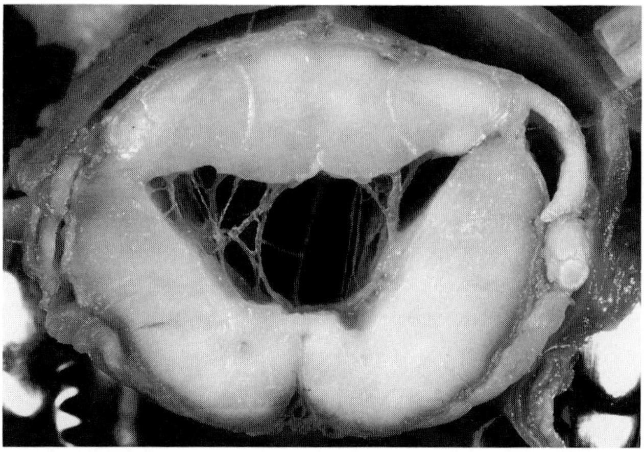

Fig. 2. Transverse section of the spinal cord showing cavitation with numerous "septa." (Different cases in Figs. 1 and 2)

Comparison between the various reports regarding the results of operative treatment of syringomyelia is difficult because of different patient groups, indications for operative treatment, and criteria for postoperative evaluation. Overall about 60%–70% of the patients treated by different shunt procedures have a stable/unchanged course or improvement and are classified in the literature as "good" and "excellent" respectively [1, 9–12]. Some reservation is necessary when counting an unchanged and stable course as a favorable result. Studies of the natural history of syringomyelia show that in untreated cases there may be very slow progression of symptoms and even stable periods [8, 10]. Therefore only cases with progressive deterioration preoperatively and long-term follow-up over several years can prove the effectiveness of operative treatment.

Our indications for operative treatment are: the patient must show significant neurological deficit, there has to be a clear progressive course, and by means of radiological methods a space-occupying spinal cord cavitation must be demonstrable (Figs. 1, 2). Using magnetic resonance imaging early diagnosis of syringomyelia is possible. So in our opinion early surgical treatment is advisable as neurological disturbances, once established, may at best regress only partially after operation.

Summary

Different aspects of the pathogenesis of syringomyelia are discussed. Twenty-four of our own patients suffering from syringomyelia were treated operatively. Different shunt procedures were performed. Most often syringosubarachnoid shunts (seven cases) and syringopleural shunts (nine cases) were used. Postoperative improvement was observed in five patients, 12 were stable/unchanged, six showed further deterioration, and one died. Indications for operative treatment were suggested.

Acknowledgment. The authors thank Prof. Dr. Kühne, Krupp Krankenhaus, Essen, for the magnetic resonance image.

References

1. Barbaro NM, Wilson CB, Gutin PH, Edwards MSB (1984) Surgical treatment of syringomyelia. Favorable results with syringoperitoneal shunting. J Neurosurg 61:531–538
2. Barnett HJM, Botterell EH, Jousse AT, Wynn-Jones M (1966) Progressive myelopathy as a sequel to traumatic paraplegia. Brain 89:159–174
3. Barnett HJM, Foster JB, Hudgson P (1973) Syringomyelia. W.B. Saunders, London
4. Benini A, Krayenbühl H (1969) Ein neuer chirurgischer Weg zur Behandlung der Hydro- und Syringomyelie. Schweiz Med Wochenschr 99:1137–1142
5. Bischof W, Frowein RA (1967) Zur Spätmyelomalazie nach Kompressionsfraktur. Zentralbl Neurochir 28:61–66
6. Gardner WJ (1965) Hydrodynamic mechanism of syringomyelia: its relationship to myelocele. J Neurol Neurosurg Psychiatry 28:247–259
7. Gardner WJ, Bell HS, Poolos PN, Dohn DF, Steinberg MJ (1977) Terminal ventriculostomy for syringomyelia. Neurosurgery 46:609–617
8. Hertel G, Kramer S, Placzek E (1973) Die Syringomyelie. Nervenarzt 44:1–13
9. Lesoin F, Petit H, Thomas CE, Viaud C, Baleriaux D, Jomin M (1986) Use of syringoperitoneal shunt in the treatment of syringomyelia. Surg Neurol 25:131–136
10. McIlroy WJ, Richardson JC (1965) Syringomyelia: a clinical review of 75 cases. Can Med Assoc J 93:731–734
11. Suzuki M, Davis C, Symon L, Gentili F (1985) Syringoperitoneal shunt for treatment of cord cavitation. J Neurol Neurosurg Psychiatry 48:620–627
12. Tator CH, Meguro K, Rowed DW (1982) Favorable results with syringo-subarachnoid shunts for treatment of syringomyelia. J Neurosurg 56:517–523
13. Williams B (1978) A critical appraisal of posterior fossa surgery for communicating syringomyelia. Brain 101:223–250

Magnetic Resonance Imaging of Syringomyelia: Idiopathic and Tumor-Associated Cases

P. Schubeus, W. Schörner, H. Henkes, and R. Felix[1]

Purpose

In recent years magnetic resonance (MR) imaging has proved to be a useful tool in the diagnosis of syringomyelia [1,4]. Multiplanar imaging, more pronounced tissue contrast, and the absence of bone artifacts are advantages of MR imaging over CT. However, differentiation between idiopathic syringomyelia and syringomyelic cavities associated with an intramedullary neoplasm may be difficult [2,3]. In the present study, we reviewed MR examinations of 38 patients with syringomyelic cavities, in order to establish MR criteria for differentiating tumor-associated from idiopathic cases.

Patients and Methods

Thirty patients with idiopathic syringomyelia and eight patients demonstrating syringomyelic cavities associated with an intramedullary tumor were studied. The diagnoses were based on clinical, radiological, and/or histological findings. Five intramedullary neoplasms were confirmed by surgery. Histological examinations revealed two hemangioblastomas, one ependymoma, and one astrocytoma (one examination was not conclusive).

Magnetic resonance examinations were performed on a 0.5-T imager (Magnetom, Siemens). All patients were studied with sagittal T1-weighted spin-echo (SE 400/30) or gradient-echo images (FLASH 315–500/14, 90°). Twenty-four patients were also studied with T2-weighted spin-echo images (SE 1600/70) and 33 patients were also examined with axial images. Other imaging parameters were 5–10 mm slice thickness and one to four excitations. In five cases of tumor-associated syringomyelia, T1-weighted sequences were repeated following administration of 0.1 mmol gadolinium (Gd)-DTPA/kg body weight (Magnevist, Schering).

[1] Radiologische Klinik, Universitätsklinikum Rudolf Virchow, Standort Charlottenburg, Freie Universität Berlin, Spandauer Damm 130, D-1000 Berlin 19

Advances in Neurosurgery, Vol. 18
K.-A. Bushe, M. Brock, and M. Klinger (Eds.)

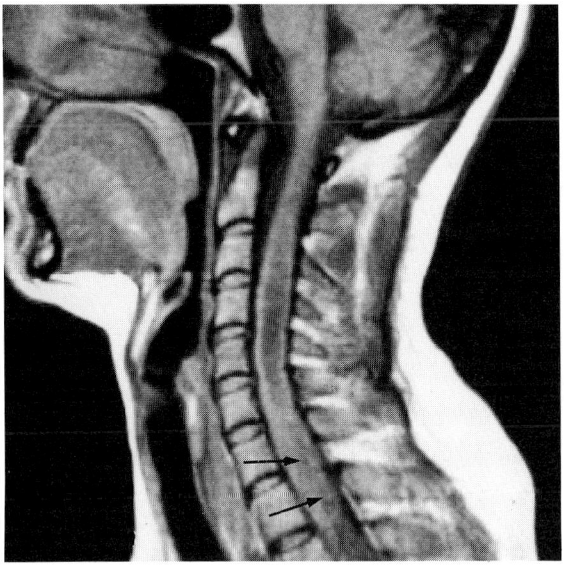

Fig. 1. Typical MR appearance of tumor-associated syringomyelia on sagittal T1-weighted image (SE 400/30): Hypointense cavity of the spinal cord from C2 to T8 (here represented to T2) with even cord contours and abrupt change of cavity diameter between C6 and T2 (*arrow*); no tonsillar ectopia. Hemangioblastoma in the lower cervical cord, histologically proven; 43-year-old woman

Results

Idiopathic and tumor-associated syringomyelic cavities were similar in localization and dimensions. Except for one tumor-associated case, all cavities involved the cervical region.

All idiopathic syringomyelias demonstrated hypointense cavities on T1-weighted images with even cord contours (30/30) and without abrupt change of cavity diameter (30/30). Most cases showed even thickness of the surrounding spinal cord (29/30) and centric cavity position (24/25). In patients with idiopathic cavities the spinal canal was frequently enlarged (8/29). On T2-weighted images idiopathic syringomyelias often showed an increased signal intensity surrounding the cavity (15/30).

Tumor-associated syringomyelic cavities were similar to idiopathic syringomyelias in most portions of the cavity. At the tumor site, however, most cases demonstrated abrupt change of cavity diameter (8/8) (Fig. 1), abrupt eccentric cavity position (7/8) (Fig. 3), uneven thickness of the surrounding spinal cord (8/8) (Fig. 3), and increased signal intensity of the spinal cord (not only surrounding the

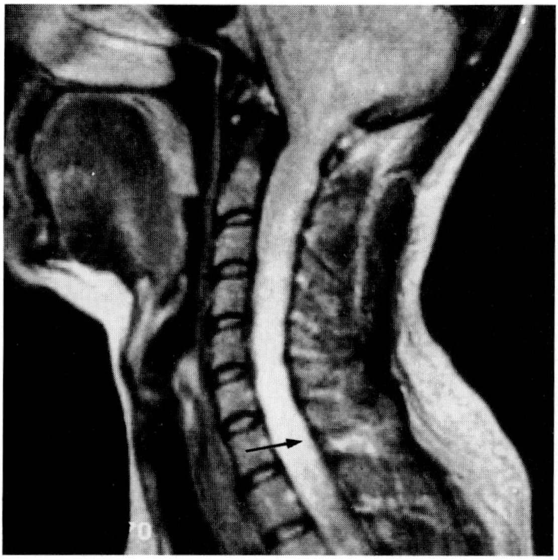

Fig. 2. Typical MR appearance of tumor-associated syringomyelia on sagittal T2-weighted image (SE 1600/70): Increased signal intensity of the spinal cord, not only surrounding the cavity (*arrow*). Hemangioblastoma (same patient as in Fig. 1)

ity) on T2-weighted images (8/8) (Fig. 2). Some tumor-associated cases showed circumscribed areas of cord enlargement (3/8), uneven cord contours (2/8), or disappearance of the cerebrospinal fluid adjacent to the tumor mass (3/7). Following administration of Gd-DTPA in tumor-associated cases all tumors showed a pathological contrast enhancement (5/5), whereas associated syringomyelic cavities did not enhance (Fig. 4). Tumor delineation from the surrounding tissue and from the syringomyelic cavity was improved.

Ectopic cerebellar tonsils were frequently found in patients with idiopathic syringomyelia (15/30). Syringomyelic cavities in patients with intramedullary tumors were not associated with any malformations.

Conclusions

In most cases MR imaging provides valuable criteria to differentiate between idiopathic and tumor-associated syringomyelia. The most typical findings in idiopathic cases were:

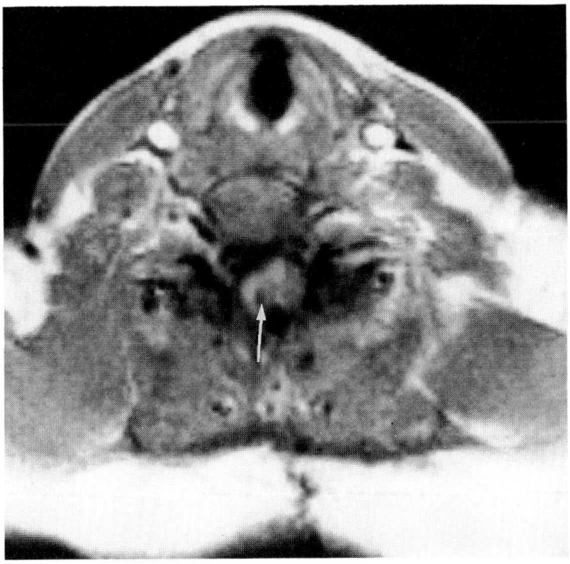

Fig. 3. Typical MR appearance of tumor-associated syringomyelia on axial T1-weighted image (FLASH 400/14) before enhancement: Eccentric cavity position (*arrow*); uneven thickness of surrounding spinal cord. Hemangioblastoma (same patient as in Fig. 1)

- Tonsillar ectopia (15/30)
- Even thickness of the surrounding spinal cord (29/30)
- No abrupt change of cavity diameter (30/30)
- Enlargement of spinal canal (8/29)

The most valuable criteria in the diagnosis of tumor-associated cases were:

- Uneven thickness of the surrounding spinal cord (8/8)
- Abrupt eccentric cavity position (7/7)
- Abrupt change of cavity diameter (8/8)
- Increased signal intensity of the spinal cord (not only surrounding the cavity) on T2-weighted images (8/8)

Administration of Gd-DTPA in tumor-associated cases may improve tumor delineation from the syringomyelic cavity.

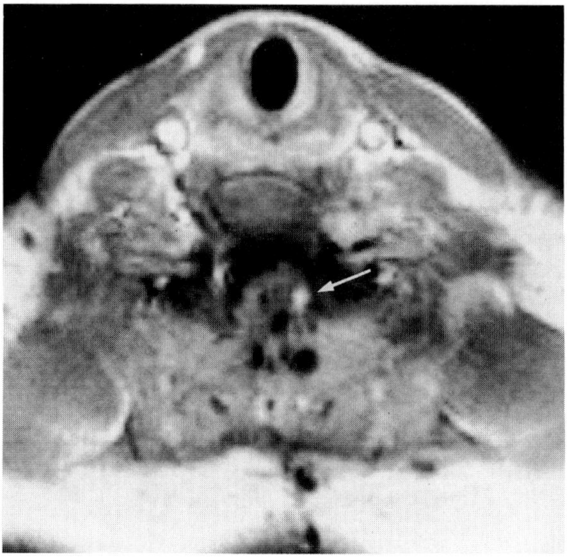

Fig. 4. Typical MR appearance of tumor-associated syringomyelia on axial T1-weighted image (FLASH 400/14) after enhancement: Pathological contrast enhancement in the spinal cord (*arrow*) corresponding with the histologically proven hemangioblastoma (same patient as in Fig. 1)

References

1. Köhler D, Treisch J, Hertel G, Schörner W, Fiegler W (1985) Die magnetische Resonanztomographie der Syringomyelie. Fortschr Roentgenstr 143:617–622
2. Sherman JL, Barkovich AJ, Citrin CM (1986) The MR appearance of syringomyelia: new observations. AJNR 7:985–995
3. Williams AL, Haughton VM, Pojunas KW, Daniels DL, Kilgore DP (1987) Differentiation of intramedullary neoplasms and cysts by MR. AJNR 8:527–532
4. Yeates A, Brant-Zawadzki M, Norman D, Kaufman L, Crooks LE, Newton TH (1983) Nuclear magnetic resonance imaging of syringomyelia. AJNR 4:234–237

Comparison of Syringoperitoneal and Syringopleural Shunting in Patients with Syringomyelia

H. Haberl, W. Zimmermann, P. Schmiedek, S. Stelzer, and F. Marguth[1]

Introduction

Since the precise pathogenesis of syringomyelia is still unknown and a matter of ongoing debate, it is not surprising that effective surgical treatment remains to be established [1,2,6]. On the other hand, the introduction of magnetic resonance imaging (MRI) has allowed much easier direct visualization of spinal cord cavitation than was possible with myelography and delayed CT scanning, which were previously used to establish the diagnosis of syringomyelia [5]. With this diagnostic improvement, it can be expected that in the future more patients will be diagnosed whose clinical symptoms are attributable to cystic changes within the spinal cord. In view of this, the question of which treatment should be offered to these patients becomes even more relevant.

Within recent years several, admittedly small, series of patients have been reported on in whom direct shunting of the intramedullary cavity has been recommended. In this study we present our own experience with two consecutive series of patients in whom either a syringoperitoneal (SPT) or a syringopleural (SPL) shunting system was employed.

Clinical Material and Methods

During the time period from 1985 to 1989 16 patients with syringomyelia underwent surgical treatment in our department. Out of the total group of 16 cases, 12 patients had either idiopathic syringomyelia or the spinal cavitation was associated with obstructive lesions of the foramen magnum as demonstrated on sagittal MRI views. In this group there were ten female and two male patients with a mean age of 48 years and an age range between 22 and 64 years. The clinical manifestation of syringomyelia in these patients disclosed the typical spectrum of combined

[1] Neurochirurgische Abteilung, Universität München, Klinikum Großhadern, Marchioninistr. 15, D–8000 München 70

Advances in Neurosurgery, Vol. 18
K.-A. Bushe, M. Brock, and M. Klinger (Eds.)

Table 1. Idiopathic and posttraumatic syringomyelia: preoperative clinical findings (n = 16)

Nr.	Age	Sex	Pain	Paresis	Sens. dis.	Veg. dis.
Idiopathic syringomyelia						
1	40	f	++	–	+	–
2	23	f	–	+	++	–
3	55	f	+	++	–	–
4	47	f	++	++	++	–
5	52	m	–	++	++	–
6	64	f	+	++	++	–
7	22	f	–	++	+	–
8	44	f	++	+	+	–
9	49	m	+	+	++	+
10	63	f	+	+	+	–
11	59	f	–	++	++	+
12	63	f	+	+	++	–
Posttraumatic syringomyelia						
1	55	m	–	++[a]	++	–
2	48	m	–	++	+	+
3	24	f	++	++[a]	++	++[a]
4	26	m	++	++	+	++

[a] Traumatic origin

sensorimotor deficits of varying degree as well as vegetative disturbances and pain syndromes as listed in detail in Table 1.

In four additional patients, three male and one female, syringomyelia had followed a spinal cord injury. The mean age of these patients was 38 years, with an age range from 24 to 55 years. Clinically, patients of this group presented with either intolerable pain and/or ascending neurological deficits [3, 4]. Only patients with severe clinical manifestation and evidence of progression of their symptoms were included in this study. Preoperatively, all patients underwent MRI studies using a spine surface coil on a 1.5-T-Siemens system. Except for the four cases of posttraumatic syringomyelia, MRI was repeated following the administration of contrast medium (gadolinium) in order to rule out the possibility of an intramedullary tumor underlying the syringeal cavity.

Postoperatively MRI was repeated early after the operation (16 cases) and during the later follow-up period (12 cases). For various reasons MRI could not be repeated subsequently in four patients. Clinical follow-up studies for evaluation of the patients' condition were routinely performed in all cases, with the follow-up period ranging from 6 months to 4 years (Fig. 1).

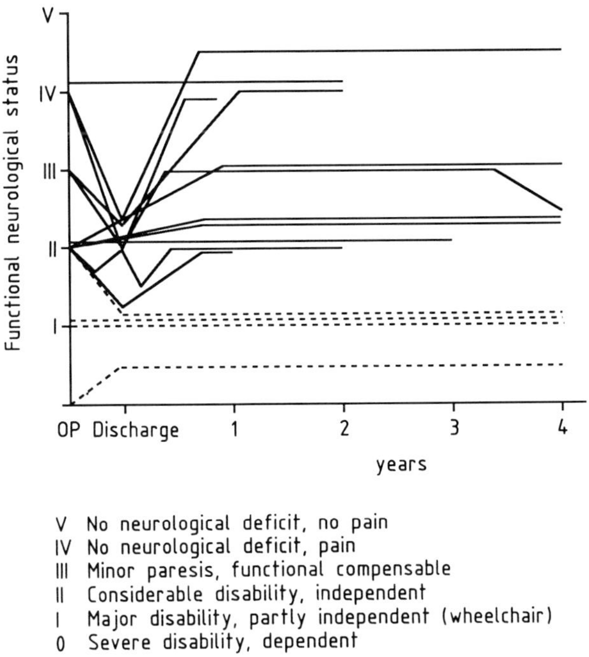

Fig. 1. Postoperative clinical course in 16 patients shunted for syringomyelia (*solid lines*, idiopathic; *broken lines*, posttraumatic)

In ten patients operated on during 1985 and 1986 an SPT shunt was used and in the remaining six patients who were operated on more recently, an SPL shunt was inserted. In all cases a Uni-Shunt (Codman) with low opening pressure was used.

Results

Clinically, in 7 out of 16 operated patients a deterioration was noted immediately following surgery; this deterioration, however, was later reversible in six patients. These new sensorimotor deficits were thought most likely to be due to the myelotomy required for insertion of the proximal shunting device into the syringeal cavity. One female patient had a thrombotic complication involving an arm vein and was treated with heparin. Another patient developed a left-sided stroke 1 week following surgery, with good functional recovery during the later follow-up period. In two patients a complete dislocation of the SPT shunting system into the peritoneal cavity was noticed, requiring endoscopic removal of the catheter which was followed by new shunt placement. One patient with an SPL shunt developed a pneumothorax postoperatively which was treated by drainage for some days. With

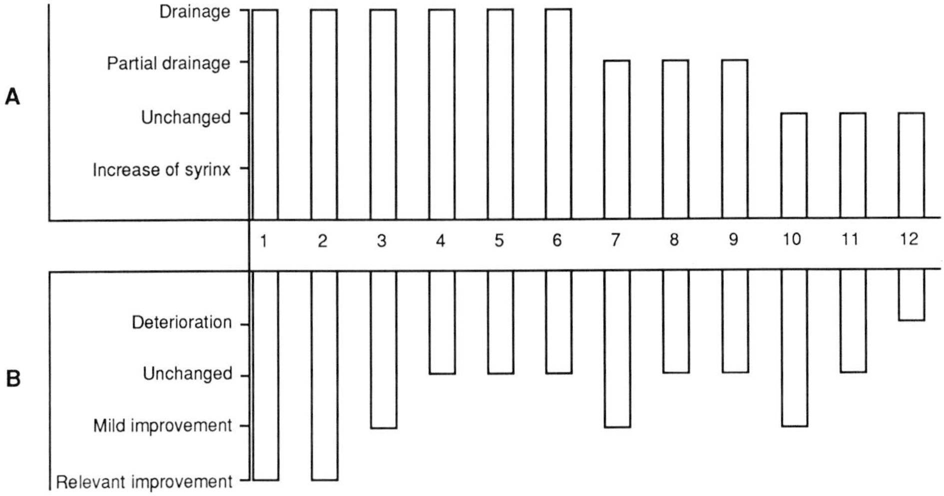

Fig. 2. Correlation of postoperative MRI findings (*A*) with postoperative clinical course (*B*) in 12 patients operated on for syringomyelia

regard to long-term follow-up results, a definite and still present improvement in neurological symptoms was noted in three patients. In 11 patients the neurological condition has shown a mild improvement or has remained stable with at least no evidence of progression of the disease. In one patient who was initially stable following surgery, new symptoms were observed recently. In this patient MRI disclosed a persisting spinal cyst, suggesting a dysfunction of the shunt (Fig. 1).

Finally, one patient deteriorated permanently following surgery. While still able to walk with assistance preoperatively, he became wheelchair bound postoperatively. One patient was operated on only recently and therefore was not taken into consideration with regard to long-term results. Postoperative MRI, which was done in 12 patients at varying intervals following surgery, revealed a complete collapse of the intramedullary cyst in six patients. A partial drainage of the cyst was noted in three patients whereas in the three remaining patients the cystic cavity was found to be unchanged, suggesting dysfunction of the shunt. Correlation of the postoperative clinical course and MRI findings revealed that only in patients with successful drainage of the entire cyst can a satisfactory clinical result be expected (Figs. 2, 3).

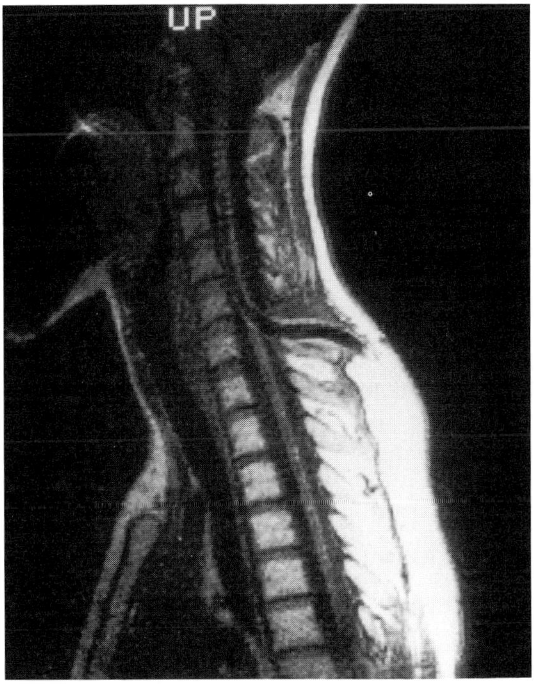

Fig. 3. Postoperative MRI study 2 years following SPL shunting showing excellent result with complete collapse of the intramedullary cavitation

Comment

Even if one concedes that the aim of operative treatment of syringomyelia is not a definite improvement in clinical condition but rather the prevention of further deterioration, the results of this study clearly indicate that a satisfactory operative treatment is still lacking. This becomes particularly evident in view of the fact that with the advent of MRI we possess a diagnostic tool to visualize the extent of the syrinx preoperatively and to evaluate the postoperative result of shunting as well. In addition, when taking into consideration the significant surgical morbidity, surgery seems to be justified only in patients with a definite progression of the disease and in the presence of a marked syringeal cavity on MRI. At present, we prefer to use the SPL shunt, primarily because it has the practical advantage of a confined operative field, in contrast to the SPT shunt, which requires repositioning of the patient for completing the procedure [7]. More satisfying operative results, however, will probably depend on a better understanding of the pathological factors causing syringomyelia.

References

1. Gardner WJ (1973) The dysraphic states. Excerpta Medica, Amsterdam
2. Peerless SJ, Quentin J, Durward MB (1982) Management of syringomyelia: a pathophysiological approach. Clin Neurosurg 30:531–576
3. Shannon N, Symon L, Logue V, Cull D, Kang J, Kendall BE (1981) Clinical features, investigation and treatment of post-traumatic syringomyelia. J Neurol Neurosurg Psychiatry 44:35–42
4. Stevens JM, Olney JS, Kendall BE (1985) Posttraumatic cystic and non-cystic myelopathy. Neuroradiology 27:48–56
5. Wilberger JE, Maroon JC, Prostko GR, Baghai P, Beckmann J, Deeb J (1987) Magnetic resonance imaging and intraoperative neurosonography in syringomyelia. Neurosurgery 20:599–606
6. Williams B (1986) Progress in syringomyelia. Neurol Res 8:130–145
7. Williams B, Page N (1987) Surgical treatment of syringomyelia with syringopleural shunting. Br J Neurosurg 1:63–80

Microsurgical Treatment of Syringomyelia: Intraoperative Findings and Results

H. Bertalanffy and H.R. Eggert[1]

Introduction

The rarity of syringomyelia, the difficulties in diagnosing the disease that were common prior to the era of computerized tomographic scanning and magnetic resonance imaging, and the lack of precise knowledge about the mechanism of fluid accumulation within the spinal cord have led to a diversity of surgical methods for treating this congenital disease.

The theory of syrinx formation as a consequence of a hindbrain malformation and the surgical technique proposed by Gardner [7, 8] have gained wide acceptance. However, the observations made by some authors have led them to doubt that the relationship between the malformation and syringomyelia is a causal one [10]. In 1980, our institution adopted the surgical technique described by Rhoton [13] to treat the Chiari I-syringomyelia complex.

Clinical Material

Fourteen patients in whom syringomyelia was associated with a Chiari I malformation were treated surgically at the Neurosurgical Department, University of Freiburg, between 1980 and 1988. The patients included seven males and seven females, whose ages ranged from 20 to 71 years (mean, 40 years).

The most common presenting symptoms were characteristic sensory changes, disturbances of motor function, and dysesthetic pain. The average duration of preoperative symptoms was 11 years, ranging from 1.5 to 38 years. All 14 patients suffered a variable degree of impaired pain, temperature, or proprioceptive sensation; 12 complained of more or less severe motor weakness, and 10 of pain. Preoperatively a progressive deterioration of the clinical condition was manifest in each patient. Computed tomography and metrizamide myelography were the diagnostic modalities used in the first years of treatment, and were performed in

[1] Neurochirurgische Universitätsklinik Freiburg, Hugstetter Str. 55, D–7800 Freiburg

Advances in Neurosurgery, Vol. 18
K.-A. Bushe, M. Brock, and M. Klinger (Eds.)

eight and six patients, respectively. Since 1984, magnetic resonance imaging has become the diagnostic tool of choice and was applied in 10 of the 14 patients.

Surgical Procedure

The surgical procedure consisted of a suboccipital craniectomy with C1 (and if necessary C2) laminectomy, removal of the posterior rim of the foramen magnum, dissection and thus mobilization of caudally displaced tonsils, splitting of arachnoid adhesions at the outlets of the fourth ventricle, plugging of the obex with a small piece of subcutaneous fat tissue fixed with fibrin glue, and insertion of a dural graft. A hemilaminectomy was then performed, usually at the level of the greatest extent of the syringomyelic cavity and on the side of symptom predominance. The cord was incised at the entry zone of the dorsal roots and a Silastic shunt tube of 1.5 mm in diameter and 6–8 cm in length was introduced to create a communication between the syringomyelic cavity and the cervical subarachnoid space. Care was taken that the lateral perforations of the catheter were placed at a distance from the dural opening in order to prevent occlusion of the catheter by scar formation.

Intraoperative Findings and Surgical Results

At the craniospinal junction, a Chiari I malformation was present in each patient, the tonsils being displaced caudally, in some cases as far as to the level of C3. The fourth ventricle was found to be displaced caudally in seven individuals. Additional findings included an adhesive velum medullare posterius with obstruction of the CSF pathways in six patients, a clearly dilated central canal of the cord at the obex in four patients, and a depressed posterior rim of the foramen magnum with compressive effect in two patients.

At the level of the posterior cervical approach, in two patients the initially bulging cord collapsed before myelotomy, immediately after the subarachnoid space was opened. In nine cases cord collapse occurred only after the syringomyelic cavity was opened. Details on the hydrodynamics were not provided in three cases.

There was no mortality associated with this operation, and no major complication was encountered in this series. Control MRI (performed in four patients between 10 days and 1 year postoperatively) disclosed a partial (Fig. 1) or total collapse of the syrinx.

All but one of the patients were contacted by phone and asked to assess the current severity of their three major complaints – pain, sensory change, and motor dysfunction – as compared with the preoperative condition. In no case was there complete relief of any of the symptoms and signs. However, intensity, incidence, and character of pain were markedly improved in 90%, the sensory deficits in

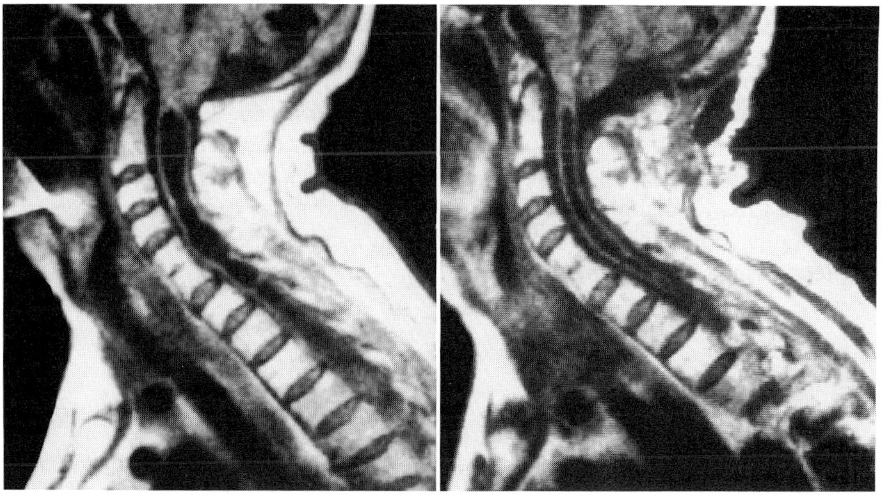

Fig. 1. MRI before and 10 days following the operation in a 71-year-old female

60%, and the motor deficits in 20% of the patients questioned. There had been no deterioration of neurological deficits. The progression of the clinical signs was arrested postoperatively in all patients. Improvement of one or more symptoms was noted in 50%, while 20% remained essentially unchanged. This accounts for an overall desirable outcome in at least 80% of surgically treated patients.

Discussion

The pathology of the spinal cord and craniospinal junction is accurately demonstrated by magnetic resonance imaging. Thus, it becomes possible to differentiate syringomyelia from other pathological entities of the spinal cord such as cystic tumors. In our opinion, each case of presumed syringomyelia without craniospinal malformation should be evaluated preoperatively by means of gadolinium-enhanced MRI in order to exclude a possible tumor.

Considerable differences exist between various published reports with regard to number of treated patients, indication for operation, surgical method, duration of follow-up, and criteria for assessing the operative outcome. With decompression of the foramen magnum alone, arrest of the progression of syringomyelia or improvement of symptoms and signs was achieved in 72%–83% [3, 5, 6, 9, 10, 14]. Other authors preferred a combined surgical treatment, including laminectomy, myelotomy, and syringostomy, leading to stabilization and improvement rates of

51%–94% [1, 4, 11–13, 15]. The long-term results in our small series concur with these results from the literature.

Despite doubts about the necessity of myelotomy and syringostomy [2], we consider this an important part of the operation. Nevertheless, it remains uncertain as to which step of the surgical method has the greatest effectiveness.

In conclusion, our experience has shown that the treatment of syringomyelia and associated Chiari I malformation with medullary and upper cervical cord decompression, plugging of the obex, and syringostomy can be expected to produce good long-term results in 80% of cases.

References

1. Alvisi C, Cerisoli M (1984) Long-term results of the surgical treatment of syringohydromyelia. Acta Neurochir (Wien) 71:133–140
2. Batzdorf U (1988) Chiari I malformation with syringomyelia. J Neurosurg 68:726–730
3. Cahan LD, Bentson JR (1982) Considerations in the diagnosis and treatment of syringomyelia and the Chiari malformation. J Neurosurg 57:24–31
4. Faulhauer K, Loew K (1978) The surgical treatment of syringomyelia. Long-term results. Acta Neurochir (Wien) 44:215–222
5. Filizzolo F, Versari P, D'Aliberti G, Arena O, Scotti G, Mariani C (1988) Foramen magnum decompression versus terminal ventriculostomy for the treatment of syringomyelia. Acta Neurochir (Wien) 93:96–99
6. Garcia-Uria J, Leunda G, Carrillo R, Bravo G (1981) Syringomyelia: long-term results after posterior fossa decompression. J Neurosurg 54:380–383
7. Gardner WJ (1965) Hydrodynamic mechanism of syringomyelia: its relationship to myelocele. J Neurol Neurosurg Psychiatry 28:247–259
8. Gardner WJ, Abdullah AF, McCormack LJ (1957) The varying expressions of embryonal atresia of fourth ventricle in adults. Arnold-Chiari malformation, Dandy-Walker syndrome, "arachnoid" cyst of cerebellum, and syringomyelia. J Neurosurg 14:591–607
9. Levy WJ, Mason L, Hahn JF (1983) Chiari malformation presenting in adults: a surgical experience in 127 cases. Neurosurgery 12:377–390
10. Logue V, Edwards MR (1981) Syringomyelia and its surgical treatment – an analysis of 75 patients. J Neurol Neurosurg Psychiatry 44:273–284
11. Love JG, Olafson RA (1966) Syringomyelia: a look at surgical therapy. J Neurosurg 24:714–718
12. Pitts FW, Groff RA (1964) Syringomyelia: current status of surgical therapy. Surgery 56:806–809
13. Rhoton AL (1976) Microsurgery of Arnold-Chiari malformation in adults with and without hydromyelia. J Neurosurg 45:473–483
14. Saez RJ, Onofrio BM, Yanagihara T (1976) Experience with Arnold-Chiari malformation, 1960 to 1970. J Neurosurg 45:416–483
15. Tator CH, Meguro K, Rowed DW (1982) Favorable results with syringosubarachnoid shunts for treatment of syringomyelia. J Neurosurg 56:517–523

SEPs and Evoked Muscle Responses After Noninvasive Magnetic Stimulation in Patients with Syringomyelia

H. Masur, S. Nedjat, C.E. Elger, and A.C. Ludolph[1]

Introduction

The clinical picture of syringomyelia is characterized by central and peripheral pareses, sensory loss, trophic disturbances, and a pain syndrome. Pathologically, the syrinx often occupies the central gray matter of the cervical portion of the spinal cord. Today, the technique of MR imaging is capable of establishing the diagnosis with great accuracy [1].

Since the introduction of the technique of noninvasive electrical stimulation of the motor cortex by Merton and Morton in 1980 and the introduction of the magnetic stimulation technique by Barker et al. in 1985, it has been possible to examine the central motor system in the awake human subject [3, 4, 12, 13].

Both methods, investigation of the SEPs and of the evoked muscle responses after noninvasive electrical and magnetic stimulation of the motor cortex, have proved to be of diagnostic value in different diseases of the central and peripheral nervous system [4–8, 11, 14]. Both methods are sensitive for the detection of morphological and functional lesions in the cervical region of the spinal cord [11, 14, 15]. A study of median and ulnar nerve evoked SEPs showed absent or reduced amplitudes and prolonged central conduction times as the prominent findings [2]. To our knowledge, evoked muscle responses after noninvasive stimulation of the motor cortex have not yet been investigated in patients with syringomyelia.

We report the results of studies of SEPs and evoked muscle responses after noninvasive magnetic stimulation in relation to clinical findings in ten patients with radiologically defined syringomyelia.

Patients and Methods

Ten patients with syringomyelia were studied (five females and five males). The mean age was 50.3 (range, 37–61 years) and the mean duration of the disease, 14.2 years (range, 3–23 years). There was no evidence of a preceding cause. In

[1] Neurochirurgische Universitätsklinik, Albert-Schweitzer-Str. 33, D–4400 Münster

Advances in Neurosurgery, Vol. 18
K.-A. Bushe, M. Brock, and M. Klinger (Eds.)

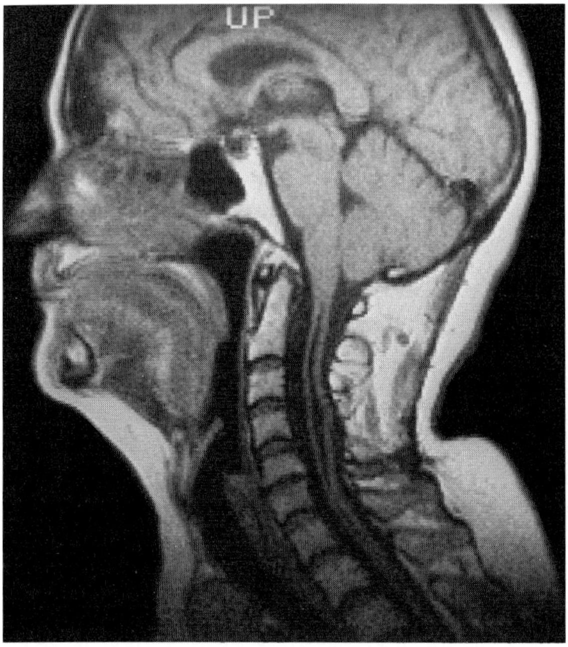

Fig. 1. Sagittal MR scan demonstrating a large cervical syrinx

two patients the diagnosis was made fortuitously. Eight patients showed typical symptoms. In all patients diagnosis was confirmed by MRI studies. A typical example is demonstrated by Fig. 1.

Each patient was individually interviewed and given a neurological examination. In this examination special attention was paid to symptoms of the central sensory and motor pathway. The long ascending (sensory) spinal tracts were assessed by median and tibial nerve SEPs and the long descending (motor) spinal tracts by the study of evoked muscle responses after noninvasive magnetic stimulation of the motor cortex.

Somatosensory evoked potentials were obtained by stimulation of the right and left median nerve at the wrist and the posterior tibial nerve at the ankle (stimuli adjusted to produce a small twitch at fingers or toes at 4 Hz and 0.2 ms duration). Evoked potentials were recorded with intradermal needle electrodes over the C7 spinous process and from the scalp at C3 and C4 (international 10/20 system) after stimulation of the median nerve, and over the L4 spinous process and from the scalp at Pz after stimulation of the tibial nerve. The reference was placed at Fpz. Filters were set at 10 Hz and 2 kHz. Typical examples of the recorded potentials are displayed in Fig. 2. The latencies of the first cortical and first spinal components were determined and a central latency (interpeak latency) was calculated. Laten-

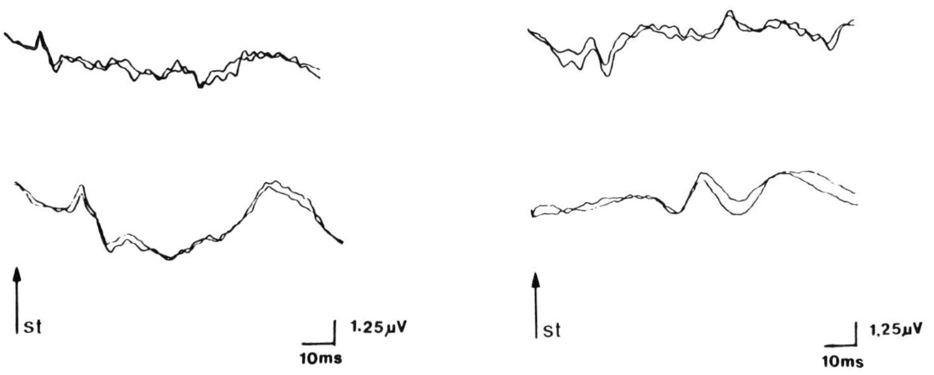

Fig. 2. Examples of median nerve SEPs (*left*) recorded from C7 (*upper trace*) and from the scalp (*lower trace*) and of tibial nerve SEPs (*right*) recorded from L4 (*upper trace*) and from the scalp (*lower trace*)

cies were subsequently compared with normative data obtained in our laboratory (n = 30).

The functional integrity of the central motor pathway was assessed by noninvasive magnetic stimulation of the motor cortex [3,4]. For central stimulation the circular coil was centered over the motor cortex (vertex), while for peripheral stimulation the coil was placed over the vertebral spine at C7 and L1. Electromyographic responses were recorded from both anterior tibial and thenar muscles. Potentials were recorded with surface electrodes. The active electrode was placed over the belly (10 cm below the inner margin of the knee joint for the anterior tibial muscle) with the reference over the tendon of the muscle.

During central stimulation the subject had to maintain a slight voluntary contraction of the target muscle. The preinnervation was monitored visually and acoustically prior to measurement. The shortest of four latencies was measured at a stimulus intensity 20% above cortical threshold.

The central (corticomotoneuronal) latency was calculated by subtraction of the peripheral latency from the total latency. Results were compared with the normative data of our laboratory [9,10]. Typical responses are shown in Fig. 3.

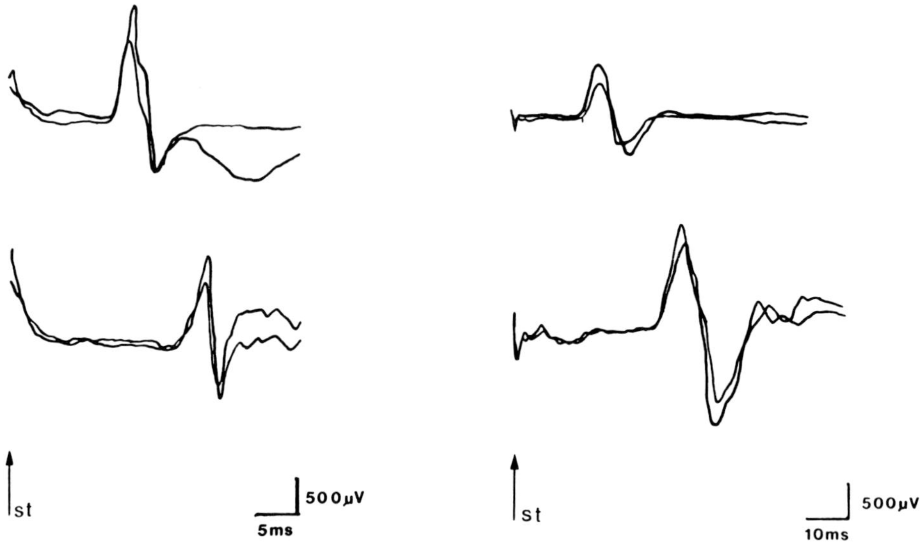

Fig. 3. Examples of evoked muscle responses after peripheral stimulation (*upper traces*) and cortical magnetic stimulation (*lower traces*) recorded from the thenar (*left*) and anterior tibial (*right*) muscles

Results

Clinical Findings

In ten patients 9 of 40 extremities showed signs of upper motor neuron dysfunction (spastic muscle tone, Babinski's sign, and sustained ankle clonus). Sensory symptoms were detected in 21 of the 40 investigated extremities (impairment of proprioception).

SEP Findings

Studies of SEPs were done in 20 median and 20 tibial nerves in ten patients. Of 40 results, 26 (65%) were abnormal. No cortical component could be identified after stimulation of 12 nerves (30%). Central conduction time (N14–N20 and N21–P40) was increased after stimulation of 14 nerves (35%). Of 20 lower extremities, 17 (85%) had a significant abnormality, whereas upper extremity testing was abnormal in 9 of 20 cases (45%). Figure 4 shows the results of the individual patients in detail.

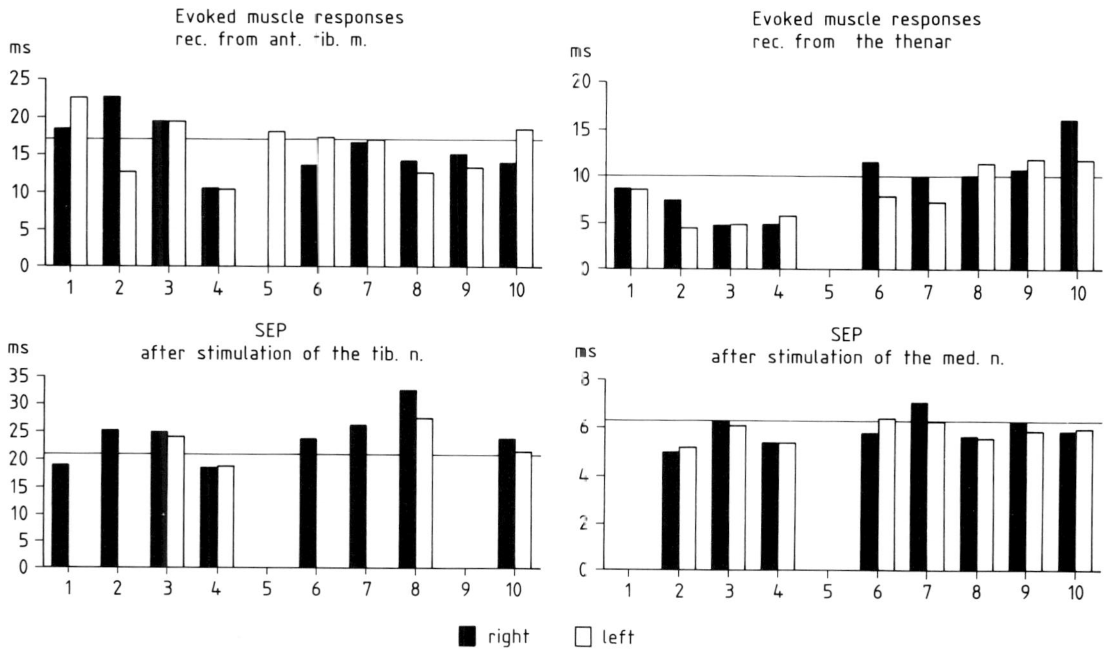

Fig. 4a, b. Central latencies of SEPs and evoked muscle responses of the patients suffering from syringomyelia (the *horizontal line* indicates the mean value in normals + 2.5 standard deviations)

Evoked Muscle Responses

Evoked muscle responses were recorded from 40 muscles in ten patients. Of 40 responses, 20 (50%) were abnormal. In three muscles (7.5%) no potential could be obtained. For 17 muscles (42.5%) central latencies were prolonged. The number of abnormal results was comparable for thenar (n = 10) and anterior tibial muscles (n = 10). The detailed results are summarized in Fig. 4.

Relations Between Findings

There was a definite relation between SEP findings and sensory symptoms. In 21 of 40 limbs sensory symptoms and abnormal SEPs were present. Normal SEP results were always associated with normal clinical examination (n = 14). In five recordings SEPs detected subclinical lesions of the ascending tracts.

In all extremities with clinical signs of upper motor neuron dysfunction, increased central motor latencies could be found (n = 9). The presence of subclinical lesions could be shown in 11 extremities. In 20, normal central latencies were observed.

Discussion

In the SEP studies the typical abnormal features were absence of cortical potentials and increase in central conduction time (interpeak latency). This confirms and extends previous results [2]. Tibial nerve SEPs more often revealed abnormal results than median nerve SEPs. This finding may be explained by a different spatial relation between fibers responsible for lower and upper limb SEP transmission and the cyst.

The main feature of abnormalities in the central motor system was a prolonged central conduction time (corticomotoneuronal latency). No response was obtained in three clinically severely affected muscles.

Both methods detected clinical and subclinical deficits. A comparison of the methods shows that SEP studies are more often abnormal than examination of the central motor system. Combined use of the two methods reveals more pathological findings than one alone.

In conclusion, our results show that both techniques can detect clinical and subclinical deficits of the spinal sensory and motor tracts in patients with syringomyelia. This may be helpful for precise definition of functional deficits and evaluation of the effects of surgical intervention.

Summary

Somatosensory evoked potentials (SEPs) and muscle responses after noninvasive stimulation of the motor cortex were investigated in ten patients with syringomyelia. With both methods, clinical and subclinical deficits of the long sensory and motor tracts could be detected. Both methods may be helpful for the definition of functional deficits and the evaluation of the effects of surgical intervention.

References

1. Aichner F, Poewe W, Rogalsky W, Wallnofer K, Willeit J, Gerstenbrand F (1985) Magnetic resonance imaging in the diagnosis of spinal cord diseases. J Neurol Neurosurg Psychiatry 48:1220–1229
2. Anderson NE, Frith RW, Synek VM (1985) Somatosensory evoked potentials in syringomyelia. J Neurol Neurosurg Psychiatry 49:1407–1410
3. Barker AT, Jalinous R, Freeston IL (1985) Non-invasive magnetic stimulation of the human motor cortex. Lancet II:1106–1107
4. Barker AT, Freeston IL, Eng B, Jalinous R, Jarratt JA (1987) Magnetic stimulation of the human brain and peripheral nervous system: an introduction and the results of an initial clinical evaluation. Neurosurgery 20:100–109
5. Cowan JMA, Dick JPR, Day BL, Rothwell JC, Thompson PD, Marsden CD (1984) Abnormalities in central motor pathway conduction in multiple sclerosis. Lancet II:304–307
6. Ingram DA, Swash M (1987) Central motor conduction is abnormal in motor neurone disease. J Neurol Neurosurg Psychiatry 50:159–166
7. Hess CW, Mills KR, Murray NMF, Schriefer TN (1987) Magnetic brain stimulation: central motor conduction studies in multiple sclerosis. Ann Neurol 22:744–752
8. Ludolph AC, Spille M, Masur H, Elger CE (1988) Befunde im peripher-motorischen System nach Stimulation der motorischen Wurzeln: Polyradikulitis, amyotrophe Lateralsklerose and Polyneuropathie. Z EEG-EMG 19:255–259
9. Ludolph AC, Wenning G, Masur H, Füratsch N, Elger CE (in press) I. Die elektromagnetische Stimulation des Nervensystems. Normalwerte im Zentralnervensystem und Vergleich mit elektrischen Stimulationsmethoden. Z EEG-EMG
10. Ludolph AC, Diekaemper S, Masur H, Elger CE (in press) II. Methodik der magnetischen Stimulation des peripheren Nervensystems – Vergleich mit elektrischen Stimulationstechniken. Z EEG-EMG
11. Masur H, Elger CE, Render K, Fahrendorf G, Ludolph AC (in press) Functional deficits of central sensory and motor pathways in patients with cervical spinal stenosis: a study of SEPs and of EMG responses to non-invasive brain stimulation. Electroencephalogr Clin Neurophysiol
12. Merton PA, Morton HB (1980) Stimulation of the cerebral cortex in the intact human subject. Nature 285:227
13. Merton PA, Hill KD, Marsden CD, Morton HB (1982) Scope of technique for electrical stimulation of the human brain, spinal cord, and muscle. Lancet II:597–600
14. Thompson PD, Dick JPR, Asselman P et al. (1987) Examination of motor function in lesions of the spinal cord by stimulation of the motor cortex. Ann Neurol 21:389–396

15. Yu YL, Jones SJ (1985) Somatosensory evoked potentials in cervical spondylosis: correlation of median, ulnar and posterior tibial nerve responses with clinical and radiological findings. Brain 108:273–300

Lumbosacral Lipomas and Lipomyelomeningoceles: Operability and Long-Term Follow-up

K.E. Richard, J. Steinhauer, P. Sanker, and R.A. Frowein[1]

Introduction

One hundred years ago, Recklinghausen [7] showed that the delayed development of neurological symptoms in patients suffering from lumbosacral lipoma is caused not by compression but by tethering of the spinal cord to the lipoma. However, the detethering of cauda as well as of conus has only been practiced over the last few years. Several authors have monitored postoperative results and a few papers have dealt with postoperative follow-up, but up to now long-term outcome has not been adequately studied.

Patients

In this study we have investigated the long-term results of 37 patients treated since 1951 at the Department of Neurosurgery, University of Cologne. At the time of surgical treatment the patients' average age was 6.5 years (range: 10 months to 65 years) for the 26 cases of lipoma and 11 months for the 11 cases of lipomyelomeningocele. The female to male ratio was 1.5:1. The median follow-up period is about 20 years, with a range from 1 to 38 years.

Results

Preoperative Status

The predominant preoperative findings in both groups of patients were gait disturbances, motor deficits, and bladder dysfunctions. The incidence of foot deformity was markedly (twofold) higher in patients with lipomyelomeningoceles.

[1] Neurochirurgische Universitätsklinik Köln, Joseph-Stelzmann-Str. 9, D–5000 Köln 41

Advances in Neurosurgery, Vol. 18
K.-A. Bushe, M. Brock, and M. Klinger (Eds.)

Surgery

The type of surgery performed was as follows (Table 1): in 22 cases the lipoma was totally or subtotally removed without adequate detethering, in six cases the lipoma was removed extradurally, and in two cases decompression was achieved by duraplasty; only in five cases was an excision of the lipoma carried out as well as complete detethering of the conus medullaris. For this reason the following neurological observations are not grouped according to different surgical procedures.

Neurological Functions During Follow-up

Gait Disturbance. Of 17 patients with gait disturbances, 8 showed a postoperative improvement, while 8 of 16 patients with a normal preoperative gait showed deterioration.

One patient suffering from a postoperative deterioration achieved a restitutio ad integrum after several years.

Motor Deficit. We observed similar results in respect of motor deficit: 7 of 14 patients showed no improvement, and 9 of 20 patients with no preoperative motor deficit had postoperative motor disturbances.

Sensory Disturbance. Postoperative improvement was generally transient. Of 22 patients with intact preoperative sensory function, 10 showed no relevant changes after surgery.

Bladder Dysfunction. Bladder dysfunction present prior to surgery in 14 patients remained unchanged after surgery in 7 patients and improved in 7 patients. Of 20 patients without preoperative disturbances, 9 showed deterioration.

Bowel Dysfunction. Apart from the usual early postoperative complaints, 30 of 34 patients retained satisfactory bowel function.

Total Evaluation of Follow-up (Figs. 1, 2)

For evaluation of the long-term results we have classified our results as follows: (a) improvement of functions, (b) unchanged good functions, (c) slight deterioration, and (d) severe deterioration.

Long-term outcome with improved (a) and with unchanged neurological function (b) was considered to be favorable, while long-term outcome with deterioration of function (c/d) was regarded as unfavorable.

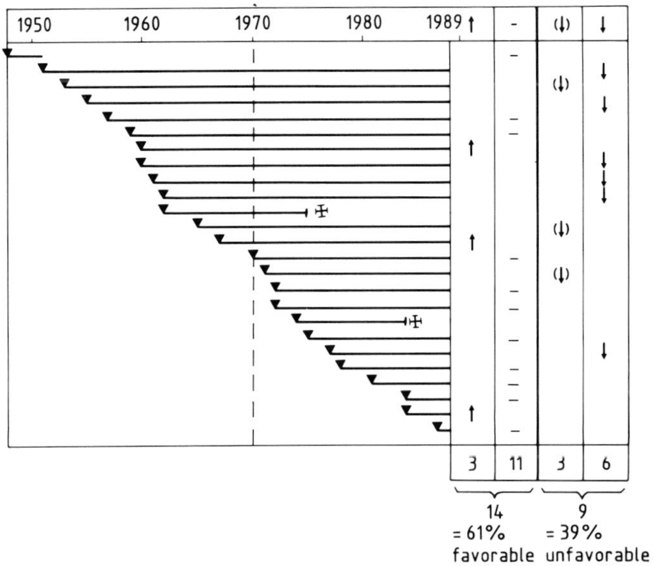

Fig. 1. Length of postoperative follow-up and long-term outcome as regards neurological function in 25 patients with lumbosacral lipoma. The following groups of long-term results were differentiated: ↑, improvement of function; –, unchanged good function; (↓), slight deterioration in function; ↓, severe deterioration in function

A satisfactory long-term result was achieved in 61% of the patients suffering from lumbosacral lipoma, while 39% of the clinical courses were not satisfactory. On the other hand, the outcome of treatment in patients with lipomyelomeningocele was only favorable in 45% of the cases and unfavorable in the remainder.

We found increasingly better results for the postoperative outcome of lipomas since 1970 (Fig. 1).

We have been able to demonstrate using magnetic resonance imaging (MRI) that several of our patients, although healthy, have a very low placement of the conus medullaris, with lipoma residues being evident in some.

Operative Procedures and Long-Term Results (Table 1)

In relating long-term results to the kind of operative management, we can see that about 45% of patients *without* detethering of the conus (Table 1; groups 2–5) showed a deterioration in neurological function, whereas patients *with* complete detethering retained satisfactory neurological function.

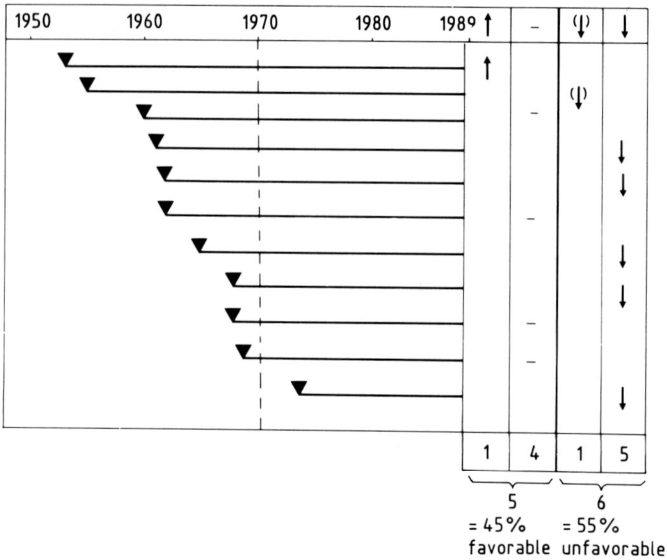

Fig. 2. Length of postoperative follow-up and long-term outcome as regards neurological function in 11 patients with lumbosacral lipomyelomeningoceles. For explanation of symbols, see Fig. 1

Table 1. Operability and long-term results in patients with lumbosacral lipoma or lipomyelomeningocele. For explanation of symbols, see Fig. 1

	Total no.	↑	–	↓
1. Removal of lipoma (totally/subtotally) + adequate detethering of conus	5	1	4	–
2. Removal of lipoma (totally/subtotally) + incomplete detethering of conus	22	1	10	11
3. Inoperable lipoma. Duraplasty only	2	–	1	1
4. Extradural removal of lipoma. Cosmetic operation	6	2	2	2
5. No surgery	2	–	1	1
	37	4	18	15

Table 2. Outcome after surgery for lumbosacral lipoma or lipomyelomeningocele. Results as described in the literature and in our own investigations. For explanation of symbols, see Fig. 1

Author	Year	n	Age (years)	$\overline{X}$	Op/ detethering	Repeat op.	Max. follow-up (years)	Outcome (%) ↑	–	↓
Bruce and Schut	1979	23	1 d-16	–	+	–	12	35	57	8
Rickwood et al.	1979	74	1 d-11.5	1.25	+	–	7	7	68	25
Chapman and Davis	1983	17	5 d-17	–	+	–	?	–	82	18
McLone et al.	1983	42	1 m-16	–	+	6	10	31	67	2
Lapras et al.	1985	57	< 2 – 15	–	+	–	?	44	53	3
Pierre-Kahn et al.	1986	64	< 2 – 52	–	+	–	?	22	75	3
Stolke et al.	1988	26	1 m-12	–	+	–	?	27	73	–
Hoffman et al.	1985	73	6 d-18.5	–	+	11	14.6	12	66	22
	1985	24	6 d-18.5	–	–	–	?	–	33	67
Cologne	1989	5	10 d-5	2.5	+	–	32	20	80	–
	1989	31	3 m-65	12	–	–	38	10	45	45

Conclusion

The pathobiomechanical concept of tethered conus medullaris in cases of lumbosacral lipoma, dating back to Recklinghausen, was confirmed by an improvement of postoperative results after meticulous detethering of conus medullaris (Table 2). Our report covers the long-term follow-up of patients with a mean clinical course of 20 years, independent of operative management. We are not able to compare our results with those of other authors because of a lack of reports with sufficient long-term follow-up after detethering of conus.

In the two large series of Hoffman et al. [3], the maximum length of follow-up was 14 years, and the mean length was only about 7 years. Therefore, the following questions still remain unanswered:

- How often does retethering of the conus occur due to scar formation in the long-term follow-up?
- If retethering occurs, what are the indications for reoperation?
- Should a patient with MRI signs of retethering be operated on despite intact neurological function?
- What is the likelihood of restoring normal function in patients with preexisting neurological deficit after reoperation?

References

1. Bruce DA, Schut L (1979) Spinal lipomas in infancy and childhood. Childs Brain 5:192–203
2. Chapman PH, Davis KR (1983) Surgical treatment and spinal lipomas in childhood. Concepts Pediatr Neurosurg 3:178–190
3. Hoffman HJ, Taecholarn C, Hendrick EB (1985) Management of lipomyelomeningoceles. Experience at the Hospital for Sick Children. J Neurosurg 62:1–8
4. Lapras CL, Patet JD, Huppert PH, Mottolese C (1985) The tethered cord syndrome. Experience of 58 cases. J Pediatr Neurosci 1:39–50
5. McLone DG, Mutluer TP, Naidich P (1983) Lipomeningoceles of the conus medullaris. Concepts Pediatr Neurosurg 3:170–177
6. Pierre-Kahn A, Lacombe J, Pichon J et al. (1986) Intraspinal lipomas with spina bifida. Prognosis and treatment in 73 cases. J Neurosurg 65:756–761
7. Recklinghausen F von (1886) Untersuchungen über die Spina bifida. Virchows Arch Path Anat Physiol 105:243–330, 373–455
8. Richard KE, Sanker P, van den Bergh P (1986) Lumbo-sacral lipoma in spina bifida occulta. In: Voth D, Glees P (eds) Spina bifida – neural tube defects. De Gruyter, Berlin
9. Rickwood AMK, Hemalatha V, Zachary RB (1979) Lipoma of the cauda equina (lumbosacral lipoma): a study of 74 cases operated in childhood. Z Kinderchir 27:159–169
10. Stolke D, Zumkeller M, Seifert V (1988) Intraspinal lipomas in infancy and childhood causing a tethered cord syndrome. Neurosurg Rev 11:59–65

The Optimal Time for Operative Treatment of the Tethered Cord Syndrome

S. Vogel, K. Neumärker, and K. Klepel[1]

The neuropathology of malformations in the rostral transition area of the neuroectoderm has been much better investigated than has that of problems in the caudal transition area (Figs. 1, 2). Therefore the methods of and indications for neurosurgical management are not based on a fetal pathology of the neural tube and the meninges [5, 6].

The tethered spinal cord is not only a special malformation of the neural tube but also an inhibition in the development of the caudal neuroectodermal segments, and an inhibition of the volume formation in the embryonic neural plate in contrast to the ectodermal somites (Fig. 3). We can definitely state today that a tethered spinal cord is a transverse caudal malformation, not an axial craniocaudal malformation with tension on nerve roots and spinal blood vessels. The ascending angle of the lumbar spinal nerves in relation to the axis of spinal cords is not the cause of the clinical findings. The clinical manifestations are secondary, resulting from the posture of the vertebral column after the first 2 years of life with an ischemic lesion of the caudal spinal segments [7].

Investigations of normal fetuses have demonstrated regression of the caudal neuroectoderm during the last 3 months of pregnancy. In pathological fetal development we can demonstrate a tethered spinal cord with ascendant spinal nerves at the end of pregnancy. A tethered spinal cord occurs in most babies with myeloceles, lipomyeloceles, congenital dermal sinus, and pilonidal sinus dimple. But the clinical symptoms vary greatly depending on the grading and the time of manifestation. In the last 10 years we have operated on:

- 52 newborns with caudal myelomeningoceles
- 4 children with epidural lipomyelomeningoceles
- 4 children with transdural lipomyeloceles
- 10 children with tethered cord syndrome

In six children we observed a secondary tethered cord syndrome following conventional surgery for myeloceles years before.

[1] Neurochirurgische Universitätsklinik, Leipziger Str. 33, DDR–3090 Magdeburg

Advances in Neurosurgery, Vol. 18
K.-A. Bushe, M. Brock, and M. Klinger (Eds.)

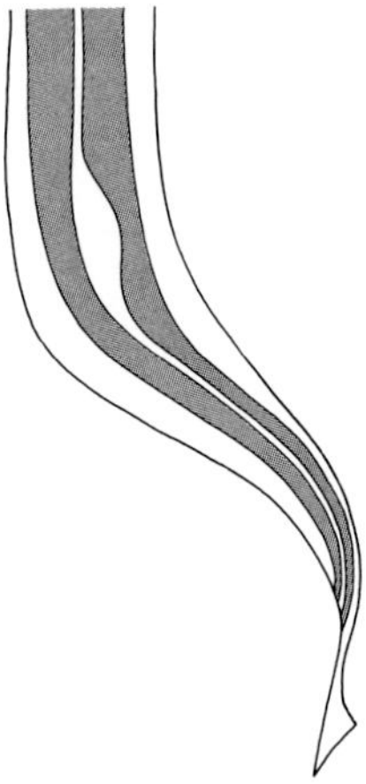

Fig. 1

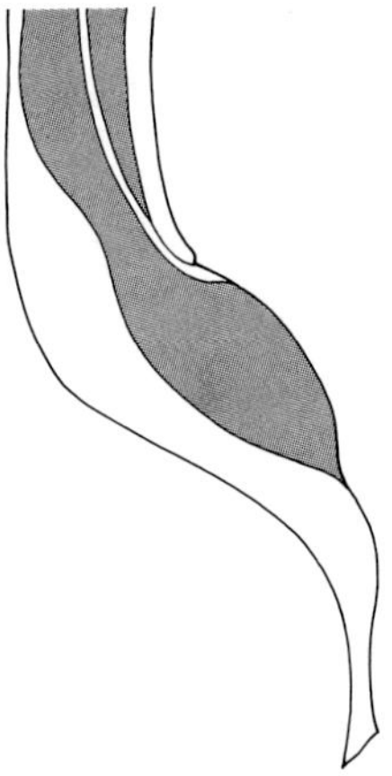

Fig. 2

Fig. 1. The lumbosacral region in the fetus after 24 weeks of pregnancy
Fig. 2. The lumbosacral region in the fetus with a myelomeningocele, with a meningeal defect and inhibition of regression in the neural plate

All newborns with myelomeningoceles were operated on immediately after delivery. The children with lipomyeloceles or lipomyelomeningoceles were operated on during the first 4 years after birth. However, the children with tethered cord syndrome were treated between the 4th and the 13th years of life. In such children we have never seen a clinical improvement following the transection of the filum terminale alone. We believe that this operation comes too late, and that an irreversible disturbance of the spinal cord is then already present.

In the children with myelomeningoceles we always employ microsurgical technique. We first dissect the end of the neural plate and then cut off the end of the neural plate close to the primitive meningeal layer. We subsequently build a neural tube with a caudal ampulla with a microsuture (7×0) at the membrana limitans gliae externa. We then cover the spinal cord only with the plain dura, without the skin. We avoid compression of the spinal cord due to a plastic skin procedure. With this method we have always been able to prevent further neurological deterioration of the motor activity of the pelvic musculature and the legs.

We employ the same method in children with lipomyeloceles. We first remove the tissue causing a compression of the nerves and then mobilize the filum terminale in the region of the meningeal defect.

In the children with minimal pathology involving regression of the filum terminale, we transect only the filum terminale, but this method is not adequate in all cases. In all children with neurological signs in the segments L2–5 we carry out a bilateral transfer of the intercostal nerves from the 11th thoracic segment to the 2nd lumbar segment for selective reinnervation of iliopsoas muscle (Fig. 4) or

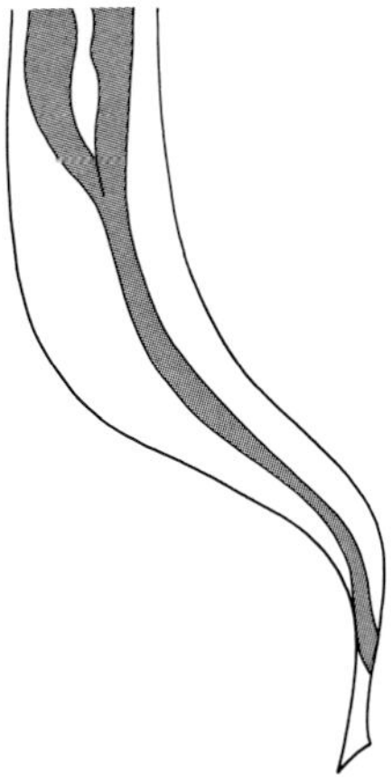

Fig. 3

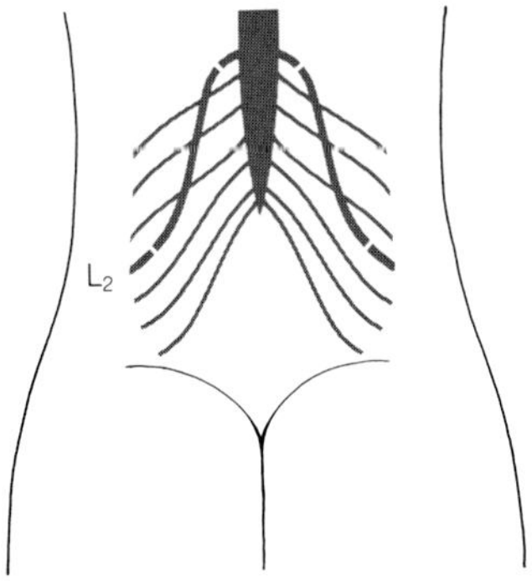

Fig. 4

Fig. 3. The lumbosacral region in the fetus with tethered cord syndrome, without a meningeal defect and with inhibition of regression in the neural tube
Fig. 4. Method of the selective motor reinnervation, with transfer of the 11th intercostal nerve end to end with the roots of the 2nd lumbar segment

to the 5th lumbar segment for selective reinnervation of m. tibialis posterior. The best results are achieved with an early operation, before irreversible neurological symptoms are present. In patients with tethered cord syndrome we are very often too late. The best time for this operation is the first year, before the child learns to walk.

References

1. Hoffman JH, Hendrick EB, Humphreys RP (1976) The tethered spinal cord. Its protean manifestations, diagnosis, and surgical correction. Childs Brain 2:145–155
2. Hoffman JH, Taecholarn C, Hendrick EB, Humphreys RP (1985) Management of lipomyelomeningoceles. J Neurosurg 62:1–8
3. Reigel DH (1983) Tethered spinal cord. Concepts Pediatr Neurosurg 4:142–164
4. Roy MW, Gilmore R, Walsh JW (1986) Evaluation of children and young adults with tethered spinal cord syndrome. Surg Neurol 26:241–248
5. Smith DW (1982) Recognizable patterns of human malformation. W.B. Saunders, Philadelphia
6. Venes JL, Stevens EA (1983) Surgical pathology in tethered spinal cord secondary to myelomeningocele repair. Concepts Pediatr Neurosurg 4:165–185
7. Yamada S, Zinke DE, Sanders D (1981) Pathophysiology of tethered cord syndrome. J Neurosurg 54:494–503

Tethered Cord Syndrome Following Spina Bifida Aperta

R.I. HogenEsch, D.J. Zeilstra, S.M.E. Breukers, G.P.A. Wiertsema, J.H. Begeer, and C.A. Ter Weeme[1]

The tethered cord syndrome is well known in patients with spina bifida occulta. It is increasingly recognized that in patients with spina bifida aperta, late onset neurological deterioration may occur, caused by a tethered cord. This report reviews the symptoms, operative findings, and results of surgery in 22 patients with tethered cord syndrome following myelomeningocele repair.

Patients and Methods

At the University Hospital of Groningen 220 patients with spina bifida aperta were checked on a regular basis from 1980 to 1989. Of these patients 88 were not able to walk or were not expected to do so on the basis of their neurological deficit, while 132 were good walkers, community or residence ambulators, or were expected to become walkers. Of the latter group 22 patients (17%) developed progressive neurological deficit. The history, neurological abnormalities, investigations, surgical observations, and results of operation were retrospectively studied. The follow-up period varied from 6 months to 8 years. Some clinical data are given in Table 1.

Table 1. Clinical data of patients with repaired myelomeningocele and tethered cord syndrome

No. of cases	22
Sex (m:f)	12:10
Delay (months)	0–240
– average	13
Age at time of operation (years)	3–37
– average	13.1

[1] Department of Neurology, University Hospital Groningen, 9700 RB Groningen, The Netherlands

Advances in Neurosurgery, Vol. 18
K.-A. Bushe, M. Brock, and M. Klinger (Eds.)

Table 2. Presenting symptoms of neurological deterioration

Symptoms	No.	%
Motor dysfunction	20	91
Increasing abnormal foot shape	10	45
Pain in the lower back or the legs	9	41
Bladder dysfunction	3	14

Signs and Symptoms

Our group included 22 patients. In most patients increasing motor dysfunction, which consisted of progressive weakness and walking problems, was an early symptom of neurological deterioration (Table 2). Increasingly abnormal foot shape was another common problem. Bladder dysfunction concerned particularly recurrent urinary infections, incontinence, or retention. Reflex changes and increasing sensory disturbances were also noted.

Neuroradiological Findings

Metrizamide myelography and combined CT-myelography were performed in 16 cases. These revealed abnormalities in a uniform pattern: The conus medullaris had a low position in the spinal canal, below the normally expected level. The conus was adherent to the scar of the former operation at the back.

In six patients magnetic resonance imaging (MRI) was performed. In all cases an abnormally low positioned conus, in the lower lumbar or the midsacral region, was found. As additional abnormalities MRI showed lipomas in three cases and a syrinx in one.

Operative Technique and Findings

A semicircular incision is made along one side of the scar, with an extension in the midline in a cranial direction. The incision reaches the fascia. On the cranial side laminectomy of the lowest intact neural arch is performed in order to reach the right cleavage plane and to identify the dura. Around the scar the subcutis is dissected from the fascia, in order to identify the laminal defects.

At the site of the first operation the dura is dissected free from the cutis. The extradural intraspinal space is identified. The dura is incised proximally in the midline and opened in a distal direction. At the site of the defect the adherent

Table 3. Operative findings

	No.	%
Scar tissue	18	82
Thickened and tight filum terminale	9	41
Lipoma	5	23
Teratoma	3	14
Diastematomyelia	2	9

structures are released from the cord. By this procedure the tethering is dissolved. The cord is now in the intradural space. The dura is closed with a lyophilized dural graft.

The most common finding at operation was scar tissue, which formed adhesions between the conus and the site of previous repair (Table 3). In nine cases a thickened and tight filum terminale was resected. Lipoma was found in five patients. In six patients early complications occurred: CSF leakage in four patients, wound infection in two. One patient needed revision of the ventriculoperitoneal shunt system shortly after the operation because of decompensation of the hydrocephalus.

Results

The follow-up period varied from 6 months to 8 years. The aim of the operation was to stop further progression of neurological deterioration. In Table 4 the results of operation are listed. Motor dysfunction and pain showed the best improvement. In most patients more than one symptom improved. Most of the amelioration appeared within 6 months postoperatively.

Unfortunately, recently three patients had to be operated on because of a retethering. In two of them the progression continued after the first operation; the third

Table 4. Results of operation

	Better	Stable	Worse
Motor dysfunction	13	7	2
Pain	7	13	2
Bladder dysfunction	5	17	0
Foot deformity	2	20	0
Overall	17	3	2

was initially stable but neurological deterioration occurred again 4 years after the first operation. The three of them are stable now, but their follow-up period is too short to draw any conclusions yet.

Discussion

Much has been written on the tethered cord syndrome in relation to spina bifida occulta [1,4,7]. Fewer reports have been published in which the tethered cord syndrome is associated with repaired myelomeningocele [2,5,6].

As in patients with spina bifida occulta, deterioration occurs in most patients in a slowly progressive way with sometimes unexpected accelerations. However, the detection of deterioration is more difficult than in patients with occult spinal dysraphism, because of the preexisting neurological deficits. Therefore, early and carefully documented neurological examination with regular follow-up examination is necessary for patients with spina bifida aperta, to detect tethered cord syndrome in an early stage. Our patients presented with increasing motor dysfunction of the legs, consisting in changes in gait and progressive weakness, increasing foot deformity, backache, or bladder dysfunction.

CT-myelography was performed in most patients. Ever since MRI was available, it has usually been used for radiological examination. MRI has several advantages over CT-myelography: It offers useful additional information for the neurosurgeon, by demonstrating good tissue differentiation and by clearly showing syringomyelia when present [3,8]. Furthermore it is noninvasive and it obviates a radiation load to a generally young population. In our opinion MRI is the screening technique of choice when a tethered cord is suspected.

In those cases in which the neuroradiological findings confirmed the diagnosis of tethered cord syndrome, the patients were operated on as soon as possible to stop further deterioration.

The untethering procedure was relatively successful; 17 of the 22 patients improved and in three patients further progression was stopped. Unfortunately one patient who initially showed improvement deteriorated after 4 years. MRI showed retethering. We have no explanation for this but it might be possible that both tetherings have been caused by the same factor. The patient was operated on for a second time and is stable up to now, but it is too early to draw any conclusions.

The fact that retethering can occur makes accurate follow-up necessary, not only after the myelomeningocele repair but also after tethered cord operations. Furthermore it seems worthwhile to search for operative procedures which can prevent postoperative tethering. Operative techniques for this purpose have been suggested [5,6], but there is no agreement so far.

At this moment careful and regular neurological follow-up with MRI, when indicated, is very important, especially in those patients who can walk.

References

1. Bakker-Niezen SH, Walder HAD, Merx JL (1984) The tethered spinal cord syndrome. Z Kinderchir 39 [Suppl II]:100–103
2. Begeer JH, Meihuizen de Regt MJ, HogenEsch RI, Ter Weeme CA, Mooij JJA, Vencken LM (1986) Progressive neurological deficit in children with spina bifida aperta. Z Kinderchir 41 [Suppl I]:13–15
3. Jaspan T, Worthington BS, Holland IM (1988) A comparative study of magnetic resonance imaging and computed tomography-assisted myelography in spinal dysraphism. Br J Radiol 61:445-453
4. Pierre-Kahn A, Lacombe J, Pichon J et al. (1986) Intraspinal lipomas with spina bifida. J Neurosurg 65:756–761
5. Tamaki N, Shirataki K, Kojima N, Shouse Y, Matsumoto S (1988) Tethered cord syndrome of delayed onset following repair of myelomeningocele. J Neurosurg 69:393–398
6. Venes JL (1985) Surgical considerations in the initial repair of meningomyelocele and the introduction of a technical modification. Neurosurgery 17:111–113
7. Villarejo FJ, Blazquez MG, Gutierrez-Diaz JA (1976) Intraspinal lipomas in children. Childs Brain 2:361–370
8. Wippold FJ, Citrin C, Barkovich AJ, Sherman JS (1987) Evaluation of MR in spinal dysraphism with lipoma: comparison with metrizamide computed tomography. Pediatr Radiol 17:184–188

Intraspinal Lipomas with Tethered Cord Syndrome – Results of Operative Treatment in 30 Children

M. Zumkeller, D. Stolke, and H. Dietz[1]

Introduction

The different growth of the vertebral column and the spinal cord leads to a relative cranial ascent of the cord and to the formation of the cauda equina during the embryonal development. After the 25th week of embryogenesis the conus has reached the L3 level, where it remains until the time of birth. During the first 5 years of life the conus migrates upwards to the interspace L1/2. Fixation of the cord by a filum terminale or by lipomatous subcutaneous tissue is the most common cause of a tethered cord syndrome. A further cause may be an intraspinal lipoma.

When severe neurological deficits are present, there will be no difficulties in the diagnosis of dysraphic disorders. But the normal or uncomplicated case will be overlooked in spite of subcutaneous lipomas or exuberant hair growth at the lumbar sacral area already present at birth.

The tension of the cord causes weakness and motor imbalance of the foot and leg which may result in foot deformities and/or scoliosis, and secondary bladder disturbances may become a leading symptom.

Patients

Thirty infants and children suffering from an occult dysraphism and tethered cord syndrome were operated on at the Medical School, Hannover, between 1973 and 1988. At the time of diagnosis the patients' age ranged from 1 month to 12 years. Seventeen patients (group 1) were below the age of 2 years. Thirteen patients (group 2) were over the age of 2 years.

[1] Neurochirurgische Klinik, Medizinische Hochschule Hannover, Konstanty-Gutschow-Str. 8, D–3000 Hannover 61

Advances in Neurosurgery, Vol. 18
K.-A. Bushe, M. Brock, and M. Klinger (Eds.)

Table 1. Pre- and postoperative clinical findings in 30 patients with tethered cord syndrome and spinal lipoma

	Preop.	Postop.	
		Improved	Unchanged
Absence of tendon reflexes	22	–	22
One side	10	–	10
Both sides	12	–	12
Bladder and bowel impairment	17	7	10
Primary	9	3	6
Secondary	8	4	4
Foot deformities	13	–	13
Motor weakness	11	–	11
Sensory disturbances	11	–	11
Muscular atrophy	11	–	11
Palpable subcutaneous lipoma	11		
Nevus	10		
Hemangioma of the skin	5		
Exuberant hair growth	6		
Back pain	1	1	–

Signs and Symptoms

Skin lesions like nevus flammeus, hyperpigmentation, hypertrichosis, or subcutaneous lipomas over the lumbosacral spine were the most common reason for admission. In 13 of the 17 patients in group 1 these signs were found at birth. Of the 13 patients in group 2, 10 also presented with similar skin lesions at birth. The reason for postponing diagnostic procedures despite these obvious dermatological signs could not be elucidated. The most frequent neurological signs were the absence of tendon reflexes (22 of the 30 patients), followed by disturbed bladder and bowel functions together with a sensory disturbance in this area. In 8 of the 13 patients in group 2 this dysfunction occurred only secondarily.

Weakness and motor deficits were equally distributed in the two groups, and usually consisted of weakness of the foot extensors and flexors in one or both legs. Muscle wasting was a common clinical sign. Four children presented with asymmetry of the legs and feet. Isolated foot deformities could be observed in 13 cases, which erroneously led to long-term orthopedic treatment just as secondary bladder impairment led to urological treatment. In the older children the progressive neurological deficits with bladder and bowel impairment were the principal signs which led to admission and to further diagnostic and finally therapeutic procedures. (A summary of the symptoms is compiled in Table 1.)

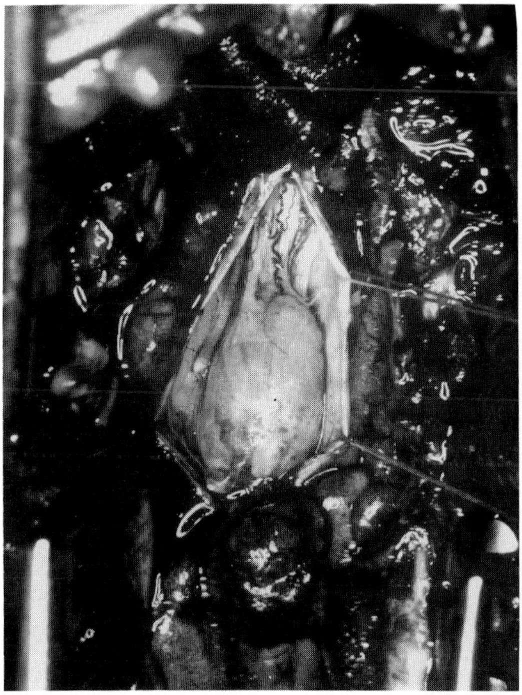

Fig. 1. Intraoperative demonstration of the lumbar sacral lipoma in situ together with the tethered conus medullaris

Diagnostic Procedures

The above-mentioned skin lesions and subcutaneous lipomas were indicative of occult spinal dysraphism, which was confirmed by plain X-ray examinations revealing anomalies of the lumbosacral arches in all of our 30 patients. In 18 cases the failed fusion of the arches was localized at the lumbar and sacral levels. In six patients open lumbar arches and sacral agenesis could be demonstrated. All our children presented scoliosis as a typical sign of spinal cord tethering.

In 25 out of 30 surgically treated children a radiological examination using a contrast medium was performed. In four cases computerized spinal tomography was added. Our latest five cases were diagnosed by MRI alone. In all of these children a conus medullaris at a low level (below L3/4) was found. Additionally in 15 cases a megacauda was found. Nerve roots running nearly horizontally to the periphery were evaluated as a typical sign of a tethered conus medullaris.

The most recent procedures, like MRI, give an excellent anatomical presentation of the structures that are to be approached surgically.

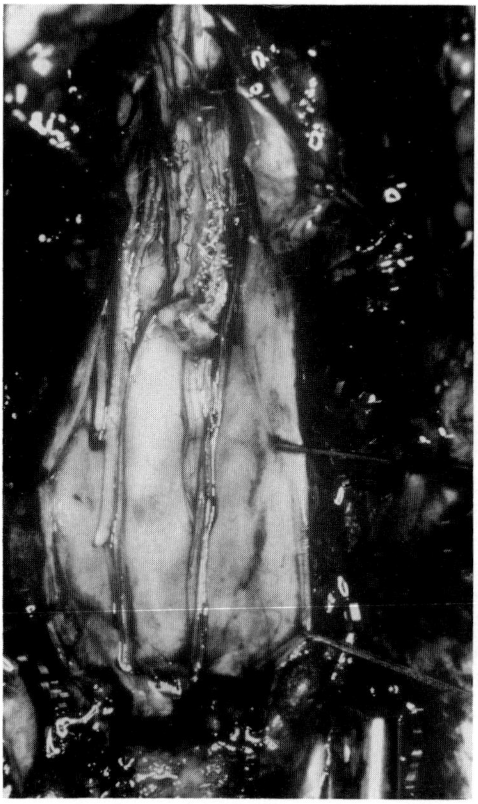

Fig. 2. Intraoperative picture of the same patient as in Fig. 1 following total extirpation of the lipoma and untethering of the conus. The thickened filum terminale is well demonstrated

Surgical Treatment

The aim of surgery is to untether the cord, to eliminate the mechanical constraint, and to release the elongated conus. The simplest way to accomplish this might be to resect the thickened filum terminale, which would release the elongated conus. Untethering is more difficult when the lipoma attachments are to be dissected and a lipomatous mass must be excised. Because of the poorly demarcated layer between lipoma and neural tissue even when microsurgical techniques are employed the lipoma bulk may often only be reduced; the dural adhesions, however, have to be dissected thoroughly, as does the filum terminale. Since CO_2 laser and Cusa are available in our clinic, they are used for dissection and reduction of the intraspinal lipoma mass (Figs. 1, 2).

The intraoperative findings can be summarized as follows: In 23 cases we found an extra- and intradural lipoma growing into the conus and into the thickened filum terminale. In seven cases an extensive intradural lipoma was found as well as adhesions to the posterior dura causing a tethering of the cord. In six cases we found an occult medullary plate directly beneath the skin and partly adherent to the lipoma. In seven cases cysts were present in the dissected conus medullaris. In four cases a diastematomyelia with bony spur and duplication of the dura was found. Excision of the bony spur was performed, as well as lysis of the surrounding adhesions. Diplomyelia and a torquated spinal cord were each found twice.

Results

In our 30 patients there were no mortalities and no postoperative complications. In none of the patients did postoperative neurological deterioration occur. Eight cases were improved. In only three cases did the bladder and bowel impairment disappear, while in four older children subjective amelioration of symptoms was reported. However, preoperative muscular atrophy, absence of the tendon reflexes, and foot deformities could not be ameliorated (Table 1).

Discussion

The disturbed relative ascent of the spinal cord within the vertebral canal caused by lipomeningoceles, a thickened filum terminale, or abnormalities like diastematomyelia is a severe disease accompanied by progressing neurological deficits as well as deformities of the lower limbs and scoliosis. These signs and symptoms worsen with the increasing age of the patient [3,4,8,9]. In addition the already tethered spinal cord is much more vulnerable to mechanical alterations and injuries to the vertebral column. The group of children beyond the age of 2 years demonstrated

Table 2. Results of surgical treatment (overview of eight series)

Author	Year	No. of operated patients	Deterioration	Improved	Unchanged
Lassman and James [7]	1967	22	2	5	15
Andersson [1]	1975	21	0	0	21
Naidich et al. [10]	1983	42	2	0	40
Chapman and Davis [3]	1983	17	0	5	12
Hoffman et al. [6]	1985	62	10	8	44
Pierre-Kahn et al. [11]	1985	55	2	26	27
Bakker-Niezzen et al. [2]	1985	25	7	1	17
Own series [12]	1988	30	0	8	22

the highest rate of progressive neurological deterioration. This is in accordance with the experience reported by other authors [3, 6, 7, 10].

The treatment of choice in diagnosed tethered cord syndrome is neurosurgical intervention. This is the opinion of all authors, especially as the operation using microsurgical techniques is no longer burdened by severe complications. Different opinions have been published concerning prophylactic surgical treatment, i.e., surgical untethering of the cord without the evidence of neurological deficits. Some authors advocate an observing attitude and monitoring of the neurological status [9]. Others tend towards prophylactic surgical treatment in order to avoid the development of neurological disturbances and deficits [7, 9]. According to our experience, as soon as dysfunctions has occurred no return to normal can be expected from the operation.

Analyzing our own patients and those of seven other authors, the results of surgery in patients without neurological findings are encouraging, and as the rate of complications is almost negligible, prophylactic early operation is to be advocated (Table 2).

In our opinion symptom-free patients will be protected against the development of neurological deficits by early operation. Observing the natural history of these patients, 60% will present symptoms and signs of a tethered cord during adolescence, whereas operative and postoperative complications range from only 0% to 8% [2, 3, 6, 10]. These facts support our preference for early operation.

References

1. Andersson FM (1975) Occult spinal dysraphism: a series of 73 cases. Pediatrics 55:826–835
2. Bakker-Niezzen SH, Walder HAP, Merx JL (1984) The tethered spinal cord syndrome. Z Kinderchir 39:100–103
3. Chapman PH, Davis KR (1983) Surgical treatment of spinal lipomas in childhood. In: Raimondi AJ (ed) Concepts in pediatric neurosurgery, vol 3. Karger, Basel
4. French BN (1982) Midline fusion defects and defects of formation. In: Youmans JR (ed) Neurological surgery, 2nd edn. Saunders, Philadelphia
5. Gerlach J, Jensen HP, Koos W, Kraus W (1967) Fehlbildungen des Rückenmarkes. Thieme, Stuttgart
6. Hoffman HJ, Taecholarn C, Hendrick HB, Humphreys HP (1985) Management of lipomeningoceles. J Neurosurg 62:1–8
7. Lassman LP, James CCM (1967) Lumbosacral lipomas: critical survey of 26 cases submitted to laminectomy. J Neurosurg Neurol Psychiatry 30:174–186
8. Lausberg G (1966) Zur Klinik und Differentialdiagnose der Diastematomyelie. Arch Kinderheilkd 175:14–25
9. Markakis E, Vogelsang H, Heyer R (1978) Neurochirurgisch-neuroradiologische Aspekte der occulten spinalen Ascensionsstörung im Kindesalter. Aktuel Neurol 5:253–263

10. Naidich TP, McLone DG, Mutluer S (1983) A new understanding of dorsal dysraphism with lipoma (lipomyeloschisis) – radiologic evaluation and surgical correction. AJNR 140:1065–1078
11. Pierre-Kahn A, Lacombe J, Pichon H et al. (1986) Intraspinal lipomas with spina bifida. J Neurosurg 65:756–761
12. Zumkeller M, Seifert V, Stoke D (1988) Intraspinale Lipome mit Malascensus des Rückenmarkes im Kindesalter. Z Kinderchir 43:384–390

The Use of Calcium Antagonists Following Subarachnoid Hemorrhage

Pharmacology of Nimodipine – A Review

A. Scriabine[1]

Among Ca^{2+} antagonists (Ca^{2+} channel antagonists, Ca^{2+} entry blockers, CAs) several drugs were reported to have anti-ischemic, anticonvulsant, or other effects on the central nervous system. One Ca^{2+} antagonist, nimodipine, was approved in the United States for prevention of neurological deficits in patients with subarachnoid hemorrhage. Some CAs are being used for the treatment of focal and global cerebral ischemia and other disorders of the central nervous system. This article reviews pharmacological properties of nimodipine which justify its use in some disorders of the central nervous system. It updates previous reviews on the same subject [28–30].

Nimodipine was synthesized at the Research Laboratories of Bayer AG by Meyer et al. [17] and evaluated pharmacologically by Hoffmeister et al. [11], Kazda et al. [13], and Towart [32]. The drug was characterized as a cerebral vasodilator, which was effective in prevention of postischemic impairment of cerebral blood flow in dogs. In addition, Kazda et al. [14] reported that nimodipine given chronically in food can prolong life in stroke-prone spontaneously hypertensive rats without any significant lowering of arterial pressure (Fig. 1).

Subsequently, nimodipine was intensively studied for anti-ischemic activity in various models of focal and global ischemia [8, 29]. In a primate model of global ischemia, Steen et al. [31] demonstrated that nimodipine can prevent neurological deficits caused by transient global ischemia (Fig. 2). Using experimental focal ischemia in rabbits, Meyer et al. [18] demonstrated that nimodipine can reverse changes in intracellular brain pH in the infarcted area (Fig. 3). Following these observations, the question was raised whether nimodipine produces its beneficial effects on the central nervous system by virtue of its cerebrovasodilator or a direct neuronal action. Although this question is still not definitively answered, recent studies in our laboratories have shown that nimodipine blocks L-type Ca^{2+} currents in rabbit dorsal root ganglia (DRG neurons) at very low concentrations ($K_i \simeq 1.3\,nM$) [6, 16]. A typical tracing showing the effect of nimodipine at 10 nM is reproduced in Fig. 4.

[1] Miles Institute for Preclinical Pharmacology, West Haven, CT 06516, USA

Advances in Neurosurgery, Vol. 18
K.-A. Bushe, M. Brock, and M. Klinger (Eds.)

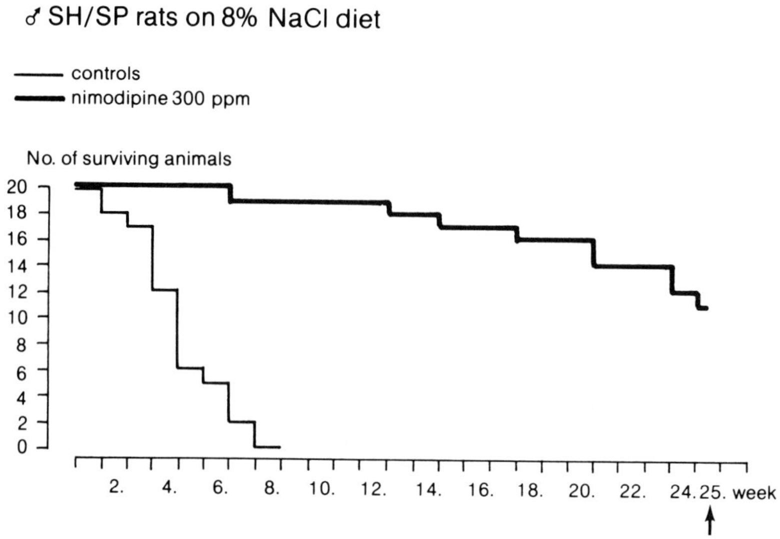

Fig. 1. Cumulative mortality of stroke-prone spontaneously hypertensive rats (SH/SP) on high salt diet. Male 5-month-old rats of 250–270 g body weight were placed on standard rat chow pellets (Sniff-Versuchsdiäten, Soet, FRG) containing 8% NaCl with or without nimodipine, 300 ppm (approximately 30 mg/kg/day) for 25 weeks. Systolic arterial pressure was measured before and 4 weeks after initiation of the study. Nimodipine had no significant effect on the arterial pressure. The systolic pressure was 210 $\pm$ 9 mmHg in control vs 220 $\pm$ 5 mmHg in animals treated for 4 weeks. (Kazda et al. [14])

Nimodipine was repeatedly shown to bind to specific dihydropyridine receptors in the brains of various species [2, 23]. Its IC_{50} in displacing a tritiated dihydropyridine from its binding sites in human brain cell membranes was found to be lower than that of nifedipine [26]. As an inhibitor of ^{45}Ca uptake in neuroblastoma cell line, nimodipine was more potent than nifedipine, flunarizine, or verapamil [5].

Physicochemical properties of nimodipine may be at least partially responsible for its beneficial effects in vivo. Nimodipine is more lipophilic than nifedipine. The higher lipophilicity can explain its higher volume of distribution in the brain of rats as compared to nifedipine [15]. As reported by Hakim et al. [10], [^{3}H]nimodipine tends to accumulate in the areas surrounding the experimental brain infarct in rats. Under conditions of relative hypoxia the binding of nimodipine to its receptor sites in vivo appears to be increased. The extent of nimodipine accumulation in the brain tissue depends on the severity as well as on the duration of the hypoperfusion, so that [^{11}C]nimodipine may represent a useful diagnostic tool in the imaging of brain infarcts.

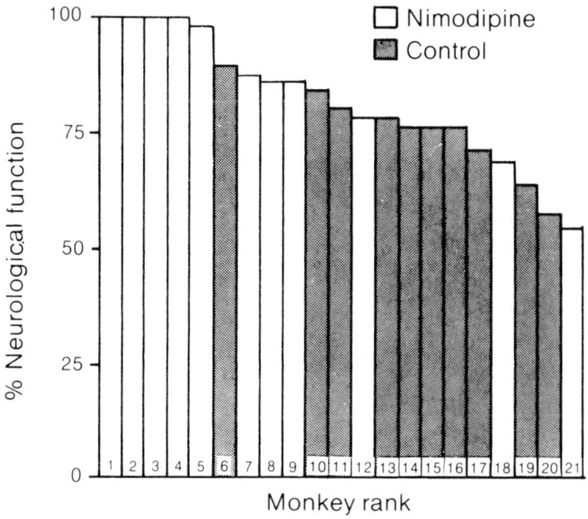

Fig. 2. Effect of nimodipine, 10 μg/kg prime, followed by infusion at 1 μg/kg/min for 10 h, on neurological function of pigtailed monkeys during recovery from transient (17 min) brain ischemia induced by hypotension and neck cuff. The animals were ranked according to neurological function at 96 h after transient ischemia. Normal animals had 100% neurological function; 0% was apparent brain death. Neurological deficit scoring system involved evaluation of consciousness, respiration, cranial nerve function, motor and sensory function, and behavior. Neurological function was significantly better in nimodipine-treated than in control animals ($P < 0.05$ by Mann-Whitney rank sum test). (Steen et al. [31])

Of practical importance for the treatment of stroke patients with a new drug are adverse effects of the therapy. The drug is expected not to increase intracranial pressure and not to facilitate brain edema. The effects of nimodipine on the intracranial pressure were studied, therefore, in baboons with occluded middle cerebral arteries (MCA) [9]. Nimodipine tended to decrease intracranial pressure in these animals. Effects of nimodipine on the brain water and electrolyte content were studied in rats with MCA occlusion [33]; nimodipine reduced water and electrolyte content in the infarcted areas of the rat brain.

By oral administration, Ca^{2+} antagonists, including nimodipine, are very safe drugs. The most commonly reported side-effect of oral nimodipine is moderate hypotension, which is dependent on the dose of nimodipine and on the blood pressure level of individual patients, since Ca^{2+} antagonists are known to lower arterial pressure in hypertensive patients to a greater extent than in normotensive individuals. By intravenous administration the doses of Ca^{2+} antagonists should be carefully controlled. Animal studies suggest that nimodipine should not be infused at doses exceeding 2 μg/kg/min.

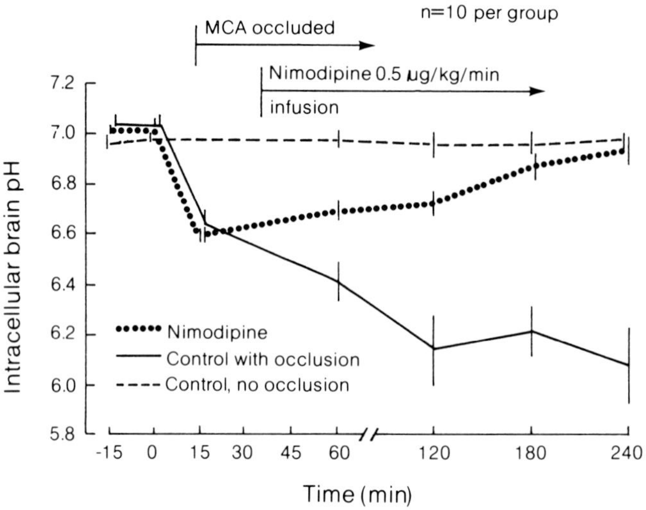

Fig. 3. Effect of nimodipine on intracellular brain pH in rabbits subjected to middle cerebral artery (MCA) occlusion. Average values for ten animals in each group. pH was measured with pH-sensitive fluorophor. Nimodipine restored brain pH to nearly normal and partially antagonized reduction in tissue perfusion (not shown). *Vertical lines* indicate standard errors of the mean. (Meyer et al. [18])

Recent experimental studies indicated that nimodipine may be useful in the treatment of age-associated memory as well as in that of functional disorders [4, 27]. Anticonvulsant effects of nimodipine in rabbits [19–21], rats [12], and man [3] were reported. The clinical efficacy of nimodipine in patients with subarachnoid hemorrhage is well documented [24, 25]. Nimodipine was also reported to reduce mortality and to improve neurological recovery in the male patients following ischemic stroke [7].

In summary, Ca^{2+} antagonists bind to specific receptors in the central nervous system and inhibit Ca^{2+} fluxes through L-type channels in the neuronal membranes. In this respect, nimodipine is more potent than nifedipine, verapamil, and many other Ca^{2+} antagonists. Nimodipine is more lipophilic than nifedipine and its volume of distribution in the brain of rats is larger than that of nifedipine. Nimodipine was found to have cerebrovasodilator and anti-ischemic effects, and to reduce neurological deficits in various models of focal and global cerebral ischemia. Its effect is associated with normalization of intracellular brain pH. In animals with experimental brain infarcts, nimodipine does not increase intracranial pressure and does not enhance brain edema. Clinical studies demonstrated that nimodipine improves neurological outcome in patients with subarachnoid hemorrhage or ischemic

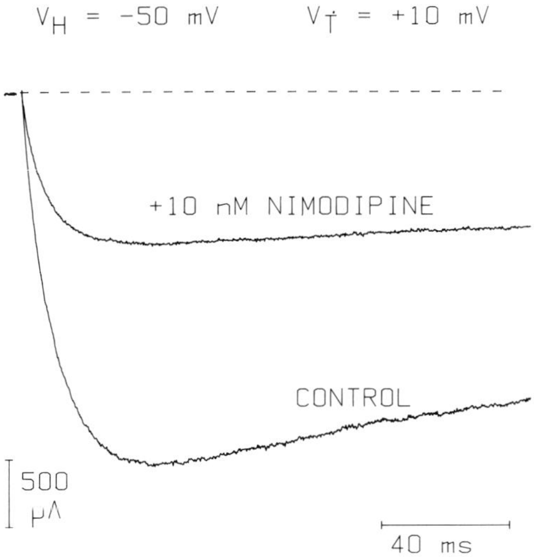

Fig. 4. Effect of nimodipine on Ca^{2+} currents through L-type Ca channels in freshly dispersed dorsal root ganglion cells of rabbits. Whole-cell current records in the presence of 10 nM nimodipine have been superimposed on control records. The −50 mV holding potential (+10 mV) was chosen to maximize L-type channel current. (Courtesy of McCarthy RT; see also [16])

stroke. New indications for nimodipine may include epilepsy and age-associated cognitive disorders.

References

1. Allen GS, Ahn HS, Preziosi TJ et al. (1983) Cerebral arterial spasm – a controlled trail of nimodipine in patients with subarachnoid hemorrhage. N Engl J Med 308:619–624
2. Bellemann P, Schade A, Towart R (1983) Dihydropyridine receptor in rat brain labeled with [^{3}H]nimodipine. Proc Natl Acad Sci USA 80:2356–2360
3. DeFalco FA, Bartiromo U, Majello I, DeGeronimo G, Mundo P (1989) Usefulness of calcium antagonist nimodipine in treatment of epilepsy in young patients with organic brain lesion. Abstracts of the 4th International Symposium on Ca^{2+} Antagonists, Florence, May 25–27, p 274
4. Deyo RA, Straube KT, Disterhoft JF (1989) Nimodipine facilitates trace conditioning of the eye-blink response in aging rabbits. Science 243:809–811
5. Freedman SB, Miller RJ (1984) Effects of nitrendipine on voltage sensitive calcium channels in brain and neuronal cultured cells. In: Scriabine A, Vanov S, Deck K (eds) Nimodipine. Urban and Schwarzenberg, Baltimore, pp 79–90
6. Fry HK, McCarthy RT (1988) Nimodipine block of L-type calcium channels in freshly dispersed rabbit DRG neurons. Soc Neurosci Abstr 14, 134

7. Gelmers HJ, Gorter K, deWeerdt CJ, Wierer HJA (1988) A controlled trial of nimodipine in acute ischemic stroke. N Engl J Med 318:203–207
8. Germano IM, Bartkowski HM, Pitts LH (1987) The therapeutic value of nimodipine in experimental cerebral ischemia: neurological outcome and histopathologic findings. J Neurosurg 67:81–87
9. Hadley MN, Spetzler RF, Fifield MS, Bichard WD, Hodak JA (1987) The effect of nimodipine on intracranial pressure – volume-pressure studies in a primate model. J Neurosurg 67:387–393
10. Hakim A, Hogan M, Berger M, Gjedde A (1989) Activation of the Ca channels in focal cerebral ischemia. J Cereb Blood Flow Metab 9 [Suppl 1]:S14
11. Hoffmeister F, Kazda S, Krause HP (1979) Influence of nimodipine (BAY E 9736) on the postischemic changes of brain function. Acta Neurol Scand 72:358–359
12. Hoffmeister F, Benz U, Heise A, Krause HP, Neuser V (1982) Behavioral effects of nimodipine in animals. Arzneimittelforschung 32:347–360
13. Kazda S, Garthoff B, Krause HP, Schlossmann K (1982) Cerebrovascular effects of the calcium antagonistic dihydropyridine derivative, nimodipine, in animal experiments. Arzneimittelforschung 32:331–338
14. Kazda S, Garthoff B, Luckhaus G, Nash G (1982) Prevention of cerebrovascular lesions and mortality in stroke-prone spontaneously hypertensive rats by the calcium antagonist nimodipine. In: Godfraind T, Albertini A, Paoletti R (eds) Calcium modulators. Elsevier Biomedical, Amsterdam, pp 155–167
15. Kerckhoff W van den, Drewes LR (1985) Transfer of the Ca-antagonists nifedipine and nimodipine across the blood-brain barrier and their regional distribution in vivo. J Cereb Blood Flow Metab 5 [Suppl I]:S459–S460
16. McCarthy RT (1989) Nimodipine block of L-type calcium channels in dorsal root ganglion cells. In: Gispen WH, Traber J (eds) Nimodipine and central nervous system function: new vistas. Schattauer, Stuttgart
17. Meyer H, Wehinger E, Bossert F, Scherling D (1983) Nimodipine: synthesis and metabolic pathway. Arzneimittelforschung 33:106–112
18. Meyer FB, Anderson RE, Yaksh TL, Sundt TM (1986) Effect of nimodipine on intracellular brain pH, cortical blood flow, and EEG in experimental focal cerebral ischemia. J Neurosurg 64:617–626
19. Meyer FB, Anderson RE, Sundt TM, Sharbrough FW (1986) Selective central nervous system calcium channel blockers – a new class of anticonvulsant agents. Mayo Clin Proc 61:239–247
20. Meyer FB, Tally PW, Anderson RE, Sundt TM Jr, Yaksh TL (1986) Inhibition of electrically induced seizures by a dihydropyridine calcium channel blocker. Brain Res 384:180–183
21. Morocutti C, Pierelli F, Sanarelli L, Stefano E, Peppe A, Mattioli GL (1986) Antiepileptic effects of calcium antagonist (nimodipine) on cefazolin-induced epileptogenic foci in rabbits. Epilepsia 27:498–503
22. Mossakowski MJ, Gadamski R (1987) Influence of the calcium channel blocker on the ischemic changes in sector CA_1 pyramidal neurons of Ammon's horn in Mongolian gerbils. Neuropatol Pol 25:439–450
23. Peroutka SJ, Allen GS (1983) Calcium channel antagonist binding sites labeled by [^{3}H]nimodipine in human brain. J Neurosurg 59:933–937
24. Philippon J, Grob R, Dagreou F, Guggiari M, Rivierez M, Vials P (1986) Prevention of vasospasm in subarachnoid hemorrhage. A controlled study with nimodipine. Acta Neurochir (Wien) 82:110–114

25. Pickard JD, Murray GD, Illingworth R et al. (1989) Effects of oral nimodipine on cerebral infarction and outcome after subarachnoid haemorrhage: British aneurysm nimodipine trial. Br Med J 298:636–642
26. Quirion R (1985) Characterization of binding sites for two classes of calcium channel antagonists in human forebrain. Eur J Pharmacol 117:139–142
27. Schuurman T, Klein H, Beneke M, Traber J (1987) Nimodipine and motor deficits in the aged rat. Neurosci Res Commun 1:9–15
28. Scriabine A, Battye R, Hoffmeister F et al. (1985) Nimodipine. In: Scriabine A (ed) New cardiovascular drugs. Raven, New York, pp 197–218
29. Scriabine A, Kerckhoff van den W (1988) Pharmacology of nimodipine, a review. Ann NY Acad Sci 522:698–706
30. Scriabine A, Schuurman T, Traber J (1989) Pharmacological basis for the use of nimodipine in central nervous system disorders. FASEB J 3:1799–1806
31. Steen PA, Gisvold SE, Milde JH, Newberg LA, Scheithauer BW, Lanier WL, Michenfelder JD (1985) Nimodipine improves outcome when given after complete cerebral ischemia in primates. Anesthesiology 62:406–411
32. Towart R (1981) The selective inhibition of serotonin-induced contractions of rabbit cerebral vascular smooth muscle by calcium-antagonistic dihydropyridines. An investigation of the mechanism of action of nimodipine. Circ Res 48:650–657
33. Young W, DeCrescito V (1989) Nimodipine reduces brain ionic and water shifts at 24 hours after MCAO. (in preparation)

The Prognostic Significance of the Concentration of Nimodipine in the CSF and Plasma in Subarachnoid Hemorrhage

W.I. Steudel, M. Carvi, S. Tiyaworabun, and R. Lorenz[1]

A major cause of mortality and morbidity in subarachnoid hemorrhage of aneurysmal origin is vasospasm. An analysis of the available placebo-controlled double-blind and open prospective studies provides statistically significant evidence that the calcium antagonist nimodipine can reduce the frequency of these spasm-induced delayed ischemic neurological deficits. However, very divergent percentages are reported in the various studies, ranging in the randomized studies from 1% with intravenous administration to 2%–16% with oral dosage [1–6, 8–11]. It is unclear what intraindividual and interindividual variations in the plasma and CSF concentration are present or are to be expected in medication with nimodipine. A high level of protein binding and the interaction with other drugs are known with regard to pharmacokinetics [12].

In the present study, the plasma and CSF levels were determined in patients with subarachnoid hemorrhage and correlated with various parameters such as grading, outcome, CSF protein, and plasma albumin.

Methods

In 1988, we measured the plasma and CSF levels postoperatively several times in a prospective study in 22 patients with subarachnoid hemorrhage. A media aneurysm was present in five cases, a carotid aneurysm in five cases, and an aneurysm of the anterior communicating artery in nine cases. The aneurysms were operated on acutely within the first 72 h after the hemorrhage in nine cases. The nimodipine treatment was started within the initial hours after the hemorrhage in 12 cases, within the first week in five cases, and later in six cases. In these patients, the CSF was sampled via cisternal drainage. In three cases there was subarachnoid hemorrhage and an intraventricular hemorrhage without detection of an aneurysm; in these, the CSF was extracted via ventricular drainage. The intravenous administration (48 mg/day) was carried out by means of a perfusor via a central line.

[1] Abteilung für Allgemeine Neurochirurgie der Johann Wolfgang Goethe-Universität, Schleusenweg 2–16, D-6000 Frankfurt 71

Advances in Neurosurgery, Vol. 18
K.-A. Bushe, M. Brock, and M. Klinger (Eds.)

Blood samples were taken in the morning, as a rule from the femoral vein. Afterwards, the blood and the CSF were centrifuged immediately, and the test tube was enclosed in a metal case and deep-frozen. The daily basic medication comprised: dexamethasone (3 × 125 mg), and hydroxyethyl starch (600 ml). Nimodipine was determined gas chromatographically in the Department of Clinical Pharmacology, Bayer Ltd., Wuppertal [12]. One hundred determinations (plasma and CSF values) were evaluated.

The statistical analysis was carried out on the basis of a data sheet comprising 87 variables. The computer analysis was made with the SPSS program. In the analysis, descriptive correlations were detected. These comprise: the Hunt and Hess classification (five grades) on admission, the Glasgow outcome scale (six grades), the time of nimodipine determination in days after SAH, operation and the beginning of treatment, the nimodipine concentration in the CSF and plasma on the 1st, 2nd and 7th postoperative days, the total protein content of the plasma and the CSF, and the daily amount of CSF and the flow rate as determined by Doppler sonography.

Results

Concentrations

The plasma concentrations of nimodipine averaged 15 ng/ml and the CSF concentrations 0.93 ng/ml with intravenous administration of 2 mg/h. The ratio of CSF to plasma concentration was about 1:15. The interindividual range of variation was considerable. It extended from 7.9 to 26.3 ng/ml in the case of the plasma values, and from 0.02 to 2.23 ng/ml for the CSF values (Table 1). The levels measured intraventricularly were about one-fourth of the cisternal concentrations (Table 1).

At a lower dosage (three cases), there was a demonstrably lower concentration of nimodipine in the plasma and in the CSF (Table 2).

Correlations

Follow-up investigations on the 2nd postoperative day did not reveal any detectable differences from the values on the 1st postoperative day. Comparing the values on the 7th postoperative day with those on the 1st postoperative day, it was shown that the nimodipine levels decreased in the plasma and CSF; however, this difference was not statistically significant (Spearman correlation coefficients). In addition, there was no statistically significant relationship between the CSF and the plasma concentration on the day of the SAH and at the beginning of the nimodipine treatment.

Table 1. Mean plasma and CSF concentrations of nimodipine after intravenous administration of the drug at 48 mg/day

	No.	Mean	Median	Minimum	Maximum	Standard deviation	P^*
Plasma	16	15.40	15.06	7.9	26.29	5.817	< 0.05
CSF, cisternal	16	0.93	0.67	0.02	2.23	0.685	
Plasma	3	16.04	13.62	12.33	22.16	5.342	< 0.05
CSF, ventricular	3	0.257	0.31	0.14	0.32	0.101	

* Mann-Whitney U-test

Table 2. Mean plasma and CSF concentrations of nimodipine after intravenous administration of the drug at lower dosage[a]

	No.	Mean	Median	Minimum	Maximum	Standard deviation
Plasma	3	5.110	3.720	2.56	9.05	3.461
CSF, cisternal	3	0.287	0.320	0.050	0.490	0.222

[a] 1 × 24 mg/day and 2 × 12 mg/day

No correlation could be detected between the plasma and the CSF concentrations and the outcome 6 months after the event as assessed using the Glasgow outcome scale.

There was a statistically significant relationship between the plasma level and the total protein in the serum: the higher the protein concentration in the serum, the lower the concentration of nimodipine measured in the plasma.

In five patients, the flow rate was measured by means of transcranial Doppler sonography. There was no statistically significant correlation between the highest measured flow values and the CSF concentration.

Discussion

There are only a few reports on measurements of the concentration of nimodipine [12, 13]. With administration of 2 mg nimodipine as a bolus injection, peak values were found after 3 min, varying between 39 and 149 ng/ml. There were also appreciable interindividual and intraindividual variations in our patients.

Our investigations indicate that the nimodipine concentration depends on the protein value in the plasma. This is explained by the high protein binding (98%)

of the nonmetabolized nimodipine [12]. However, the same correlation could not be found between the CSF concentration and the protein concentration. The CSF protein value obviously is subject to substantial interindividual and intraindividual variation after an SAH. Together with the small number of cases, this may explain the lack of correlation.

The values on the 7th day after the operation were on average less than the values on the 1st day. This unequivocally proves that there is no pharmacokinetically significant accumulation or raised elimination of nimodipine.

There is also no demonstrable correlation between the CSF levels measured and the final outcome. This may be explained by the appreciable alterations in the intraindividual concentrations. The vartiations in the CSF levels evidently do not affect the flow rate as measured by Doppler sonography. The flow pattern develops regularly with increase in the values, in particular on the operated side [7]. However, since Doppler sonographic investigations were only performed in five patients, a definitive conclusion cannot be drawn. It is possible that short-term administration of nimodipine in the initial days after the operation and a continuous low dosage over the first 2 weeks have the same effect.

References

1. Allen GS, Hyo SA, Preziosi TJ et al. (1983) Cerebral arterial spasm – a controlled trial of nimodipine in patients with subarachnoid haemorrhage. N Engl J Med 308:619–624
2. Auer LM (1984) Acute operation and preventive nimodipine improve outcome in patients with ruptured cerebral aneurysms. Neurosurgery 15:57–66
3. Auer LM, Brandt L, Ebeling U et al. (1986) Nimodipine and early aneurysm operation for good condition SAH patients. Acta Neurochir (Wien) 82:7–13
4. Auer LM, Schneider GH, Auer T (1986) Computerized tomography and prognosis in early aneurysm surgery. J Neurosurg 65:217–221
5. Gilsbach J (1988) Nimodipine in the prevention of ischaemic deficits after aneurysmal subarachnoid haemorrhage. Acta Neurochir (Wien) [Suppl] 54:41–50
6. Grotenhuis JA, Bettag W (1986) Prevention of symptomatic vasospasm after SAH by constant venous infusion of nimodipine. Neurol Res 8:243–248
7. Harders A, Gilsbach J (1988) Haemodynamic effectiveness of nimodipine on spastic brain vessels after subarachnoid haemorrhage evaluated by the transcranial Doppler method. Acta Neurochir (Wien) 45:21–28
8. Kazner E, Sprung C, Adelt D et al. (1985) Clinical experience with nimodipine in the prophylaxis of neurological deficits after subarachnoid haemorrhage. Neurochirurgia 18:110–113
9. Neil-Dwyer G, Mee E, Dorrance D, Lowe D (1987) Early intervention with nimodipine in subarachnoid haemorrhage. Eur Heart J 8:41–47
10. Öhman J, Heiskanen O (1987) The effect of nimodipine on cerebrovascular spasm after subarachnoid haemorrhage – a randomized, double-blind study, presented at the 2nd World Congress of Neuroscience, Budapest, Hungary, August 16–21, 1987
11. Philippon J, Grob R, Dagreou F, Guggiari M, Rivierez M, Viars P (1986) Prevention of vasospasm in subarachnoid haemorrhage. A controlled study with nimodipine. Acta Neurochir (Wien) 82:110–114

12. Raemsch KH, Graefe KH, Sommer J (1985) Pharmacokinetics and metabolism of nimodipine. In: Betz E, Deck K, Hoffmeister F (eds) Nimodipine, pharmacological and clinical properties. Schattauer, Stuttgart, pp 147–159
13. Scriabine A, Battye R, Hoffmeister F et al. (1985) Nimodipine. In: Scriabine A (ed) New drugs annual: cardiovascular drugs, vol 3. Raven, New York, pp 197–218

Is High Dose Nimodipine Therapy Necessary Following Subarachnoid Hemorrhage? A Retrospective Study of 80 Patients with Aneurysmal Subarachnoid Hemorrhage over the Past 2 Years

G. Birn and M. Stallmach[1]

Introduction

The protective role of the calcium channel blocker nimodipine has been verified by a considerable number of open or double-blind studies over the past few years. In general a dose of 2 mg/h i.v. or 240 mg/day p.o has been used. Here we present a retrospective analysis of our treatment results with varying intravenous nimodipine dosages during the past 2 years.

Material and Methods

Between November 1986 and November 1988, 80 patients with aneurysmal subarachnoid hemorrhage (SAH) were treated in our department. Fourteen patients receiving no or only oral nimodipine therapy were excluded from the protocol. The continuous intravenous nimodipine treatment given to the remaining 66 patients did not strictly follow a preset pattern, but was adjusted according to defined blood pressure levels in order to guarantee a steady cerebral perfusion. The systolic blood pressure levels for normotensive patients were usually set between 120 and 140 mmHg, and for hypertensive patients between 140 and 160 mmHg. Based on the clinical course and the hemodynamic changes in the basal cerebral arteries as monitored by transcranial Doppler sonography (TCD), the blood pressure was raised by catecholamines if necessary.

The 66 patients were divided into two groups. The first group (A) comprised 43 patients whose average dosage during the treatment period (mean 12.2 days) stayed below 1 mg/h (mean 0.62 ± 0.2 mg/h). The second group (B) consisted of 23 patients with a dosage equal to or more than 1 mg/h (mean 1.45 ± 0.42 mg/h) over 14.2 days. With $P < 0.01$, the nimodipine dosage of the two groups showed a significant difference; the treatment interval, however, did not.

[1] Neurochirurgische Abteilung, Katharinenkrankenhaus, Kriegsbergstr. 60, D-7000 Stuttgart 1

Advances in Neurosurgery, Vol. 18
K.-A. Bushe, M. Brock, and M. Klinger (Eds.)

There were no significant differences in respect of the following parameters: age (A:46.1 ± 14.8 years; B: 48.6 ± 12.1), sex (A/B: male 32.6%/30.4%, female 67.4%/69.4%), the amount of blood seen on the initial CT scan, localization of the aneurysms, timing of the operation (A/B within 3 days 66.7%/52.2%, within 7 days 81%/73.9%, within 14 days 88.1%/95.7%), complications (e.g., rebleeding, hydrocephalus, cardiopulmonary), risk factors (e.g., arterial hypertension, diabetes mellitus), and supplementary medical treatment (dexamethasone, osmotherapy, antibiotics, anticonvulsive drugs).

Evaluation of hemodynamic changes in the basal cerebral arteries by TCD is part of our standard management after SAH. For this study the time course of the mean flow velocity in the middle cerebral artery both ipsi- and contralateral to the side of the surgical approach was selected as the parameter for comparison. The measured data were summarized in 2-day cycles.

Statistical analysis was performed using the χ^2 test and the nonparametric U-test according to Mann and Whitney (significance level, $P < 0.05$).

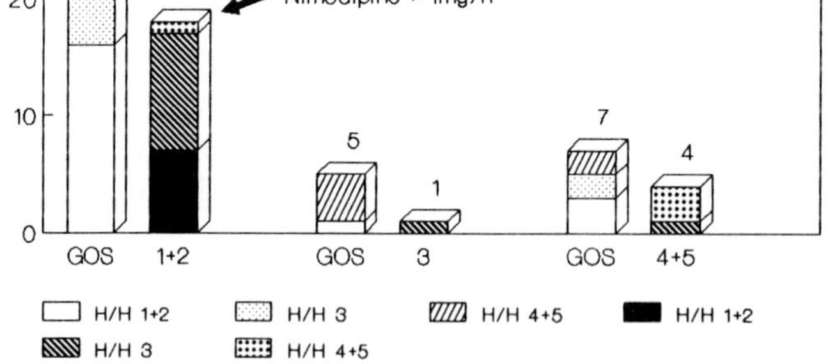

Fig. 1. Glasgow outcome scale (*GOS*) as related to the neurological status at admission according to Hunt and Hess (*H/H*). χ^2 test: $P = 0.58$ (not significant)

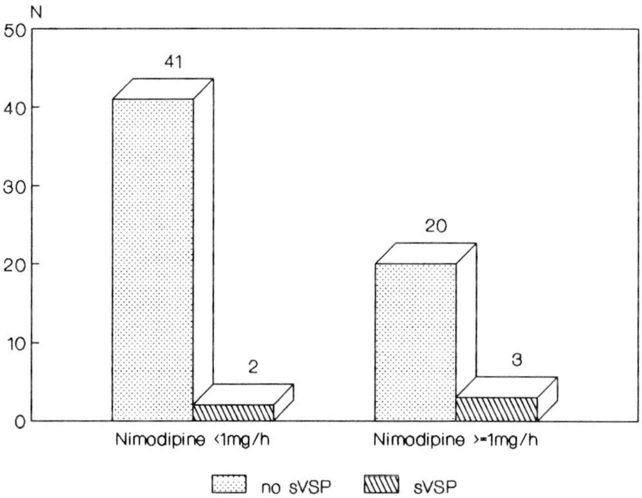

Fig. 2. Occurrence of symptomatic vasospasm (*sVSP*) after SAH in both groups (n = 66). χ^2 test: $P = 0.22$ (not significant)

Results

Figure 1 shows the score on the Glasgow Outcome Scale upon reexamination as related to the classification of Hunt and Hess established at admission.

Symptomatic vasospasm occurred in two patients in group A and in three patients in group B (Fig. 2). One patient from each group made a complete recovery.

The hemodynamic changes after SAH in the middle cerebral artery as monitored by TCD both ipsi- and contralateral to the side of the surgical approach were similar in both groups (Figs. 3, 4). Patients with a higher nimodipine dosage showed a tendency toward lower mean flow velocities.

For the above-mentioned parameters no statistically significant differences could be found.

Discussion

Since its introduction various authors have been able to prove a positive influence of the calcium channel blocker nimodipine in comparison to placebo [1, 6, 9, 10]. The standard dosage for intravenous administration reported in the literature is 2 mg/h [4, 6–8, 11]. There are no comparative studies concerning a lower i.v. dosage. Regarding the clinical outcome, the frequency of symptomatic vasospasm, and the time course of hemodynamic changes monitored by TCD in the presented sub-

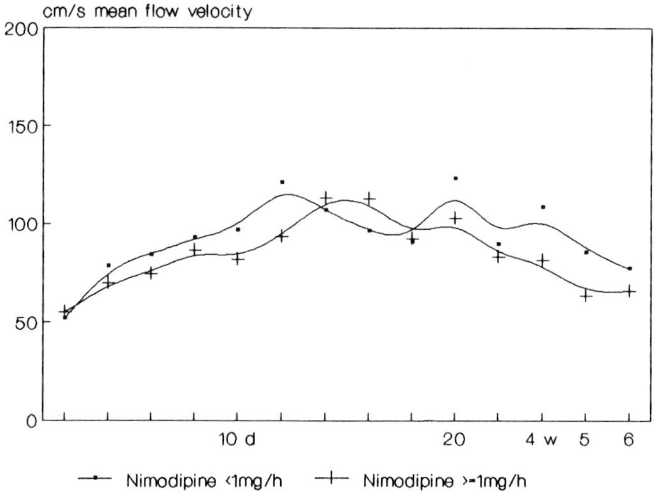

Fig. 3. Hemodynamic changes in the middle cerebral artery after SAH monitored by TCD ipsilateral to the surgical approach. Nonparametric U-test of Mann and Whitney: P = not significant

groups and the entire group of 80 patients, our results are comparable to those of other authors [2, 3, 5, 11–13].

As we did not find any advantage of the higher nimodipine dosage, we suppose that a dosage lower than 1 mg/h could be sufficient. However, this is a retrospective evaluation, and a prospective double-blind study with standardized varying dosage to confirm our results would be desirable.

Summary

Out of 80 patients treated for aneurysmal SAH in our department during the last 2 years, 66 could be evaluated retrospectively with regard to the effectiveness of different intravenous nimodipine dosages. Fourty-three patients received less than 1 mg/h nimodipine (mean, 0.62 mg/h), and 23 a dosage of more than or equal to 1 mg/h (mean, 1.45 mg/h).

Comparative criteria employed were: (a) the Glasgow Outcome Scale in relation to the clinical condition at admission according to the classification of Hunt and Hess, (b) the occurrence of delayed ischemic deficits, and (c) the hemodynamic changes in the middle cerebral arteries as monitored by transcranial Doppler sono-

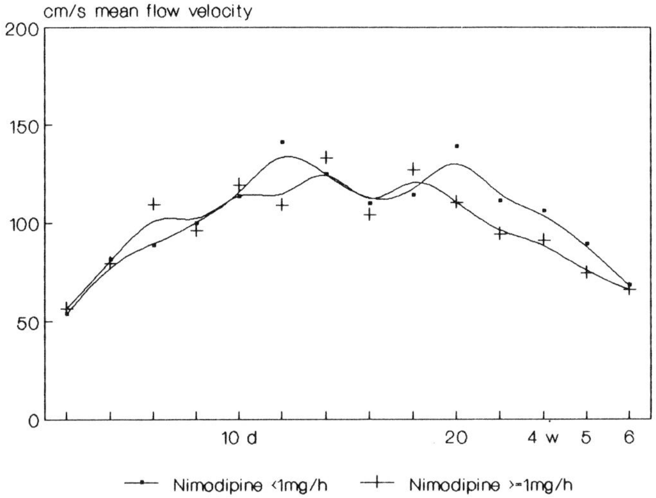

Fig. 4. Hemodynamic changes in the middle cerebral artery after SAH monitored by TCD contralateral to the surgical approach. Nonparametric U-test of Mann and Whitney: P = not significant

graphy. No statistically significant difference was found between the two groups in respect of these parameters.

Acknowledgment. We are indebted to Mrs. C. Diggett and Miss I. Schöckel for the English translation.

References

1. Allen GS, Ahn HS, Preziosi TJ et al. (1983) Cerebral arterial spasm – a controlled trial of nimodipine in patients with subarachnoid hemorrhage. N Engl J Med 308:619–624
2. Auer LM (1984) Acute operation and preventive nimodipine improve outcome in patients with ruptured cerebral aneurysms. Neurosurgery 15:57–66
3. Gilsbach JM, Harders AG, Eggert HR et al. (1988) Early aneurysm surgery: a 7 year clinical practice report. Acta Neurochir (Wien) 90:91–102
4. Grotenhuis JA, Bettag W (1986) Prevention of symptomatic vasospasm after SAH by constant venous infusion of nimodipine. Neurol Res 8:243–249
5. Harders A (1986) Neurosurgical applications of transcranial Doppler sonography. Springer, Vienna New York
6. Jan M, Bucheit F, Tremoulet M (1988) Therapeutic trial of intravenous nimodipine in patients with established cerebral vasospasm after rupture of intracranial aneurysms. Neurosurgery 23:154–157
7. Kazner E, Sprung C, Adelt D et al. (1985) Clinical experience with nimodipine in the prophylaxis of neurological deficits after subarachnoid hemorrhage. Neurochirurgia 28:114–117

8. Ljunggren B, Brandt L, Saeveland H et al. (1984) Outcome in 60 consecutive patients treated with early aneurysm operation and intravenous nimodipine. J Neurosurg 61:864–873
9. Mee E, Dorrance D, Lowe D et al. (1988) Controlled study of nimodipine in aneurysm patients treated early after subarachnoid hemorrhage. Neurosurgery 22:484–491
10. Philippon J, Grob R, Dagreou F et al. (1986) Prevention of vasospasm in subarachnoid hemorrhage. A controlled study with nimodipine. Acta Neurochir (Wien) 82:110–114
11. Seiler RW, Grolimund P, Zurbruegg HR (1987) Evaluation of the calcium-antagonist nimodipine for the prevention of vasospasm after subarachnoidal hemorrhage. Acta Neurochir (Wien) 85:7–16
12. Seiler RW, Reulen HJ, Huber P et al. (1988) Outcome of aneurysmal subarachnoid hemorrhage in a hospital population: a prospective study including early operation, intravenous nimodipine, and transcranial Doppler ultrasound. Neurosurgery 23:598–604
13. Stolke D, Seifert V (1988) Früh- oder Spätoperation des rupturierten Aneurysmas? Eine Analyse anhand von 356 Fällen. Neurochirurgia 31:81–87

Interactions Between Nimodipine and Anesthetics Used in Neurosurgical Procedures

H. Müller and J. Zierski[1]

Introduction

In order to reveal interactions between the calcium antagonist nimodipine and the drugs used in anesthesia during neurosurgical procedures, we have performed hemodynamic, respiratory, metabolic, and endocrine investigations to address the following questions:

1. What is the role of nimodipine as well as alcohol, the solvent of its preparation, in achieving intraoperative hypotension in normo- and hypertensive patients?
2. Where is the site of action of nimodipine in producing hemodynamic alterations, and are there differences from other vasodilative drugs, such as nitroglycerin?
3. Apart from its vasoactive properties, does nimodipine also influence pulmonary circulation, gas exchange, metabolism, and endocrine responses?

Methods

To answer these questions a total of 102 patients, divided into three different groups, were studied. All investigations were carried out under standardized general anesthesia for neurosurgical operations, with thiopental for induction, fentanyl for analgesia, pancuronium for relaxation, and artificial ventilation with oxygen and nitrous oxide. The following parameters were monitored or calculated: systolic, diastolic, and mean arterial pressure, heart rate, cardiac output and index, stroke volume and index, systolic, diastolic, and mean pulmonary arterial pressure, pulmonary capillary wedge pressure, central venous pressure, systemic and pulmonary vascular resistance, left and right ventricular stroke work or cardiac work indices, rate-pressure product, triple index, arterial and venous oxygen and carbon dioxide pressures, hemoglobin, fraction of inspired oxygen, arterial-venous oxygen difference, oxygen consumption, availability and extraction, pulmonary shunting,

[1] Institut für Anästhesiologie und Intensivmedizin, Krankenhaus Kemperhof, D–5400 Koblenz

Advances in Neurosurgery, Vol. 18
K.-A. Bushe, M. Brock, and M. Klinger (Eds.)

arterial and venous oxygen concentration, glucose, lactate, free glycerol, albumin, antidiuretic hormone, adrenocorticotropic hormone, and cortisol. In this paper only the parameters undergoing significant changes during the investigations will be presented.

Group 1: Nimodipine was infused in a group of normotensive (n = 17) and hypertensive patients (as ascertained from their clinical history, pretreated with antihypertensive drugs, n = 14). An additional group of 11 patients received a placebo infusion containing alcohol in the same concentration as in the nimodipine preparation. Measurements were made during anesthesia without PEEP ventilation at intervals of 30 min, first without infusion (I), then after 30 min with 0.4 μg/kg/min (II), after another 30 min with 0.7 μg/kg/min (or the same volume of placebo) (III), and 30 min after the infusion had been stopped (IV).

Group 2: Nimodipine was given in the same doses and at the same intervals (I–IV) during continuous administration of inhaled anesthetics [halothane (n = 10) or isoflurane (n = 10) in equipotent concentration of MAC 1]. Besides these groups with inhaled anesthetics, a control group received nimodipine only during the already mentioned standardized opiate anesthesia, this time providing PEEP of 5 cmH_2O (n = 10).

Group 3: A dosage of 1.0 μg/kg/min of nimodipine (n = 10) was compared to nitroglycerin in an equipotent dosage of 2.4 μg/kg/min (n = 10) during anesthesia with PEEP ventilation. A control group (n = 10) did not receive either nitroglycerin or nimodipine. Monitoring was done at intervals of 30 min for a period of 3 h (I = after start of operation, without hypotensive medication; II–VI: with hypotensive medication; VII: at the end of operation without hypotensive medication).

Results

In the first group of patients the hypotensive effect of nimodipine was compared in normo- and hypertensive patients. Nimodipine reduced blood pressure especially in hypertensive patients, i.e., similarly reduced blood pressure was obtained in both nimodipine groups (systolic arterial pressure of almost 110 mmHg at measuring point III), although the initial values before infusion of the calcium antagonist were completely different (170 mmHg in the hypertensive and 130 mmHg in the normotensive group). Stroke and cardiac indices were increased with nimodipine as a consequence of the reduction of vascular resistance, once again starting from a higher level in the group of hypertensive patients. A slight decrease was seen in pulmonary vascular resistance. Cardiac work indices demonstrated improved conditions for cardiac contraction with less energy consumption. There was a slight decrease in arterial oxygen pressure due to an increase in pulmonary shunting (maximum of about 25% instead of the normal value of up to 10%). There was an increase in glucose, lactate, antidiuretic hormone, adrenocorticotropic hormone, and cortisol during the course of hypotension in both groups. There were no changes in any of the parameters with placebo, i.e., alcohol.

In the second group nimodipine was combined with inhaled anesthetics and the fall in blood pressure in the control group, receiving only nimodipine, was similar to that in the first group. In the halothane and isoflurane groups systolic arterial blood pressure reached 90–100 mmHg at measuring point III. The increase in cardiac index, still present in the control group, was almost not demonstrable during the combination of nimodipine and inhaled anesthetics. There was, however, no difference from the control group (and group 1) with regard to reduction of systemic vascular resistance. Pulmonary shunting and arterial oxygen pressure were unchanged, but the decrease in oxygen extraction and increase in oxygen availability, also seen with all other nimodipine-receiving groups, were still present. The increase in hormone levels was prevented by the addition of inhaled anesthetics to the opiate anesthesia during infusion of nimodipine.

In the third group, which compared nimodipine and nitroglycerin in equipotent dosages, the reductions in blood pressure were similar but the cardiac changes were different. Nimodipine enhanced cardiac and stroke indices, while nitroglycerin decreased them. Systemic vascular resistance was diminished with nimodipine and increased with nitroglycerin. The enhancement of pulmonary shunting seen in group 1 was not present. Decrease in oxygen extraction and increase in oxygen availability were combined with an increase in the anaerobic metabolic derivate lactate (maximum of 3.5 mmol/l), which became especially obvious in this group with long-term measurement for more than 3 h and which was not present with nitroglycerin or in the control group. The increase in stress hormones was similar in both hypotensive drug groups and not seen in the control group.

Discussion

According to our investigations the hypotensive action of nimodipine is especially obvious when there is preexisting vasoconstriction, e.g., in hypertensive patients (group 1). This particular effect may even be used to achieve controlled hypotension in surgery for cerebral aneurysms [2] and is enhanced in combination with other vasodilator drugs, such as sodium nitroprusside [1]. The hypotensive effect of nimodipine is different from that of nitroglycerin (as well as sodium nitroprusside [1]) and due to arterial vasodilation, which leads to a decrease in systemic vascular resistance and an increase in cardiac output. Nitroglycerin acts by venous pooling, thus reducing cardiac output. Both mechanisms lead to a more economic use of cardiac work and contractility. With alcohol vasodilation may occur with high concentrations, but it was not seen in the placebo group (group 1).

Inhaled anesthetics are known to produce a negative inotropic effect by impairing cardiac calcium influx. Therefore in our combination with nimodipine (group 2) we saw a marked drop in blood pressure and the increase of cardiac index, usually present during nimodipine infusion, could no longer be observed.

Pharmacological vasodilation is present in systemic and pulmonary circulation and the use of vasodilators leads to an increase in pulmonary shunting, especially in artificial ventilation. Nimodipine is no exception and PEEP ventilation, which produces an adequate relation between ventilation and circulation, may be used to reduce this side-effect (comparison of groups 2 and 3 to group 1).

Calcium antagonists are known to reduce oxygen consumption and mitochondrial oxidative phosphorylation in muscle and brain. In our patients we saw a decrease in oxygen extraction and an increase in oxygen availability which was accompanied by a slow rise in lactate. The clinical importance of this possibly unfavorable action remains to be established.

In some reports an analgesic effect of nimodipine was discussed according to changes in the endocrine reaction. We were unable to demonstrate indirect changes of the stress hormones related to analgesia. On the contrary, during hypotension (with nimodipine as well as with nitroglycerin, see group 3) endocrine factors were increased and this increase was only reduced if the basic opiate anesthesia was combined and therefore deepened by addition of inhaled anesthetics.

In general, our studies have demonstrated that nimodipine, in spite of its so-called cerebrovascular specificity, is also characterized by the typical qualities of calcium antagonists, which have to be taken into account whenever nimodipine is used.

References

1. Müller H, Kafurke H, Marck P, Zierski J, Hempelmann G (1988) Interactions between nimodipine and general anaesthesia – clinical investigations in 124 patients during neurosurgical operations. Acta Neurochir (Wien) [Suppl] 45:29–35
2. Müller H, Marck P, Otto O, Gips H, Börner U, Hempelmann G (1986) Intraoperative Anwendung von Nimodipin bei Operation von zerebralen Aneurysmen. In: Kazner E (ed) Nimotop in der Prophylaxe and Therapie neurologischer Ausfälle durch zerebrale Ischämie. G. Braun, Karlsruhe, pp 35–47

Influence of Nimodipine on Pulmonary Function

J. Nadstawek, M. Födisch, D. Schmalohr, and D.-K. Böker[1]

Introduction

The calcium antagonist nimodipine has been in use for 8 years in patients with subarachnoid hemorrhage to prevent ischemic neurological deficits caused by cerebral vasospasm. Many studies have shown a benefit from preventive nimodipine treatment. Nimodipine has proved to be a predominantly cerebroarterial dilating drug that can be administered over a long period and in sufficiently high doses without severe arterial hypotension. Until now, nothing is known about a possible influence of nimodipine on pulmonary function, although clinical observations indicate that continuous administration of nimodipine may cause pulmonary venous congestion and, finally, development of pulmonary edema. A retrospective study was performed to verify our observations by comparing the initial chest X-ray of patients with subarachnoid hemorrhage with the subsequent chest X-rays under continuous administration of nimodipine.

Patients and Methods

One hundred and seventy patients suffering from subarachnoid hemorrhage were included in a retrospective study. Of these patients, 123 received continuous nimodipine at a dosage of 2 mg/h. In 47 patients no nimodipine was administered. In the two groups, the initial chest X-ray was compared to the subsequent chest X-rays taken during the treatment course. The development of pulmonary venous congestion was classified according to three phases. Phase 1a was defined as discrete pulmonary venous congestion with enlargement of the apical veins of the lung, phase 1b as a further enlargement of the veins in addition to broadened hili, and phase 1c as diffuse shadowing of the lung; the ultimate stage in development was interstitial and intra-alveolar edema. Other parameters recorded include changes in the mechanical ventilation, such as positive end-expiratory pressure and increase in

[1] Institut für Anästhesiologie der Rheinischen Friedrich-Wilhelms-Universität Bonn, Sigmund-Freud-Str. 25, D-5300 Bonn 1

Advances in Neurosurgery, Vol. 18
K.-A. Bushe, M. Brock, and M. Klinger (Eds.)

the inspiratory oxygen fraction, and administration of diuretics and catecholamines. Deterioration of chest X-rays was rated according to the following scheme. Grade 1 corresponded to patients with a normal chest X-ray on admission who developed pulmonary congestion as described for phase 1b and patients who had initial pulmonary congestion as described for phase 1a with further progression to phase 1c. Grade 2 included patients who had a normal chest X-ray on admission and deteriorated to phase 1c as well as patients in phase 1b on admission developing pulmonary edema during the treatment. All patients with a normal chest X-ray on admission progressing to pulmonary edema were classified as grade 3. It should be mentioned that progression to phase 1a was not considered as significant; thus these patients were classified as having an unchanged chest X-ray. The statistical analysis was carried out with the Wilcoxon test. Significance was set at $P < 0.05$.

Results

On admission 4.8% of the patients treated with nimodipine had pulmonary edema, 37.3% had slight pulmonary congestion, and 57.7% had no abnormalities on the chest X-rays. During continuous administration of nimodipine, 34% of these patients did not reveal any changes on chest X-rays, while 66% progressed to a worse phase (Fig. 1). Of these latter patients, 44% were classified as grade 1, 49% as grade 2, and 7% as grade 3 (Fig. 2). In 57 patients the mechanical ventilation had to be changed because of deterioration of ventilation parameters; 16 patients needed a higher oxygen fraction (Fig. 3). In 45 patients catecholamines were administered and in 31 patients diuretics. A reduction of nimodipine dosage was necessary in 24 patients. In three patients nimodipine had to be discontinued (Fig. 4).

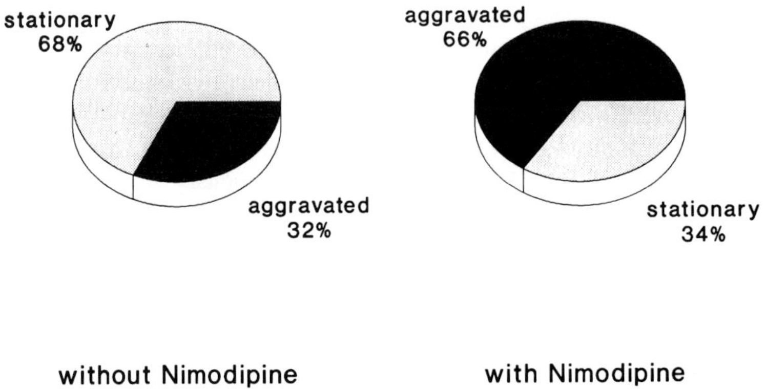

Fig. 1. Progression in chest X-rays in patients with subarachnoid hemorrhage treated with and without nimodipine

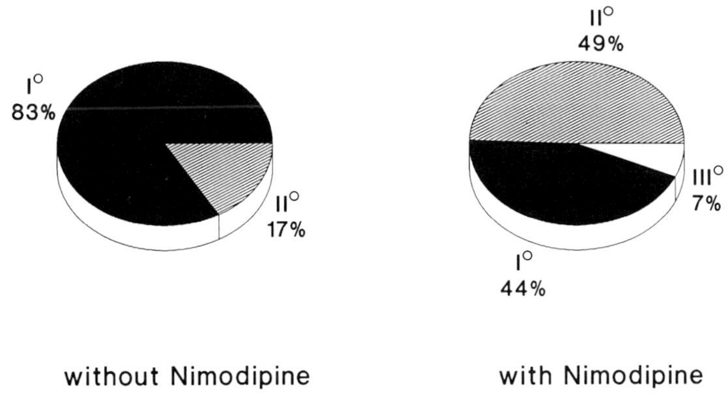

Fig. 2. Deterioration of chest X-rays divided into three grades in patients treated with and without nimodipine

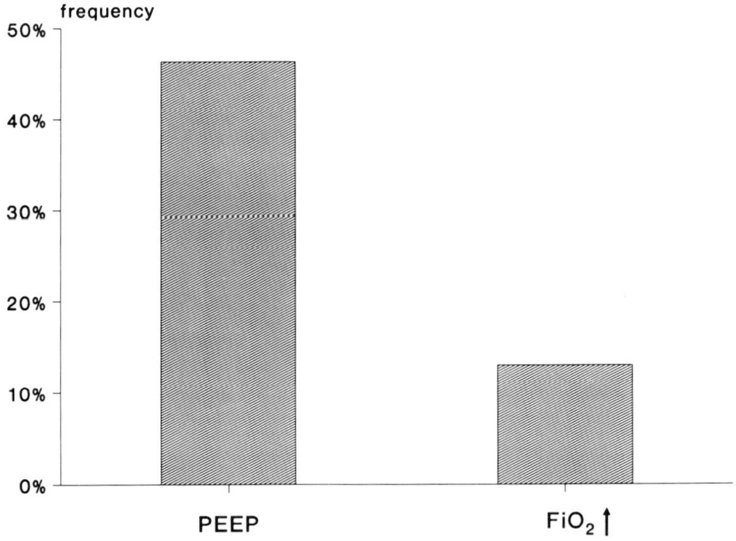

Fig. 3. Changes in ventilation parameters and oxygen fraction in patients treated with nimodipine

Of the patients who did not receive any nimodipine, two had pulmonary edema on admission, 22 revealed a slight congestion, and 23 had no abnormalities in the chest X-ray. Fifteen patients had a deterioration during the course of treatment. The chest X-ray did not change in the other 32 patients (Fig. 1). Concerning the three grades of progression, five patients were grade 1, only one patient grade

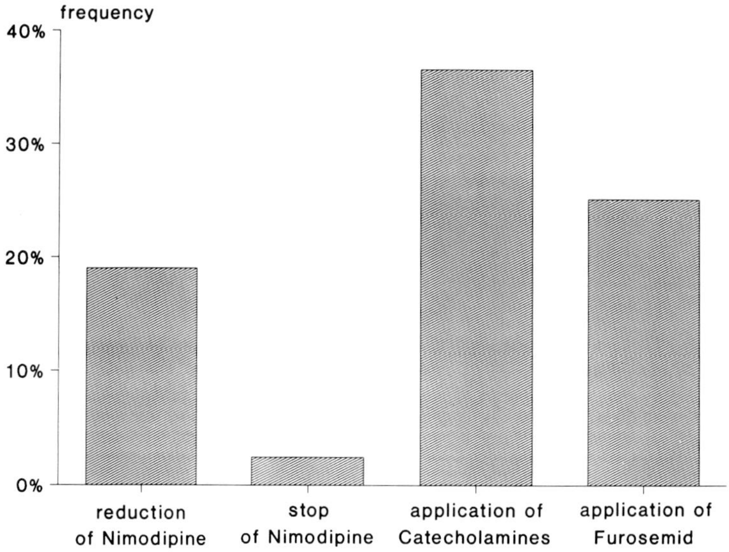

Fig. 4. Changes in medication in patients with subarachnoid hemorrhage treated with nimodipine

2, and no patients grade 3 (Fig. 2). None of the patients required a change in mechanical ventilation or medication such as catecholamines or diuretics, except for two patients who were given catecholamines for a short period.

Discussion

The results indicate a direct effect of nimodipine on lung circulation that may promote development of lung congestion and lung edema. In dosages used for the prevention of cerebral vasospasms in subarachnoid hemorrhage, the lung function may decrease, thus affecting oxygen supply. A significantly higher percentage of patients treated with nimodipine, compared with patients who did not receive nimodipine, required either changes in mechanical ventilation for deterioration of arterial blood gases or medication to support hemodynamics. The mechanism of this effect, however, remains unclear. A decrease in total peripheral as well as in pulmonary resistance, as described by Bormann et al. [2] in patients undergoing aortocoronary bypass grafting with nimodipine dosages similar to those used in the treatment of vasospasm in patients with subarachnoid hemorrhage, could be responsible. Furthermore, an increase in the intrapulmonary right-to-left shunt, described for nimodipine, may contribute to this phenomenon [1,3,4]. Effects of subarachnoid hemorrhage itself on the development of pulmonary congestion and are possible. Thus, the mechanism of the described observation has to be investigated as it is important in any patient presenting with subarachnoid hemorrhage to prevent a decrease in lung function that could further impair cerebral oxygen supply.

References

1. Boldt J, Bormann B, Kling D, Ratthey K, Hempelmann G (1987) Influence of nimodipine and nifedipine on intrapulmonary shunting – a comparison to other vasoactive drugs. Intensive Care Med 13:52
2. Bormann v. B, Boldt J, Kling D, Mulch J, Weidler B, Hempelmann G (1985) Influence of nimodipine on cardiovascular parameters during coronary surgery. Neurochirurgia 28:79
3. German P, Aloy A, Zabloudil B, Richling B, Zimpfer M (1987) Treatment of subarachnoid hemorrhage with the calcium entry blocker nimodipine affects pulmonary gas exchange in patients requiring controlled mechanical ventilation. Anesthesiology 67, 3A:A115
4. Guggenberger H, Kottler B, Heuser D (1986) Cardiorespiratorische Problematik beim Einsatz eines Calciumantagonisten zur Vasospasmusprophylaxe bei Subarachnoidalblutung. Anaesthesist 35:429

Tolerability of Calcium Antagonists

D. Tettenborn, J. Dycka, L. Porto, and N. Kurtz[1]

Introduction

Among all classes of calcium antagonists, probably the most important groups are the dihydropyridines (nifedipine, nimodipine, nicardipine, felodipine, and others), the papaverine derivatives, verapamil and its analogues, and the benzodiazepine derivative diltiazem.

As a group the calcium antagonists share several side-effects which can be related to their pharmacological mechanisms of action, although the intensity and frequency of these adverse effects differ between the various agents.

In recent years various calcium antagonists have been widely studied in the prevention and/or treatment of neurological disorders such as ischemic complications of subarachnoid hemorrhage (SAH), ischemic stroke, and migraine. Nimodipine has now been licensed in many countries for the prevention or treatment of ischemia following SAH.

Here, we report on the spectrum of side-effects of nimodipine as derived from clinical studies in patients with SAH. Data from studies on approximately 2600 patients, including uncontrolled, prospective studies with intravenous followed by oral nimodipine, one placebo-controlled randomized double-blind intravenous study, four placebo-controlled randomized double-blind studies, and one double-blind dose ranging study using oral nimodipine, were analyzed to evaluate the safety profile of the drug.

[1] Medizinische Abteilung, Pharmazeutisches Forschungsinstitut, BAYER AG, D–5600 Wuppertal 1

Advances in Neurosurgery, Vol. 18
K.-A. Bushe, M. Brock, and M. Klinger (Eds.)

Table 1. Adverse drug reactions (ADRs) in intravenous studies

Sign/symptom	No. of all events	%	No. of events possibly or probably related to nimodipine	%
Hypotension	54	4.1	44	3.4
Headache	29	2.2	16	1.2
Bradycardia	29	2.1	20	1.5
Tachycardia	9	0.7	8	0.6
Extrasystoles	10	0.8	8	0.6
Arrhythmias	7	0.6	4	0.3
Hypertension	6	0.5	4	0.3
Perspiration	2	0.2	2	0.2
BUN increase	20	1.5	4	0.3
Transaminases ↑	31	2.4	28	2.1
Hyperglycemia	10	0.8	4	0.3
Thrombophlebitis	3	0.3	3	0.2
Vomiting	3	0.2	2	0.2
Hyperkalemia	9	0.7		
Laboratory test abnormal	4	0.3		
Lactic acidosis	2	0.2		
Acute renal failure	2	0.2		
Pancreatitis	2	0.2		
Rash	2	0.2		
Vascular disorder	2	0.2		
Single occurrences	21	2.1	15	1.1
Total no. of ADRs	255		162	
No. of patients with ADRs	178	13.6	127	9.7
Total no. of patients	1309		1309	

Results

Intravenous Administration

In 178 of the 1309 patients (13.6%), 255 adverse reactions were reported. The type and incidence of these reactions are shown in Table 1. Only 162 reactions which occurred in 127 patients (9.7%) were considered by the investigators to be either probably or possibly related to nimodipine; the remaining reactions were considered either doubtful, undetermined, or of no relationship to the drug. However, for the purpose of this survey all reported adverse reactions will be discussed.

Mild or moderate decreases in blood pressure were the most frequent reported side-effects, being reported in 54 of the 1309 patients (4.1%). The effect of ni-

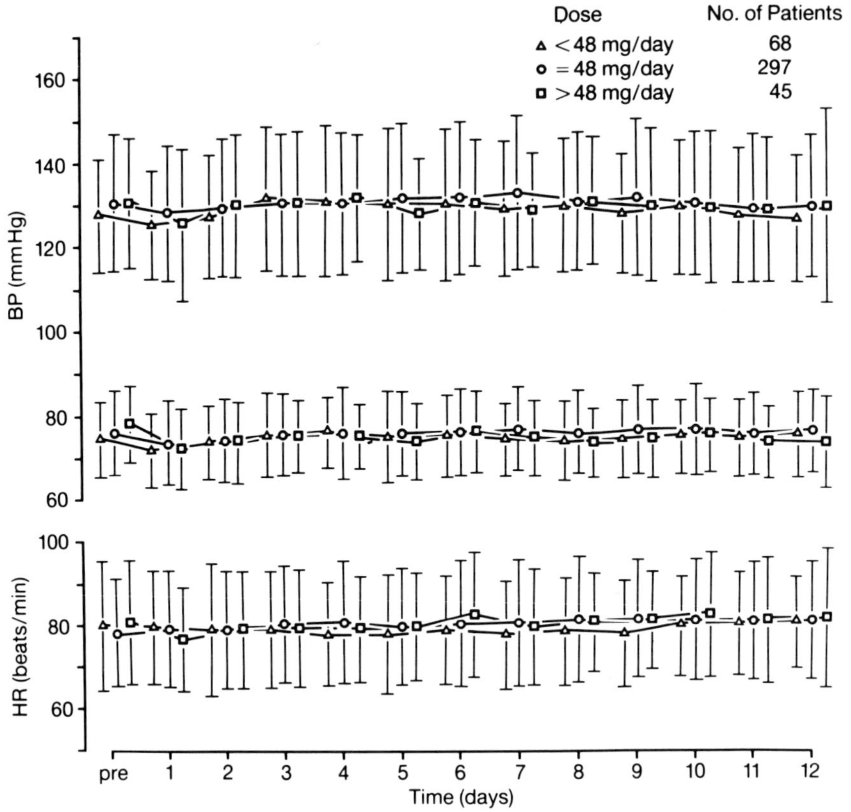

Fig. 1. Blood pressure (*BP*) and heart rate (*HR*) in normotensive SAH patients (n = 410) during intravenous treatment with nimodipine

modipine on blood pressure appears to be dependent in part on whether or not the patient is normotensive when therapy is initiated. For normotensive patients, there was no clear dose-response relationship in the incidence of hypotensive reactions. The mean changes in both diastolic and systolic recordings were indistinguishable from baseline (Fig. 1). The conclusion therefore is that in the recommended dose of 48 mg/h, intravenous nimodipine has a negligible effect on blood pressure in normotensive individuals. However, in some patients with elevated blood pressure, due to either preexisting hypertension or the underlying disease, nimodipine appears to exert a hypotensive effect. In hypertensive patients, intravenous nimodipine decreased both the mean systolic and the mean diastolic blood pressure recordings (Fig. 2). This decrease was most pronounced on the 1st day of therapy and then leveled off, remaining stable throughout the remainder of therapy. Heart rate was not changed in either the normotensive or hypertensive group of patients.

Electrocardiographic abnormalities – bradycardia, tachycardia, arrhythmias, or extrasystoles – were reported as side-effects in 53 of the 1309 patients (4.1%). There was no dose-response relationship in the incidence of these events either. Since the association of ECG abnormalities with SAH is well known [3], it is likely that these are changes occurring during the natural course of the disease, presumably due to stimulation of the parasympathetic/sympathetic systems by the hemorrhage [11, 17]. This interpretation is supported by the results of placebo-controlled, double-blind studies using oral nimodipine where the type and frequency of electrocardiographic changes were identical in nimodipine- and placebo-treated patients [12].

Headache, which is another symptom that is difficult to assess in patients with SAH, was reported in 29 patients (2.2%). Abnormal liver function tests were reported in 31 patients (2.4%).

Other more rarely reported adverse drug experiences are shown in Table 1. Most of them were believed by the investigators to have either a remote or no relation to the study drug treatment.

In a placebo-controlled, double-blind study in France, no significant differences in the incidence and type of adverse reactions were reported, hypotensive episodes being noted in both patient groups with similar rates [10]. This shows the importance of a control group for assessing adverse reactions.

Oral Administration

In randomized, double-blind studies, a total of 823 patients were treated with doses which ranged from 180 to 540 mg per day [1, 12, 14–16; data on file]. In 92 patients (11.2%), 124 side-effects were reported (Table 2). The most frequent adverse drug reactions were hypotension (4.4%), followed by nausea (1.1%), headache (0.8%), rash (0.8%), and bradycardia (0.7%). Almost all of the other adverse reactions occurred with a similar incidence to the placebo group or with very few incidences only in either treatment group. It may be worth noting that more patients in the placebo group experienced abnormal liver function tests than in the active treatment group (1.5% vs 0.5%). This, once again, shows the importance of a control group when assessing the incidence of adverse drug effects.

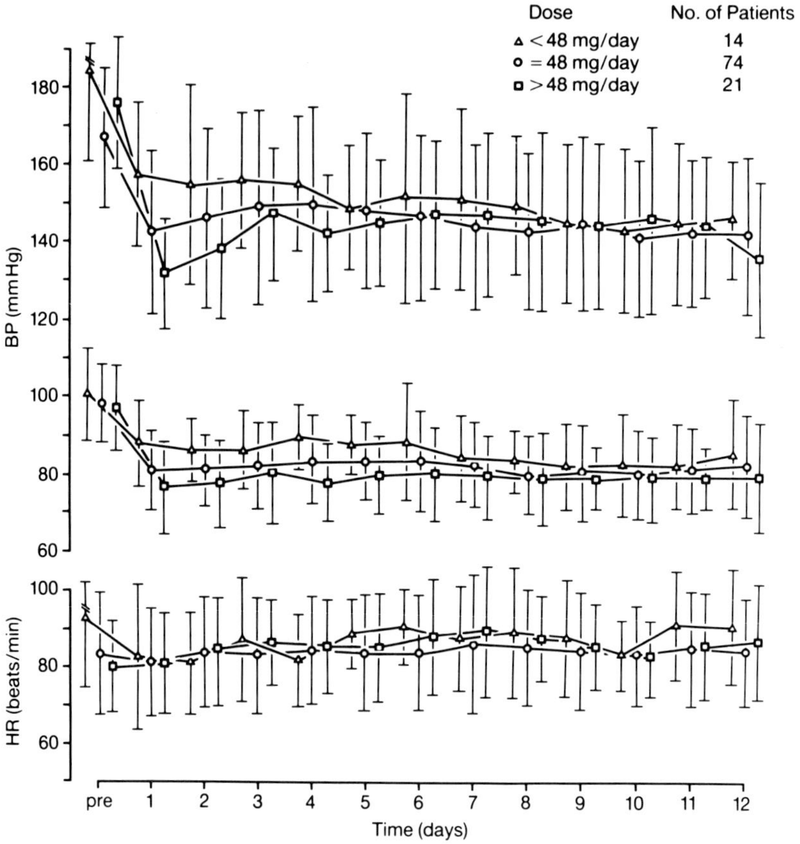

Fig. 2. Blood pressure (*BP*) and heart rate (*HR*) in hypertensive SAH patients (n = 109) during intravenous treatment with nimodipine

Spontaneous Reports of Adverse Drug Experiences

There are a few reports on increased intrapulmonary shunting allegedly related to nimodipine [5, 7]. Some authors [e.g., 13] believe this to be a general response to vasoactive drugs, similar changes having been found with nifedipine, nitroglycerin, and dopamine.

Table 2. Adverse drug reactions (ADRs) in oral studies

Sign/symptom	Nimodipine No. of events	Nimodipine %	Placebo No. of events	Placebo %
Hypotension	36	4.4	6	1.2
Nausea	9	1.1		
Rash	7	0.8	3	0.6
Headache	7	0.8	1	0.2
Bradycardia	6	0.7		
Diarrhea	6	0.6	3	0.6
Gl symptoms	4	0.5		
Gl hemorrhage	2	0.2	1	0.2
Edema	4	0.5	3	0.6
Transaminases ↑	4	0.5	7	1.5
Thrombocytopenia	3	0.4	1	0.2
Itching	2	0.2	1	0.2
Diaphoresis	2	0.2		
Flushing	2	0.2		
Vomiting	2	0.2		
Palpitations	2	0.2		
Muscle cramp	2	0.2		
Jaundice	2	0.2		
Single occurrences	22	2.7	13	2.7
Total no. of ADRs	124		39	
No. of patients with ADRs	92	11.2	29	6.1
Total no. of patients	823		479	

Effects on Intracranial Pressure

Intracranial pressure has not been measured regularly in clinical trials. There are a few reports of slight increases in intracranial pressure [2,4], whereas others did not find any evidence of change [9]. In animal experiments even decreases have been described [8]. In patients resuscitated following cardiac arrest, in whom intracranial hypertension frequently occurs, nimodipine did not cause significant changes in intracranial pressure [6].

Conclusions

The overall frequency of side-effects during treatment with nimodipine is relatively modest, with hypotension being the most frequent, occurring in 4.1% of patients on intravenous and 4.4% of patients on oral treatment. There is no clear dose-response relationship in the reported incidence of this reaction, although regular blood pressure measurements indicated that higher doses of nimodipine (i.e., 48 mg/day i.v.) also caused more pronounced decreases in blood pressure in hypertensive patients. Drops in blood pressure can be reduced or avoided by stepping up the dose more slowly to the desired level. Most of the other side-effects were minor and only a few appeared to have more than a remote relation to the study drug. However, a full picture of incidence, as for all pharmacological agents, will have to await further long-term use and evaluation.

Acknowledgment. The secretarial assistance of Mrs. M. Wilke is gratefully acknowledged.

References

1. Allen GS, Ahn HYOS, Preziosi TJ et al. (1983) Cerebral arterial spasm – a controlled trail of nimodipine in patients with subarachnoid hemorrhage. N Engl J Med 308:619–624
2. Brinker TH, Spring A, Brendel R (1987) Risiken der intravenösen Nimodipin Therapie. Anaesthesist 36 [Suppl]:279
3. Eisalo A, Präsalo J, Halonen PJ (1972) Electrocardiographic abnormalities and some laboratory findings in patients with subarachnoid haemorrhage. Br Heart J 34:217–226
4. Gaab MR, Haubitz I, Brawanski A, Korn A, Czech T (1985) Acute effects of nimodipine on the cerebral blood flow and intracranial pressure. Neurochirurgia 28:93–99
5. Germann P, Aloy A, Zahloudil H, Richling B, Zimpfer M (1987) Treatment of subarachnoid hemorrhage with the calcium entry blocker nimodipine affects pulmonary gas exchange in patients requiring controlled mechanical ventilation. Anesthesiology 67, 3A:A155
6. Gueugniaud P-Y, personal communication
7. Guggenberger H, Kottler B, Heuser B (1986) Cardio-respiratorische Problematik beim Einsatz eines Calciumantagonisten zur Vasospasmusprophylaxe bei Subarachnoidalblutung. Anaesthesist 35:429–432
8. Hadley MN, Spetzler RF, Fifield MS, Bichard WD, Hodak JA (1987) The effect of nimodipine on intracranial pressure. J Neurosurg 67:387–393
9. Hoffmann P (1987) Auswirkungen von Calcium-Antagonisten auf den normalen und auf den erhöhten intrakraniellen Druck. Anaesthesist 36 [Suppl]:281
10. Jan M, Buchheit F, Tremoulet M (1988) Therapeutic trial of intravenous nimodipine in patients with established cerebral vasospasm after rupture of intracranial aneurysms. Neurosurgery 23:154–157
11. Marion DW, Segal R, Thompson ME (1986) Subarachnoid hemorrhage and the heart. Neurosurgery 18:101–106
12. Mee E, Dorrance D, Lowe D, Neil-Dwyer G (1988) Controlled study of nimodipine in aneurysm patients treated early after subarachnoid hemorrhage. Neurosurgery 22:484–491

13. Müller H, Marck P, Gips H, Börner U, Otto O, Adams H-A, Hempelmann G (1987) Der Einfluß des Kalzium-Antagonisten Nimodipin auf Hämodynamik, Gasaustausch und Endokrinium während einer Opiatnarkose. Anaesthesist 36:561–569
14. Petruck CK, West M, Mohr G et al. (1988) Nimodipine treatment in poor-grade aneurysm patients. J Neurosurg 68:505–517
15. Philippon J, Grob R, Dagreou F, Guggiari M, Rivierez M, Viars P (1986) Prevention of vasospasm in subarachnoid hemorrhage. A controlled study with nimodipine. Acta Neurochir (Wien) 82:110–114
16. Pickard JD, Murray GD, Illingworth R et al. (1989) Effect of oral nimodipine on cerebral infarction and outcome after subarachnoid hemorrhage: British aneurysm nimodipine trial. Br Med J 298:636–642
17. Stober T, Kunze K (1982) Electrocardiographic alterations in subarachnoid hemorrhage. J Neurol 227:99–113

Nimodipine in the Treatment of Subarachnoid Hemorrhage Secondary to Aneurysm Rupture: Five Years' Experience

L. Cristante, N. Freckmann, D. Winkler, and H.-D. Herrmann[1]

Introduction

The prevention of vasospasm has always been a key issue in the treatment of subarachnoid hemorrhage (SAH) secondary to aneurysmal rupture. According to the Iowa Cooperative Study [8], this complication is present in 38% of patients who have suffered an SAH. The subsequent cerebral ischemia has been reported to be clinically relevant in about 25%–30% of those cases [7, 9]. In the last few years the therapeutic approach to this problem as well as the management of ruptured aneurysms in general has undergone substantial changes [1–6, 11]. Some of these recent advances were included in a therapeutic protocol we have been using since March 1984. The following is a brief summary of our new therapeutic strategy:

- Intravenous and intrathecal application of nimodipine
- Early operation except for patients in grades IV and V of the Hunt and Hess classification without a relevant intra- or extra-axial hematoma, or patients harboring aneurysms of the upper vertebrobasilar circulation or with angiospasm
- Use of intraoperative "brain protection" (isoflurane and Brevimytal)
- Use of external ventricular drainage
- Induction of hypervolemia/hypertonia after aneurysm clipping and in the period thereafter
- Perioperative Doppler sonographic monitoring

Clinical Material

During the period from January 1981 to December 1988, 444 consecutive nonselected patients who suffered an SAH secondary to the rupture of an intracranial aneurysm were admitted to our department. Their mean age was 47.2 years; there was a slight preponderance of female patients (male/female ratio = 1:1.6).

[1] Universitätskrankenhaus Hamburg-Eppendorf, Abteilung für Neurochirurgie, Martinistr. 52, D–2000 Hamburg 20

Advances in Neurosurgery, Vol. 18
K.-A. Bushe, M. Brock, and M. Klinger (Eds.)

Table 1. Selected preoperative characteristics

	Group A (n = 135)	Group B (n = 309)
Re. SAH	34 (25%)	59 (19%)
Vasospasm	30 (22%)	52 (17%)
Re. SAH + vasospasm	9 (26%)[a]	17 (29%)[a]
Preop. deaths	25 (18%)	30 (10%)
Fatal vasospasm	6	4

Re. SAH, multiple subarachnoid hemorrhage
[a] Percentage figure related to total number of Re. SAH cases

Table 2. Surgical modalities and mortality

	Group A (n = 135)	Group B (n = 309)
Surgery	103	256
Early surg.	36 (35%)	183 (71%)
Delayed surg.	67 (65%)	73 (29%)
Op. mortality	12 (12%)	28 (11%)
Early surg. mortality	8	26
Del. surg. mortality	4	2
Overall mortality	37 (27%)	58 (19%)

For the purposes of the study and in accordance with the modification of our therapeutic strategy we distinguished two groups:

- Group A (Jan. 1981 – Feb. 1984) = 135 patients
- Group B (March 1984 – Dec. 1988) = 309 patients

Results

The relevant clinical data are briefly summarized in Tables 1–4. In group B there was a 6% decrease in patients suffering multiple SAH. In the preoperative phase, angiographically demonstrated vasospasm – either local or generalized – was also less frequent (17%). The preoperative death rate went down by 8%, with a parallel decrease in fatalities due to severe cerebral ischemia. Further, the ratio of early vs delayed surgery was reversed. The operative mortality in the early surgical group went down to 11%; that of "good-risk patients" (grades 1–3) was now 8.5%. Delayed surgery carried a mortality of less than 1%. In the last 5 years, there was

Table 3. Outcome after early surgery

Preoperative grade	Outcome (GOS) 1	2	3	4	5	Death
Group A (n = 36)						
1 (n = 17)	7	5	2	2		1
2 (n = 9)	3	2	3			1
3 (n = 3)		1		1		1
4 (n = 5)			1		1	3
5 (n = 2)						2
Group B (n = 181)						
1 (n = 71)	15	29	15	6		6
2 (n = 36)	4	15	7	6	1	3
3 (n = 47)	3	14	19	5	2	4
4 (n = 19)		2	4	4	2	7
5 (n = 8)			1	1		6

Preop. grade according to Hunt and Hess
GOS, Glasgow Outcome Score

Table 4. Postoperative neurological deficits related to temporary clipping/vasospasm

	Group A	Group B
Temporary clipping	20 (19%)	78 (30%)
Related neurol. deficits	11 (55%)	21 (27%)
Transient neurol. deficits	6	9
Permanent neurol. deficits	5	12
Postop. vasospasm	11 (9%)	37 (14%)
No neurol. deficits	2	10
Transient neurol. deficits	1	17
Permanent neurol. deficits	8	10

more liberal use of temporary clipping, while the occlusion time did not exceed 10 min in most of the cases. Related neurological deficits were now present in only 27% of the patients. The incidence of postoperative vasospasm, as assessed by transcranial Doppler sonography, was higher in group B. Its clinical relevance, in terms of permanent neurological deficits, was much reduced (25%).

Discussion

A retrospective analysis of our clinical data shows that our present therapeutic strategy has had a positive impact on morbidity and mortality. The objective advantages of this regimen were particularly striking in grade 3 patients, a substantial group making up 25% of our population. The preoperative mortality in this group was as high as 50% in the period 1981–1983. A further improvement in the mortality, as far as "good-risk patients" are concerned, seems feasible since four out of the nine patients in grades 1 and 2 who died did so because of complications which were unrelated to SAH and surgery (pneumonia and kidney failure). On the other hand, we do not presently see any possibility of improving the outcome in patients in grades 4 and 5 upon admission.

Since 1984 a decrease in preoperative vasospasm has been paralleled by a similar increase in postoperative vasospasm. The incidence of this complication was therefore not changed by our new policy. The more aggressive treatment of this complication in the postoperative phase made its clinical course undeniably more benign. An evaluation of the therapeutic impact of nimodipine is difficult because of the limits imposed by the design of this study. Nonetheless, we feel able to state that this impact seems relatively limited. In the preoperative period there was, in fact, a higher incidence of vasospasm in patients who suffered multiple SAH in spite of the administration of nimodipine. In the postoperative phase we had the opportunity to verify in single cases a sometimes clinically relevant change in blood flow related to manipulations of the infusion rate of nimodipine. On the other hand, in cases of hypotension, a side-effect of nimodipine we saw in approximately 25% of our patients, we did not hesitate to cut down drastically the administration of the calcium blocker and counteract vasospasm by increasing that of plasma expanders and catecholamines. The clinical outcome of these patients was by no means worse than that of others receiving higher doses of nimodipine.

Temporary clipping was better tolerated in the second group. We tend to believe that this increased tolerance to focal ischemia is related not only to nimodipine administration, as already demonstrated in experimental models of global and partial ischemia [10, 12], but also to the intraoperative induction of hypertension and the administration of 100% O_2 during that maneuver [11].

References

1. Allen GS, Ahn HS, Preziosi TJ et al. (1983) Cerebral arterial spasm – a controlled trial of nimodipine in patients with subarachnoid hemorrhage. N Engl J Med 308:619–624
2. Arnold H, Freckmann N, Herrmann H-D (1985) Is early operation for ruptured cerebral aneurysm justified? In: Auer LM (ed) Timing of aneurysm surgery. de Gruyter, Berlin, pp 283–286

3. Finn SS, Stephensen SA, Miller CA (1986) Observations on the perioperative management of aneurysmal subarachnoid hemorrhage. J Neurosurg 65:48–62
4. Freckmann N, Renz D, Müchler H-C et al. (1985) Hirnprotektive Maßnahmen aus der Sicht des Neurochirurgen. In: Heuser D, Freckmann N, Renz D et al. (eds) Indikation und Praxis cerebroprotektiver Maßnahmen in der Neurochirurgie. Springer, Berlin Heidelberg New York
5. Harders A, Gilsbach J (1985) Time course of cerebrovascular spasm in early aneurysm operation: transcranial Doppler findings. In: Auer LM (ed) Timing of aneurysm surgery. de Gruyter, Berlin, pp 437–444
6. Ljunggren B, Brandt L, Säveland H et al. (1984) Outcome in 60 consecutive patients treated with early aneurysm operation and intravenous nimodipine. J Neurosurg 61:864–873
7. Millikan CH (1975) Cerebral vasospasm and ruptured intracranial aneurysm. Arch Neurol 32:433–449
8. Sahs AL, Nibbelink DW, Torner JC (1981) Aneurysmal subarachnoid hemorrhage. Urban & Schwarzenberg, Baltimore, pp 249–272
9. Schneck SA, Kriceff II (1964) Intracranial aneurysm rupture, vasospasm and infarction. Arch Neurol 11:668–680
10. Steen PA, Gisvold SE, Milde JH et al. (1985) Nimodipine improves outcome when given after complete cerebral ischemia in primates. Anesthesiology 62:406–414
11. Symon L (1985) Thesholds of ischemia applied to aneurysm surgery. Acta Neurochir (Wien) 77:1–7
12. Tally PW, Sundt TM, Anderson RE (1989) Improvement of cortical perfusion, intracellular pH and electrocorticography by nimodipine during transient focal cerebral ischemia. Neurosurgery 24:80–87

Experience with Deliberate Early Aneurysm Surgery and Nimodipine

A. Brawanski, W. Ullrich, M. Holzschuh, and J. Meixensberger[1]

Introduction

During the last decade two innovations have been introduced in the treatment of aneurysms:

1. Early surgery
2. The administration of nimodipine for the prevention and treatment of vasospasm

However, there is still discussion about several issues: What is the real effect of nimodipine and how significant are the beneficial effects of calcium antagonists in combination with early surgery? On this issue several controlled double-blind studies have been reported with positive results [5, 7, 8]. In addition to these publications it seems interesting to consider the results in a series of patients treated in clinical routine and not under the ideal circumstances of a controlled study. The treatment results in such a group of patients are reported here, with special reference to postoperative vasospasm.

Patients and Method

Retrospectively the outcome in 88 patients operated on in our institution between 1986 and 1988 was analyzed 3 months after the operation according to the Glasgow Outcome Scale (GOS). The clinical condition before the operation was graded according to the scale of Hunt and Hess (HH) and the extension of the subarachnoid hemorrhage was scored according to its appearance on CAT scan after Davis [2].

In general patients were operated on, irrespective of the day of admission, as early as possible. However, there were the following exceptions: We preferred a delayed operation in patients over 60 years of age with multiple risk factors and in patients in grades HH4 and HH5. Postoperatively all patients were subjected to hypervolemia and slight hypertension. Pre-, intra-, and postoperatively

[1] Neurochirurgische Klinik und Poliklinik der Universität Würzburg, Josef-Schneider-Str. 11, D–8700 Würzburg

Advances in Neurosurgery, Vol. 18
K.-A. Bushe, M. Brock, and M. Klinger (Eds.)

all were treated with nimodipine. Postoperative vasospasm was defined as a clinical deterioration not earlier than 2 days after the operation, whereby other causes like intraoperative problems, hydrocephalus, and intracerebral hematoma were excluded. In addition to the clinical symptoms, vasospasm was confirmed by rCBF studies and/or transcranial Doppler studies in nearly all cases.

Results

Patient Group Description

Of the 88 patients, 57 were female and 31 male. The mean age of the group was 44.3 ± 15.2 years. The preoperative condition according to the Hunt and Hess scale is given in Fig. 1. Thirty-nine of the aneurysms were located at the anterior cerebral artery or the anterior communicating artery, 27 at the internal carotid artery, 17 at the middle cerebral artery, and 5 in the posterior circulation. Thirty-seven patients (42%) were operated on during the first 3 days after the SAH, 14 patients (16%) between the 4th and 7th days, 16 patients (18%) between the 8th and 14th days, and 21 patients (24%) later than 14 days.

General Outcome

Sixty-nine patients (78%) achieved an excellent or favorable outcome (GOS 4/5), 12 (14%) had a GOS of 3, and 7 (8%) had a GOS of 1/2. The outcome according to the preoperative HH grade is shown in Fig. 2. Of the patient of groups HH1–3, 81% achieved a GOS of 4/5, whereas the same good outcome was achieved in only 20% of the HH4 patients.

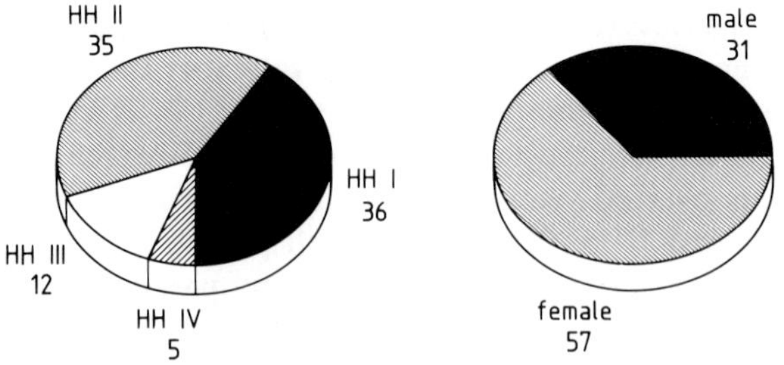

Fig. 1. Patient group description (n = 88). Classification according to Hunt and Hess scale

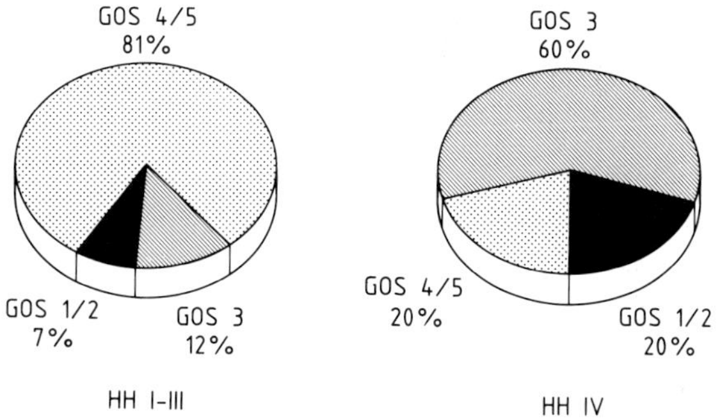

Fig. 2. Outcome according to the HH grade

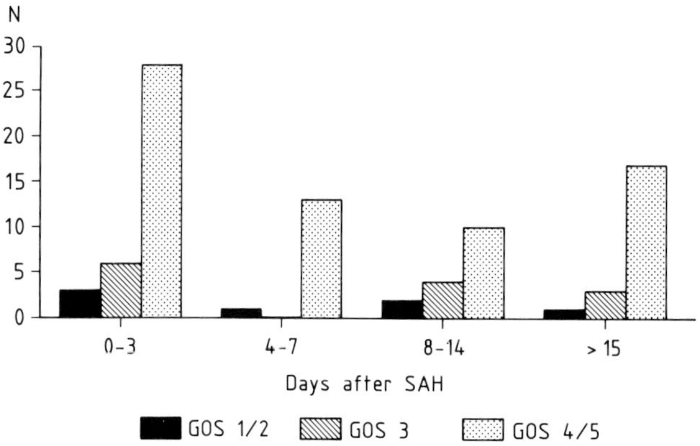

Fig. 3. Outcome according to day of operation

Outcome According to the Day of Operation

As Fig. 3 shows, at all time intervals mostly good postoperative results were achieved, and no interval with especially bad results could be observed. However, a GOS of 1/2 was seen in some cases in each subgroup.

Incidence of Postoperative Vasospasm

The overall incidence of postoperative vasospasm amounted to 11%. The distribution of vasospasm according to the time of operation is given in Fig. 4. In general we found that the later the time of operation, the lower was the incidence of postoperative vasospasm. Most commonly it occurred in those patients who had

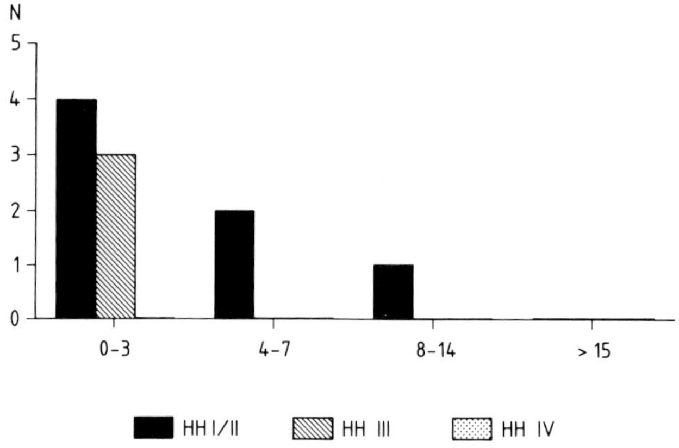

Fig. 4. Incidence of postoperative vasospasm according to day of operation and HH grade

been operated on within 72 h after the SAH: here, its incidence amounted to 17%. Interestingly, most of the patients in this subgroup were preoperatively in grade HH1/2.

Despite the administration of nimodipine the outcome of patients with symptomatic vasospasm was poor: 60% presented with an unfavorable clinical course (three died from the sequelae of vasospasm) whereas only 40% achieved a GOS of 4/5. In all cases with vasospasm its incidence was closely correlated to an extensive SAH as seen on CAT scan (Davis grade 3/4).

Discussion

With microsurgical procedures and proper timing of the operation good results can be achieved. This has been proven by several recent publications [e.g., 3, 9]. Especially the risk of rehemorrhage is reduced to a minimum by early operation. In addition, the administration of calcium antagonists further improves outcome by reducing the incidence and the sequelae of vasospasm [6, 8]. Similar beneficial effects are attributed to induced postoperative hypertension and hypervolemia [10]. The overall outcome of our patients corresponds to the data published in other studies [1, 4]. We want to emphasize that our results were obtained in an uncontrolled study, whereby this was mainly due to the different times of patient admission. However, in normal clinical daily life good outcomes can be achieved using the above-described therapeutic regime.

In contrast, the incidence of vasospasm under the administration of nimodipine is higher in our patient population than reported by others [5]. We found an incidence of 17.2% in good grade patients operated on within 72 h. Nearly all of them had an extended SAH on CAT scan. On the other hand no correlation of the preoperative clinical condition and the incidence of vasospasm was found in our patients. Furthermore the administration of calcium antagonists generally seems not to improve the prognosis once vasospasm has become symptomatic. Summing up, the administration of calcium antagonists improves the outcome after SAH in combination with other treatment modalities. However, vasospasm still occurs and symptomatic vasospasm is still a severe threat to the patient. Further work has to be done to cope with these specific issues.

References

1. Auer LM, Brandt L, Ebeling U et al. (1986) Nimodipine and early aneurysm operation in good condition SAH patients. Acta Neurochir (Wien) 82:7–13
2. Davis JM, Davis KR, Crowell RM (1980) Subarachnoid haemorrhage secondary to ruptured intracranial aneurysm: prognostic significance of cranial CT. AJR 134:711–715
3. Disney L, Weir B, Petruk K (1987) Effect on management mortality of a deliberate policy of early operation on supratentorial aneurysms. Neurosrugery 20:695–701
4. Gilsbach JM (1988) Nimodipine in the prevention of ischaemic deficits after aneurysmal subarachnoid hemorrhage. Acta Neurochir (Wien) 45:41–50
5. Gilsbach JM, Harders AG (1989) Morbidity and mortality after early aneurysm surgery – a prospective study with nimodipine prevention. Acta Neurochir (Wien) 96:1–7
6. Ljunggren B, Brandt L, Säveland H, Nilsson PE, Cronquist S, Andersson KE, Vinge E (1984) Outcome in 60 consecutive patients treated with early aneurysm operation and intravenous nimodipine. J Neurosurg 61:864–873
7. Mee E, Dorrance D, Lowe D, Neil-Dwyer G (1988) Controlled study of nimodipine in aneurysm patients: treatment early after subarachnoid hemorrhage. Neurosurgery 22:484–491
8. Pickard JD, Murray GD, Illingworth R et al. (1989) Effect of oral nimodipine on cerebral infarction and outcome after subarachnoid haemorrhage: British aneurysm nimodipine trial. Br Med J 298:636–642
9. Sano K (1985) Timing of surgery, operative mortality, and follow-up results in cases with subarachnoid hemorrhage due to aneurysm rupture. In: Auer LM (ed) Timing of aneurysm surgery. de Gruyter, Berlin, pp 227–236
10. Solomon RA, Post KD, McMurtry JG III (1984) Depression of circulating blood volume in patients after subarachnoid hemorrhage: implications for the management of symptomatic vasospasm. Neurosurgery 15:354–361

Effect of Nimodipine on the Outcome Following Operation for Ruptured Intracranial Aneurysms

W. Köning, K. Lukow, and J. Menzel[1]

The outcome of patients treated surgically for ruptured intracranial aneurysms is influenced by the severity of the initial subarachnoid hemorrhage (SAH) [4, 10], the onset of delayed ischemic deterioration (DID) [9, 11], and rebleeding [1]. Several studies have suggested a protective anti-ischemic effect of nimodipine for DID [2, 3, 5–7]. Therefore we have changed our regime of operative management to early surgery combined with intravenous nimodipine medication. Retrospectively we have reviewed the outcome of our patients treated with and without nimodipine.

Clinical Material and Method

During the period 1983–1988, 108 patients with intracranial ruptured aneurysms (43 males, 65 females) admitted to the neurosurgical department in Cologne were treated by the direct clipping method. Their ages ranged from 12 to 73 years with an average of 46.8. The most frequent site of the aneurysms was the anterior communicating artery (55 patients); the middle cerebral artery was the site in 28 patients, and the internal carotid artery in 18. Seven patients had multiple aneurysms.

On the day of surgery 36 patients were in the Hunt-Hess grade I, 34 in grade II, 30 in grade III, and 8 in grades IV and V. Early surgery (within 3 days after SAH) was performed in 42 patients and delayed surgery in 66 patients (Table 1). Intravenous nimodipine treatment at a dosage of 2 mg/h was administered to 63 patients for 7–10 days after the SAH and continued by oral medication for a total of 21 days. Forty-five patients had no nimodipine treatment. The neurological deficit or deterioration of the clinical state was attributed postoperatively to one of the following causes:

- Effect of SAH, if the preoperative neurological deficit remained unchanged or deteriorated progressively
- Complication of surgery, if immediate deterioration occurred postoperatively as a direct effect of surgery

[1] Neurochirurgische Klinik der Städt. Krankenanstalten Köln-Merheim, Ostmerheimer Str. 200, D-5000 Köln 91

Advances in Neurosurgery, Vol. 18
K.-A. Bushe, M. Brock, and M. Klinger (Eds.)

Table 1. Timing of surgery

Operation groups	Hunt-Hess grade	Nimodipine (n)	Non-nimodipine (n)	Total
Acute surgery	I – II	22	10	32
(0–3 days post-SAH)	III – V	7	3	10
Delayed surgery	I – II	22	16	38
(≥4 days post-SAH)	III – V	12	16	28
Total		63	45	108

- DID, if deterioration occurred secondary to an uncomplicated postoperative interval
- Extracranial complication

The operative outcome 3 months later was categorized as: excellent/good = favorable, fair/poor = unfavorable, and dead.

Results

The 3-month operative outcome revealed no significant difference in the mortality rates between the nimodipine group (14%) and the non-nimodipine group (16%) (Tables 2 and 3). In both groups the mortality ranged from 2% to 3% in the good condition patients of grades I and II and from 32% to 42% in the poor condition patients of grades III–V.

In the nimodipine group 46 patients (73%) were classified as being in good condition at the 3-month outcome as compared to 23 (51%) in the non-nimodipine group. This difference is caused by the different outcomes in the grade III–V patients: in the nimodipine group four of them (21%) had a good outcome compared with only one (5%) in the non-nimodipine group. By contrast, in the good-risk patients of grades I and II no difference was seen, whether in the mortality rate or in the good outcome rate.

Among the 69 patients with a good condition at the 3-month outcome (Table 4), 66 had an uneventful postoperative course – 45 (71%) of the nimodipine group and 21 (47%) of the non-nimodipine group. Transient ischemic deficits occurred in one patient (2%) in the nimodipine group and in two patients (5%) in the non-nimodipine group.

The causes of permanent neurological deficits or deterioration were analyzed for the 39 patients who had a poor outcome or died. The rate of surgical and medical complications was about 5% in both the nimodipine and the non-nimodipine group. The rate of DID with fixed neurological deficit and the effect of initial SAH were

Table 2. Three-month outcome of patients treated without nimodipine

Hunt-Hess grade	Total cases	Favorable		Unfavorable		Dead	
		n	%	n	%	n	%
I – II	26	22	85	3	12	1	3
III – V	19	1	5	12	63	6	32
Total	45	23	51	15	33	7	16

Table 3. Three-month outcome of patients treated with nimodipine

Hunt-Hess grade	Total cases	Favorable		Unfavorable		Dead	
		n	%	n	%	n	%
I – II	44	42	95	1	2	1	2
III – V	19	4	21	7	37	8	42
Total	63	46	73	8	13	9	14

Table 4. Causes of transient or permanent deterioration

Outcome groups	Causes	Nimodipine		Non-nimodipine	
		n	%	n	%
Total cases		63		45	
Favorable		46	73	23	51
	Uneventful	45	71	21	47
	Transient DID	1	2	2	5
Unfavorable or dead		17	27	22	48
	Permanent DID	3	4.7	7	16
	SAH	7	11	10	22
	Surgery	3	4.7	2	4.4
	Other	4	6.3	3	6.6

SAH = effect of subarachnoid hemorrhage
DID = delayed ischemic deterioration
Other = medical complications

4.7% and 11% respectively in the nimodipine group and rose to 16% and 22% respectively in the non-nimodipine group.

Discussion

The rate of DID with permanent neurological deficit causing an unfavorable outcome or death in the present retrospective study corresponds to the results reported in recent studies [5, 7, 8]. The favorable outcome rates for the nimodipine and non-nimodipine groups were quite different for the high-risk patients of Hunt-Hess grades III–V (21% vs 5% respectively), whereas the rates did not differ markedly for the lower risk patients of grades I and II (95% vs 85% respectively). These results confirm that nimodipine has a beneficial effect on the favorable outcome rate especially in high risk patients.

References

1. Adams HP, Kassell NF, Torner JC, Nibbelink DW, Sahs AL (1981) Early management of aneurysmal subarachnoid hemorrhage. J Neurosurg 54:141–145
2. Allen GS, Ahn HS, Preziosi TJ et al. (1983) Cerebral arterial spasm – a controlled trial of nimodipine in patients with subarachnoid hemorrhage. N Engl J Med 308:619–624
3. Auer LM, Brandt L, Ebeling U, Gilsbach J, Groeger U, Reulen HJ, Saeveland H (1986) Nimodipine and early aneurysm operation in good condition SAH patients. Acta Neurochir (Wien) 82:7–13
4. Hunt WE, Hess RM (1968) Surgical risk as related to time of intervention in the repair of intracranial aneurysms. J Neurosurg 28:14–20
5. Ljunggren B, Brandt L, Säveland H et al. (1984) Outcome in 60 consecutive patients treated with early aneurysm operation and intravenous nimodipine. J Neurosurg 61:864–873
6. Öhman J, Heiskanen O (1988) Effect of nimodipine on the outcome of patients after aneurysmal subarachnoid hemorrhage and surgery. J Neurosurg 69:683–686
7. Säveland H, Ljunggren B, Brandt L, Messeter K (1986) Delayed ischemic deterioration in patients with early aneurysm operation and intravenous nimodipine. Neurosurgery 18:146–150
8. Seiler RW, Reulen HJ, Huber P, Grolimund P, Ebeling U, Steiger HJ (1988) Outcome of aneurysmal subarachnoid hemorrhage in a hospital population: a prospective study including early operation, intravenous nimodipine, and transcranial Doppler ultrasound. Neurosurgery 23:598–604
9. Sundt TM (1977) Cerebral vasospasm following subarachnoid hemorrhage: evolution, management, and relationship to timing of surgery. Clin Neurosurg 24:228–239
10. Sundt TM, Kobayashi S, Fode NC, Whisnant JP (1982) Results and complications of surgical management of 809 intracranial aneurysms in 722 cases. J Neurosurg 56:753–765
11. Weir B, Aronyk K (1981) Management mortality and the timing of surgery for supratentorial aneurysms. J Neurosurg 54:146–150

Parenteral Nimodipine Therapy After Subarachnoid Hemorrhage: A Randomized Clinical Study

W. Reif, M.R. Gaab, D. Stolke, H.G. Höllerhage, and H. Dietz[1]

Introduction

With the advent of common clinical use of the calcium antagonist nimodipine, controversy arose over its efficiency in the treatment of vasospasm after subarachnoid hemorrhage (SAH). Great hopes were placed in nimodipine, but on the other hand the possibility of negative effects, e.g., the ability of nimodipine to decrease blood pressure, to augment right-to-left shunts, and to disturb the blood-brain barrier, could not be neglected.

To us, the variety of opinions about nimodipine was the motive for undertaking a randomized clinical study which was to be concerned especially with the effectiveness of nimodipine when administered intravenously. The question to be investigated in our study was: Does nimodipine have a positive influence on the course of patients with SAH? Prior to our study few data were available from randomized studies on the i.v. administration of nimodipine.

Patient Selection Criteria

Three criteria had to be met by patients for admission to the study: an age of 14–65 years, a Hunt-Hess (HH) grade I–IV, and an SAH dating back no longer than 72 h.

Under these conditions 120 patients were accepted into the study between January 1985 and August 1988, 99 of whom were operated on for aneurysms. About 80 patients did not fulfil the above-mentioned criteria, most because of their age. In patients who were to be treated with nimodipine, administration was commenced as soon as possible, beginning with 0.5 mg i.v./h. After 2 h the dose was raised to 1 mg i.v./h, and after 6 more hours, to 2 mg i.v./h. We assumed a blood pressure of at least 100/60 mmHg under nimodipine therapy.

In the evaluation of our study the most important data were the HH grade at the time of admission to our department, the rating on the Glasgow Coma

[1] Neurochirurgische Klinik, Medizinische Hochschule Hannover, Konstanty-Gutschow-Str. 8, D-3000 Hannover 61

Advances in Neurosurgery, Vol. 18
K.-A. Bushe, M. Brock, and M. Klinger (Eds.)

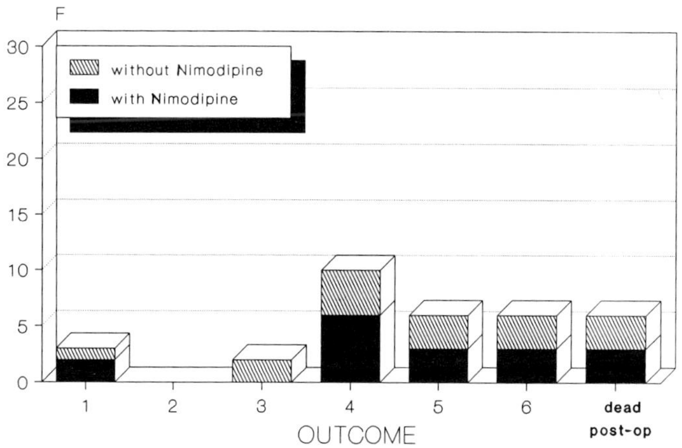

Fig. 1. Outcome after at least 12 months in 27 hypertensive patients who underwent surgery

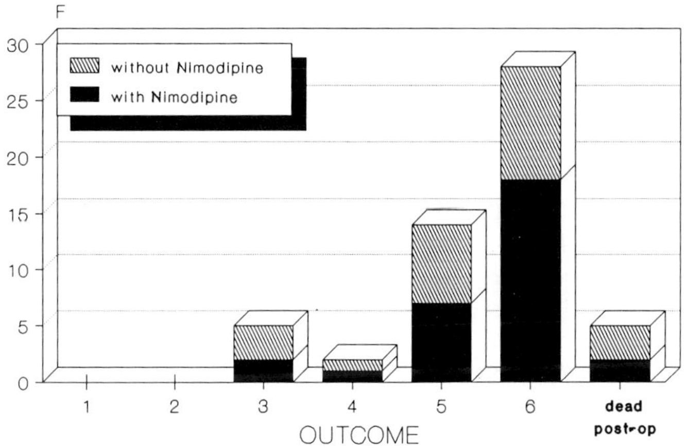

Fig. 2. Outcome after at least 12 months in 54 normotensive patients who underwent surgery

Scale (GCS) within the first 2 weeks after SAH, and the outcome of patients at the time of discharge as well as at least 1 year after SAH. Further criteria were the frequency of infarctions (as proved by CT), of infections, and of re-rupture (included intraoperative), and the blood pressure levels. For the evaluation of the outcome we used the modified Glasgow Outcome Scale (GOS), ranging from 6 (very good) to 1 (dead).

Results

The data for the 120 patients, separated into groups with and without nimodipine, showed the well-known age distribution: most of our patients with SAH were between 40 and 60 years old, with a peak between 50 and 55 years. No difference between the groups was evident in respect of the location of malformations of intracerebral blood vessels. Aneurysms of the internal carotid artery were underrepresented, while aneurysms of the anterior and middle cerebral arteries were seen a little more frequently than usual. The outcome of our patients, operated on or not, was 5 or 6 (good/very good) in about three-quarters of all cases; 24 patients died, and about 10% had a poor outcome. The distribution of the outcomes within the two groups (with/without nimodipine) was almost equal. The reliability of these figures, however, is restricted by the presence of patients in whom no aneurysm could be found or who were in too poor a condition to be operated on. So, from here on, the data will be related to the 99 patients who were operated on for aneurysms.

Evaluation of the HH grade at the time of admission showed that most of our patients were admitted with an HH grade II or III. Grade IV was rather rare; one patient was in grade I and one in grade V. The patients operated on showed a lower mortality, and again almost three-quarters had a good or very good outcome. Important differences between the two groups were not found.

The important intraoperative complications were compared with the complications in the postoperative course; more than one complication in a single patient was possible. In more than two-thirds of the operations there were no complications at all, the figure being the same in both groups. There were no differences between the two groups in the distribution of intraoperative complications, but among the postoperative complications, re-ruptures were more frequent in the group with nimodipine while infections and especially infarctions were more frequent in the group without nimodipine.

Comparing the outcomes of patients with increased or normal blood pressure, we saw a significantly worse outcome in hypertensive patients. One-third of our patients suffered from hypertension, but they accounted for 50% of our mortality; a good or very good outcome was less frequent in these patients.

We followed up 100 patients for at least 1 year after SAH. Three-quarters of them had a good or very good outcome; three patients had died after discharge. A trend for or against nimodipine did not appear.

Comparison of early operations (within 72 h after SAH) with delayed operations was not possible, because there were only 12 delayed operations. It is important, however, that during the period of delay, five patients died due to rebleeding.

After 12 months, differences in outcome in patients with and without hypertension were evident: The outcome of patients with hypertension was much worse; for example, three more patients had died after discharge. The normotensive patients seemed to benefit from nimodipine administration (Figs. 1, 2).

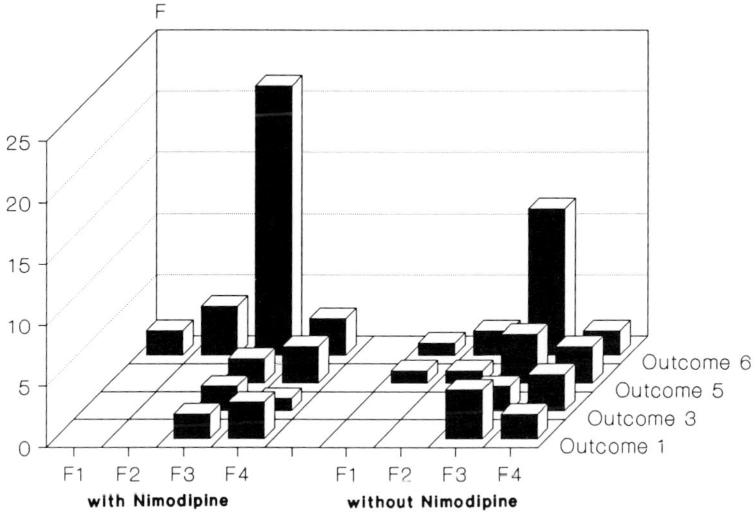

Fig. 3. Severity of SAH (according to Fisher) and outcome in patients who underwent surgery (n = 81)

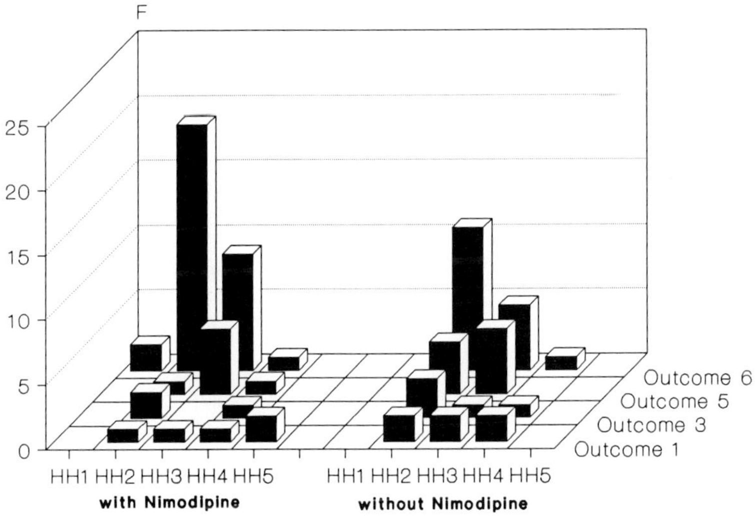

Fig. 4. HH grade at operation and outcome (n = 81)

When outcome was related to the Hunt-Hess grading at the time of operation and the severity of SAH according to Fisher, it was seen that the better the condition of the patient at the time of operation and the less severe the SAH, the better the outcome. Mortality increased with poorer HH grading and the severity of SAH. Patients with favorable prognoses (HH grades I–III) had a postoperative mortality of 8%, despite the fact that most of our patients were admitted with severe SAH (Fisher 3 or 4).

Comparing the groups with and without nimodipine, the number of good outcomes was higher in the nimodipine group and the number of poor outcomes was lower, but a difference in mortality was not found (Figs. 3, 4).

Postoperative Long-Term Results After Aneurysm Surgery Among Patients Treated with Nimodipine

J. Hampl, H. Berger, M. Degenhard, E. Irle, J. Kunert, and S. Kunze[1]

Introduction

With the introduction of the calcium channel blocker nimodipine, the incidence of permanent neurological deficits (PNDs) was significantly reduced [1,4]. In our present study 9.8% of the patients had PNDs at a follow-up evaluation; 2.4% of these neurological deficits were due to postoperative vasospasm. Despite this success in respect of neurological outcome, several studies [11, 14, 15] have shown severe cognitive impairment of patients after subarachnoid hemorrhage (SAH).

Questions

1. Does the calcium channel blocker nimodipine prevent cognitive impairment after aneurysm surgery?
2. Does the quality of cognitive impairment vary in relation to the location of the ruptured aneurysm?
3. What impact does postoperative rehabilitation have on the occupational reintegration of patients after aneurysm surgery?
4. What happens to patients with no postoperative rehabilitation as regards occupational reintegration and neuropsychological sequelae?

Patient Random Samples

Between 1983 and 1987 248 patients with the diagnosis of SAH were admitted (113 males, 135 females). Seventy-four patients had no visible aneurysm on angiography; 20 patients died before the follow-up. Of the remaining population, 135 patients (88%) have undergone follow-up evaluation. Their mean age was 47.4

[1] Neurochirurgische Klinik der Universität Heidelberg, Im Neuenheimer Feld 400, D-6900 Heidelberg

Advances in Neurosurgery, Vol. 18
K.-A. Bushe, M. Brock, and M. Klinger (Eds.)

years, ranging from 21 to 77 years. Mean time since operation was 3.1 years, ranging from 335 days to 2029 days.

All patients received nimodipine intravenously, at a dosage of 2 mg/h, immediately after the diagnosis of SAH. Times between SAH and aneurysm surgery were as follows:

< 72 h: 67 patients (53.2%)
> 72 h and < 9 days: 20 patients (16.6%)
$>$ 9 days: 38 patients (30.2%)

The locations of the ruptured aneurysms were:
- Anterior cerebral artery: 59 patients (46.8%)
- Internal carotid artery: 36 patients (28.6%)
- Medial cerebral artery: 31 patients (24.6%)

The grading of SAH on admission and at discharge is shown in Table 1.

The Psychological Methods

1. WIP (reduced intelligence test for adults) with the parameters full-scale IQ, verbal IQ, and performing IQ [19].
2. MMPI (Minnesota Multiple Choice Personality Inventory) with the dimensions depression, hypomania, psychasthenia, schizophrenia, and psychopathic deviation [8].
3. POMS (Profile of Mood States) including the parameters vigor, fatigue, depression, and anger-hostility [12].
4. Stroop Test: a psychomotor speed test which measures the speed of information processing, vigilance, interference, and concentration [5].
5. TMT (Trail Making Test): again a speed test which focuses on perceptual speed, review, and accuracy [2].

Table 1. Grading of SAH (Hunt and Hess)

Hunt and Hess	On admission	At discharge
Grade I	13 (10.3%)	79 (62.7%)
Grade II	72 (57.1%)	35 27.8%)
Grade III	30 (23.8%)	12 (9.5%)
Grade IV	10 (7.9%)	–
Grade V	1 (0.8%)	–
	126 (100%)	126 (100%)

6. Selective Reminding Test: memory test for verbal short-term and long-term memory [7].
7. Paired Associate Test: memory test for verbal long-term memory with the focus on interference [20].
8. BVRT (Benton Visual Retention Test): test to investigate visual short-term memory dysfunctions [6].
9. Gollin Test: test to establish disturbance of visual long-term memory and activation of perceptual concepts [9, 17, 18].

Analyses of variance and discriminant analyses were used to investigate differences between the sample group and a control group of patients with lumbar disk operation.

Results

Ad 1. With the exception of the paired associate test the sample's gain for each test was significantly lower than for the control group (Table 2).

Ad 2. No difference could be found between the groups as regards the site of the ruptured aneurysm. Looking at the subtests, patients with a ruptured aneurysm of the anterior cerebral artery had a higher score on the depression and schizophrenia scales of the MMPI and a lower score on the vigor dimension of the POMS.

Ad 3. Only 46.3% of the patients employed before the SAH returned to a full-time job postoperatively and out of 126 patients only 47 (37.3%) went through a rehabilitation program. Despite the significantly lower Hunt and Hess grades,

Table 2. Sample's gain for each test as compared with the control group

Test	*P*	Reclassification
WIP	0.01	72.2%
MMPI	0.02	60.3%
POMS	0.04	61.0%
Stroop Test	0.03	59.2%
Trail Making Test	0.08	53.6%
Selective Reminding Test	0.001	90.9%
Paired Associate Test	NS	72.3%
Benton Visual Retention Test	0.04	67.4%
Gollin Test	0.03	70.7%

46.8% of the patients in the rehabilitation group returned to work compared with 38.4% in the nonrehabilitation group.

Ad 4. When patients without rehabilitation were compared with the control group (Table 3), significantly lower scores could again be found in the patient group. Compared to the total sample the amount of cognitive impairment was less in the nonrehabilitation group. This can be explained by the better Hunt and Hess grades in this group of patients.

Discussion

The frequency of PNDs can be significantly reduced by the calcium channel blocker nimodipine. Nevertheless, cognitive impairment can be found in a high percentage of patients after SAH [11, 14, 15]. These results were confirmed in this study. Interestingly, no difference could be found between groups based on different aneurysm locations. The conclusion from this finding is that an SAH has an impact on the whole brain and not only on the area of the ruptured aneurysm. This supports the theory of Grote and Hassler, that just after the rupture of a cerebral aneurysm a complete circulatory arrest with concomitant hypoxia will occur.

In accordance with other studies [3, 13, 16], only 46.3% of the patients returned to full-time work. The results regarding occupational reintegration were better in the group of patients who went through rehabilitation therapy than in the nonrehabilitation group. It is remarkable that only 37.3% of the patients went through a rehabilitation program. Comparison of the nonrehabilitation group with the healthy control group as regards cognitive impairment showed a significant difference disfavoring the nonrehabilitation group.

Table 3. Comparison between patients without rehabilitation and the control group

Test	*P*	Reclassification
WIP	0.07	61.1%
MMPI	0.09	62.5%
POMS	0.02	56.9%
Stroop Test	0.08	47.9%
Trail Making Test	NS	45.5%
Selective Reminding Test	0.01	75.4%
Paired Associate Test	NS	75.0%
Benton Visual Retention Test	NS	59.7%
Gollin Test	0.03	60.9%

Summary

1. The early administration of the calcium channel blocker nimodipine can significantly reduce the incidence of permanent neurological deficits after SAH.
2. Despite these good neurological results a high percentage of patients demonstrated cognitive dysfunction on neurophysiological test batteries.
3. These deficits are independent of the location of the ruptured aneurysm.
4. After rehabilitation a higher percentage of patients could be reintegrated in full- or part-time employment. The referral to a rehabilitation program should be initiated by the neurosurgeon.

Refences

1. Allen GS, Ahn HS, Preziosi TH et al. (1983) Cerebral arterial spasm – a controlled trial of nimodipine in patients with subarachnoid hemorrhage. N Engl J Med 308:619–624
2. Army Individual Test Battery (1944) Manual of directions and scoring. War Department, Adjutant Generals Office, Washington, D.C.
3. Artiola I, Fortuny L, Prieto-Valiente L (1981) Long term prognosis in surgically treated intracranial aneurysms. Part 2: Morbidity. J Neurosurg 54:35–43
4. Auer LM, Brandt L, Ebeling U et al. (1986) Nimodipine and early aneurysm surgery in good condition SAH patients. Acta Neurochir (Wien) 82:7–13
5. Bäumler G (1985) Farbe-Wort-Interferenztest (FWIT) nach J.R. Stroop, 1st edn. Verlag für Psychologie Dr. C.J. Hogrefe, Göttingen
6. Benton AL (1981) Der Benton-Test, 5th edn. Hans Huber, Bern
7. Buschke H (1973) Selective reminding for analysis of memory and learning. J Verbal Learning Verbal Behav 12:543–550
8. Gehring A, Blaser A (1982) MMPI Minnesota Multiphasic Personality Inventory, deutsche Kurzfassung für Handauswertung, Handbuch, 1st edn. Hans Huber, Bern
9. Gollin ES (1960) Developmental studies of visual recognition of incomplete objects. Percept Mot Skills 11:289–298
10. Grote E, Hassler W (1988) The critical first minutes after subarachnoid hemorrhage. Neurosurgery 22:654–661
11. Ljunggren B, Sonesson B, Säverland H, Brandt L (1985) Cognitive impairment and adjustment in patients without neurological deficits after aneurysmal SAH and early operation. J Neurosurg 62:673–679
12. McNair D, Iorr M, Droppleman LF (1981) Eits manual for the Profile of Mood States (POMS), 1st edn. Educational and Industrial Testing Service, pp 1–25
13. Ropper A, Zervas NT (1984) Outcome 1 year after SAH from cerebral aneurysm. J Neurosurg 60:909–916
14. Sonesson B, Ljunggren B, Säbeland H, Brandt L (1987) Cognition and adjustment after late and early operation for ruptured aneurysm. Neurosurgery 21:279–287
15. Stolke D, Hoppe B, Seifert V, Rückert N, Dietz H (1987) How good are the good results in aneurysm surgery? Psychological aspects of rehabilitation of early and late operated patients with cerebral aneurysm. In: Wüllenweber R, Klinger M, Brock M (eds) Advances in neurosurgery, vol 15. Springer, Berlin Heidelberg New York, pp 297–301

16. Sundt TM Jr, Kobayashi S, Fode NC et al. (1982) Results and complications of surgical management of 809 intracranial aneurysms in 722 cases. Related and unrelated to grade of patient, type of aneurysm, and timing of surgery. J Neurosurg 56:753–765
17. Warrington EK, Weiskrantz I (1968) New methods of testing long-term retention with special reference to amnesic patients. Nature 217:972–974
18. Warrington EK, Weiskrantz L (1970) Amnesic syndrome: consolidation or retrieval? Nature 228:628–630
19. Wechsler D (1982) Handanweisung zum Hamburg-Wechsler-Intelligenztest für Erwachsene (HAWIE). Hans Huber, Bern
20. Winocur G, Weiskrantz L (1976) An investigation of paired-associate learning in amnesic patients. Neuropsychologie 14:97–110

The Acute Effect of Nimodipine on Intracranial Pressure and Cerebral Blood Flow

J.F. Schmidt, G. Waldemar, M. Albeck, O.B. Paulson, and F. Gjerris[1]

In acute stroke it has been shown that the calcium antagonist PY 108-068 induces a slight increase in mean cerebral blood flow (CBF) in some cases with a concomitant flow reduction in focal ischemic areas [8]. In experimental studies in rat, nimodipine increases the CBF and impairs the autoregulation [3, 5]. Different calcium antagonists might, however, have different effects on the CBF. Whether the increase in CBF exerts any adverse effect on intracranial pressure (ICP) has not been studied. In the present study we have investigated this using nimodipine.

Method

Twelve patients with normal pressure hydrocephalus (NPH) were studied. ICP and mean arterial blood pressure (MABP) were continuously measured intraventricularly and intraarterially, respectively. In all patients CBF was estimated by the arteriovenous method. The relative CBF was determined from $CMRO_2 = (aO_2 - vO_2) \times CBF$. On the assumption of a constant cerebral metabolic rate of oxygen ($CMRO_2$), this is a good estimate of changes in CBF [6].

In eight patients intravenous nimodipine was administered via an indwelling catheter at a dosage of 15 μg/kg/h during the first 2 h, after which the dosage was increased to 30 μg/kg/h for the next 2 h. CBF was measured by the arteriovenous method before and every 10 min during the nimodipine infusion.

In four other patients regional CBF (rCBF) was measured by xenon 133 inhalation and single photon emission computerized tomography (SPECT) using the Tomomatic 64 [1, 7] before and 1 h after oral administration of 60 mg nimodipine. Hypo- and hypertension were induced in the control period and 1 h after the oral dose of nimodipine. CBF was measured for every 5 mmHg change in MABP by the arteriovenous method. The lower limit of CBF autoregulation was determined according to a computer program.

[1] Department of Anesthesiology, Rigshospitalet, Blegdamsvej 9, 2100 Copenhagen Ø, Denmark

Advances in Neurosurgery, Vol. 18
K.-A. Bushe, M. Brock, and M. Klinger (Eds.)

Results

Following intravenous nimodipine MABP was reduced by 23 mmHg (4–47 mmHg) ($P < 0.05$). Median CBF was unchanged, but in the three patients with the greatest reduction in perfusion pressure the CBF decreased by 11%, 23%, and 34%, respectively.

Following oral nimodipine MABP was reduced slightly, by 7 mmHg (0–16 mmHg). There was no significant change in ICP and CBF, and no changes in the regional distribution of CBF as measured with SPECT. The lower limit of CBF autoregulation did not change significantly. However, the lower limit was not reached despite a decrease in perfusion pressure of 29 mmHg (from 86 to 57 mmHg) in one patient, and in one patient CBF increased despite a decrease in the cerebral perfusion pressure, suggesting an abolished autoregulation of CBF.

Comments

A slight but significant increase in ICP was found after intravenous nimodipine. After oral nimodipine there was no change in ICP. The data from the literature are conflicting. Guggiari et al. [4] found an increase in ICP after nimodipine treatment. In contrast Gaab et al. [2] found no change in ICP. Both studies dealt with patients suffering from an acute cerebral disease.

In our study no significant change in CBF was found, although a major decrease in CBF was found in the patients with the greatest decrease in MABP. The patients were elderly outpatients without acute diseases. This finding stresses the importance of maintaining the perfusion pressure under treatment with nimodipine.

In conclusion, a slight increase in ICP and no significant changes in CBF after nimodipine administration were found. However, major falls in perfusion pressure should be avoided using this potent vasodilator in elderly people and in patients with cardiovascular instability.

References

1. Celsis P, Goldman T, Henriksen L, Lassen NA (1981) A method for calculating regional cerebral blood flow from emission computed tomography of inert gas concentration. J Comput Assist Tomogr 5:641–645
2. Gaab MR, Haubitz I, Brawanski A, Korn A, Czech T (1985) Acute effects of nimodipine on the cerebral blood flow and intracranial pressure. Neurochirurgia 28:93–99
3. Gaab MR, Høllerhage HG, Zumkeller M, Trost HA (1987) The effect of the Ca-antagonist nimodipine on cerebral blood flow autoregulation. J Cereb Blood Flow Metab 7 [Suppl 1]:170

4. Guggiari M, Guillaume A, Dagreou F, Philippon J, Viars P (1983) Intracranial pressure and hemodynamic effects of a new calcium channel blocking agent: nimodipine. Anesthesiology 59:A357
5. van der Kerckhoff W, Kazda S (1987) The autoregulation of cerebral blood flow is influenced by calcium antagonists. J Cereb Blood Flow Metab [Suppl 1]:169
6. Lassen NA (1959) Cerebral blood flow and oxygen consumption in man. Physiol Rev 39:183–238
7. Stokely EM, Sveinsdottir E, Lassen NA, Rommer PA (1980) Single photon dynamic computer assisted tomography (DCAT) for imaging brain function in multiple cross sections. J Comput Assist Tomogr 4:230–240
8. Vorstrup S, Andersen A, Blegvad N, Paulson OB (1986) Calcium antagonist (PY 108-068) treatment may further decrease flow in ischemic areas in acute stroke. J Cereb Blood Flow Metab 6:222–229

Clinical Significance of TCD Pulsatility Indices in Subarachnoid Hemorrhage and Arteriovenous Malformations

R. Steinmeier, R. Laumer, K. Appelbaum, M. Mück, and R. Priem[1]

The first transcranial Doppler sonography (TCDS) examinations were all performed in an era without the standard clinical use of the calcium antagonist nimodipine. All these examinations had shown a strong correlation between signs of radiological vasospasm, elevated flow velocity (FV) measured by TCDS, and clinical signs of delayed ischemic (= neurological) deficits (DIDs). Best prediction of impending DID was possible in the area of the middle cerebral artery (MCA), and limits of absolute FV for predicting the impending risk of DIDs were published. In addition, the importance of the daily increase in FV (delta FV) was stressed. Since recent publications [7] on the therapeutic influence of nimodipine on cerebral vasospasm show a statistical decrease in the clinical DID without any remarkable change in the radiological vasospasm, it was the aim of this study to verify the above-mentioned data in a group of patients treated routinely with nimodipine and to examine the clinical significance of different pulsatility indices (PIs). Different authors [6,9] have suspected a significant change to occur in cases of DID. In cases of proximally located vasospasm accessible to TCDS, increased PIs measured proximally to an area of increased FV (= vasospasm) would be expected. In cases of DID without any alteration of FV, where a peripheral vasospasm could be assumed to lie outside the TCDS range, examination of the PIs might be of great interest, because an elevation of the peripheral resistance might cause an elevation of the PIs. As a well-known example of PI sensitivity caused by an increase in resistance, pre- and postoperative hemodynamic changes in arteriovenous malformations (AVMs) are precisely documented [5].

Material and Methods

In a series of 66 patients with SAH, 455 TCDS follow-up examinations were performed. SAH was demonstrated by computed tomography, and one or multiple aneurysms were shown by angiography to be the cause of the SAH in every case.

[1] Neurochirurgische Klinik der Universität Erlangen-Nürnberg, Schwabachanlage 6, D–8520 Erlangen

Advances in Neurosurgery, Vol. 18
K.-A. Bushe, M. Brock, and M. Klinger (Eds.)

Each patient was treated with nimodipine intravenously from the beginning of the clinical course, and the therapy regimen lasted for at least 10 days when no clinical signs of vasospasm were seen. Cases of SAH without aneurysm or with intracerebral hematoma were excluded, as were cases with less than three TCDS follow-up investigations. "Delayed ischemic = neurological deficit" was defined as a neurological disturbance (paresis, aphasia) occurring at least 3 days following the initial bleeding. As a consequence, three different groups could be distinguished: group I without any neurological signs during the clinical course (n = 35), group II with neurological deficits occurring during the first 3 days following bleeding or the day following aneurysm surgery (n = 19), and group III with DIDs of the definition given above (n = 12).

If possible, TCDS examinations were performed at least every second day. All examinations were done using a three-dimensional transcranial Doppler scanner (Transscan, EME, D–7770 Überlingen, FRG) according to internationally accepted standards. On both sides of the patient's head, five measuring points were taken routinely for the statistical analysis: one point for the internal carotid artery bifurcation (ICA), one point for the anterior cerebral artery (ACA), and three points for the MCA (one point before and one distal to the point of the maximum value). Depths of measurement, systolic (V_{sys}), mean (V_{mean}), and diastolic (V_{diast}) FVs were stored semiautomatically in a PC database. The following transformations were performed: For each measurement point, the Fourier pulsatility index 1 (FDI_1, relative amplitude of the first harmonic), the Gosling index (GI; V_{sys}-V_{diast}/V_{mean}), and the Purcelot resistance index (RI; V_{sys}-V_{diast}/V_{sys}) were calculated [1, 3]. Statistical analysis and graphic output were done on a standard PC statistical system[2]. As a control group for the sensitivity of PIs, six angiographically proven AVMs with one main feeding artery accessible to TCDS examination without any clinical signs of NPPB ("normal perfusion pressure breakthrough" [10]) were taken, comparing FV and PIs pre- and postoperatively. AVMs might be an ideal model for testing the sensitivity of PIs because of the generally accepted rise in peripheral resistance pre- and postoperatively when examining the main feeder.

[2] STATGRAPHICS V. 3.0, STSC Plus Ware, Rockville, USA: median, 25% and 75% quantile, 2SD limit; curve fitting by polynomial smoothing of order 3 (Fig. 1); means and 95% confidence intervals of means (Fig. 2); regression of an exponential model ($y = e^{a+bx}$) with standard error of estimation (inner dotted line) and 95% confidence and prediction limits (outer dotted line), significance testing for the model by analysis of variance (Fig. 3); Spearman rank correlation coefficients (Table 1); bounds of the P values: *, $P < 0.05$; **, $P < 0.01$; ***, $P < 0.001$

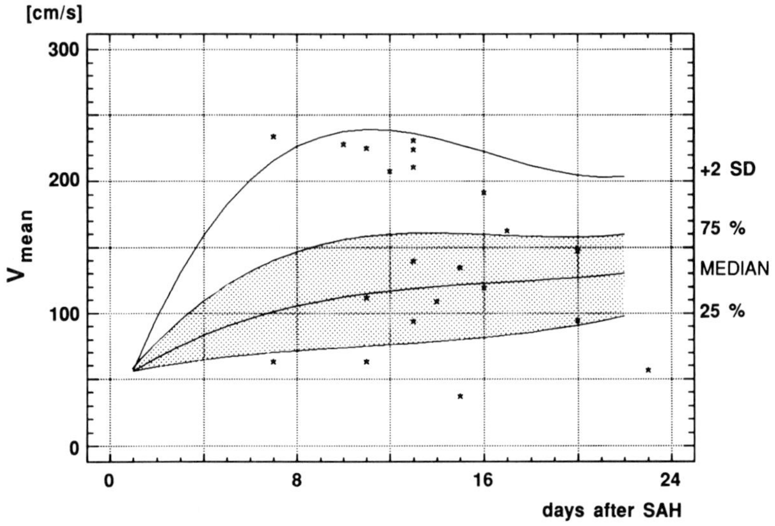

Fig. 1. Comparison between FV of non-DID and DID cases (right side): median, 25% and 75% quantiles, and +2SD of mean for group I (non-DID) (*lines*) and group III (DID (*asterisks*)

Results

To examine the influence of SAH on FV, the Fisher CT grade [2] of every patient was analyzed according to the FV of this patient. A statistically significant difference between the amount of blood in the initial CT and the subsequent rise in FV was found, underlining the fact that vasospasm also occurs in patients treated with nimodipine (*t*-test for independent samples ***: Fisher grade I – Fisher grade III, both sites).

To test the sensitivity of FV by TCDS examinations in distinguishing between patients with and without DID, FV in the group of patients never showing any neurological deficits (group 1) was compared with FV in the patients showing DID (group III; Fig. 1). Although some data points were found to be in the upper range (above the 75% quantile), most cases with DID could not be distinguished from the non-DID group, and even patients with DID and high FV did not reach the +2SD limit of the non-DID group. To test the predictive capacities of changes in FV, TCDS data of patients with DIDs were synchronized according to their date of DID onset (day "0"). According to the literature, a significant increase in velocities before the onset of DID should be assumed, but the graphic analysis did *not* show markedly elevated FV in the range before or at the beginning of the DID (Fig. 2).

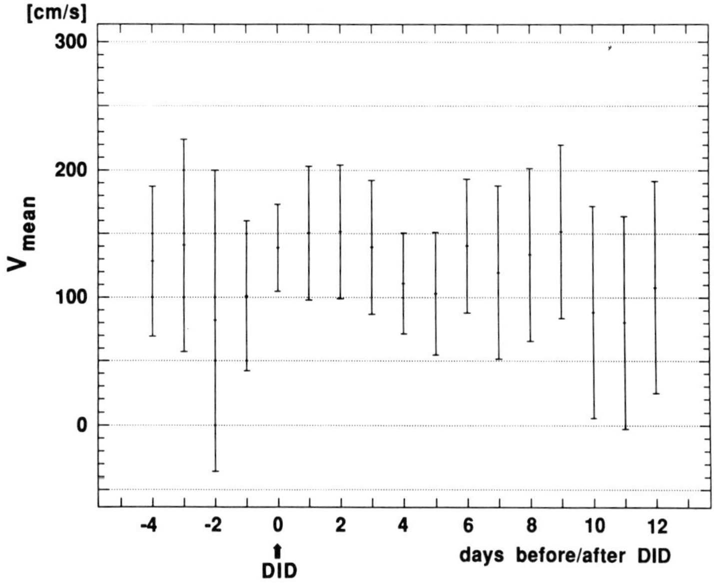

Fig. 2. Changes in FV (mean and 95% confidence intervals) for synchronized onset of DID (group III; "side of interest" contralateral to neurological deficit)

To analyze the predictive capacities of PIs, FV as a marker of vasospasm ($V_{mean\ MCA}$) was correlated with different PIs measured at a point located proximally (ICA bifurcation). Although a significant correlation was found (Table 1), an inverse relationship was observed between the parameters; the higher the MCA velocity, the lower the ICA PIs. To examine whether increased peripheral resistance would act as a possible cause of DIDs without increased FV, the dependency of RI_{MCA} on FV_{MCA} was tested independently for group I and group III cases (Fig. 3). DID cases (group III) were compared with cases of group I: No significant difference in PIs between the two groups could be found, and even DID cases with normal FV did not differ from cases without DID of the same velocity range.

To express the sensitivity of PIs in AVMs, the results of PI measurement (same feeding artery before and following surgery) are demonstrated in Fig. 4.

Discussion

Despite routinely applied nimodipine therapy in our patients, a significant rise in FVs was seen, depending on the amount of subarachnoid blood in the initial CT. As reported by other authors [4, 8], there might be a reduction of absolute FV in

Table 1. Correlation between FV in the MCA and different pulsatility indices of ICA bifurcation (right and left side)

	$V_{mean\ ICA}$	$FPI_{1\ ICA}$	RI_{ICA}	GI_{ICA}
$V_{mean\ MCA}$ right	+ 0.51***	−0.60***	−0.57***	−0.58***
RI_{ICA} right		+0.82***		+0.96***
GI_{ICA} right		+0.80***	+0.98***	
$V_{mean\ MCA}$ left	+ 0.54***	−0.56***	−0.57***	−0.58***
RI_{ICA} left		+0.75***		+0.97***
GI_{ICA} left		+0.71***	+0.97***	

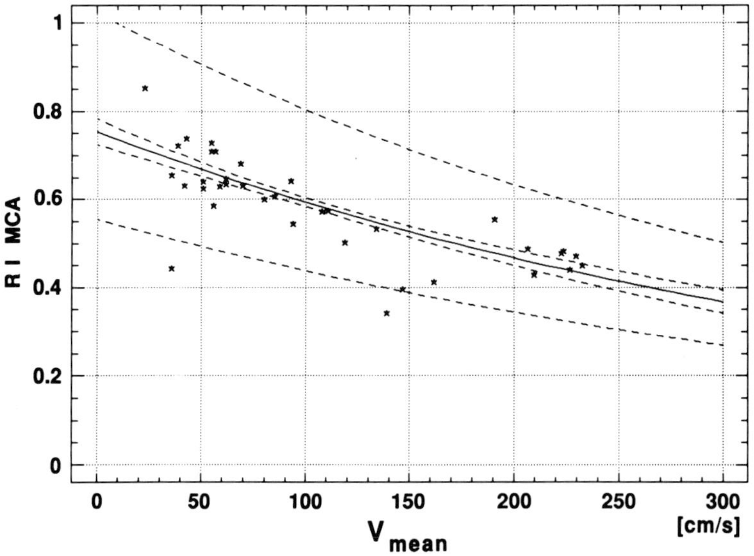

Fig. 3. Comparison between PIs of non-DID and DID cases (*right side*): exponential regression ($y = e^{-0.308 + (-0.0022x)}$; ***) with 95% confidence intervals between $V_{mean\ MCA}$ and RI_{MCA} of group I (no DID) (*lines*); RI_{MCA} of group III (DID) (*asterisks*)

SAH under nimodipine treatment, but the level of the increase in FV despite nimodipine is striking. These data correspond with the findings of persisting radiological vasospasm despite nimodipine treatment [6].

Several mechanisms are discussed to explain the clinical benefits – the decrease in DIDs – in these patients:

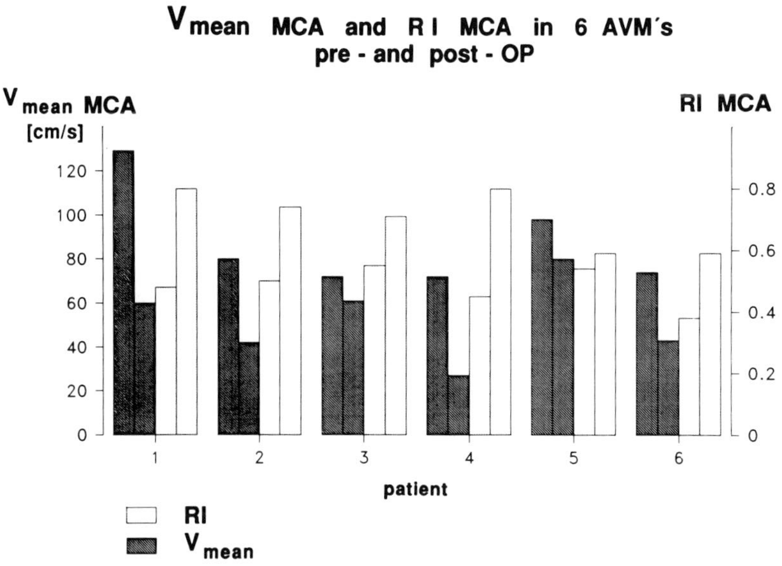

Fig. 4. FV and Pourcelot resistance index pre- and postoperatively in six AVM cases

- Decrease in vasospasm in basal cerebral arteries (= radiological vasospasm)
- Decrease in peripheral resistance and improvement of collateral flow
- Increase in O_2 deficiency states and ischemic tolerance on a cellular level

We are not convinced that the benefits of nimodipine rely primarily on reducing vasospasm in basal cerebral arteries: Several patients had FVs higher than 200 cm/s without any neurological deficits.

The idea of this study was to examine the influence and possible change of peripheral resistance under nimodipine treatment. In view of the different PIs, it should be possible to examine peripheral resistance with TCDS investigations when considering the model of hemodynamics in AVMs. Presuming vasospasm in cases of increased FV, PIs proximally located to vasospasm should show a direct proportionality to FV. Since we found a statistically significant inverted proportionality, some of the assumptions in the theory of pulsatility indices might not be correct: PIs are not parameters for peripheral resistance in every hemodynamic model. The possibility of increased blood flow instead of vasospasm in cases of increased FV might not be too high according to data obtained by other blood flow measuring techniques in SAH. As could be derived from Fig. 3, no further parameter for discriminating between DID and non-DID cases evolves from the examination of PIs: DID PIs are completely within the range of non-DID cases.

The value of TCDS in determining the risk of impending DID under nimodipine therapy should be investigated further.

References

1. Aaslid R, Lindegaard KF (1986) Cerebral hemodynamics. In: Aaslid R (ed) Transcranial Doppler sonography. Springer, Wien New York, pp 60–85
2. Fisher LM, Kistler JP, Davis JM (1980) Relation of cerebral vasospasm to SAH visualized by computerized tomographic scanning. Neurosurgery 6:1–9
3. Gosling RG, King DH (1974) Arterial assessment by Doppler-shift ultrasound. Proc R Soc Med 67:447–449
4. Harders AG, Gilsbach JM (1987) Time course of blood velocity changes related to vasospasm in the circle of Willis measured by transcranial Doppler ultrasound. J Neurosurg 66:718–728
5. Hassler W (1986) Hemodynamic aspects of cerebral angiomas. Acta Neurochir [Suppl] (Wien) 37:1–136
6. Pasqualin A, Acerbi G, Licata C et al. (1988) Transcranial Doppler findings in the early stage of subarachnoid hemorrhage: relation to the amount of cisternal blood deposition and modality of treatment. In: Wilkins RH (ed) Cerebral vasospasm. Raven, New York, pp 25–32
7. Petruk KC, West M, Mohr G et al. (1988) Nimodipine treatment in poor-grade aneurysm patients. Results of a multicenter double-blind placebo-controlled trial. J Neurosurg 68:505–517
8. Seiler RW, Grolimund P, Zurbruegg HR (1987) Evaluation of the calcium-antagonist nimodipine for the prevention of vasospasm after aneurysmal subarachnoid haemorrhage. Acta Neurochir (Wien) 85:7–16
9. Seiler RW, Grolimund P (1988) Transcranial Doppler evaluation of cerebral vasospasm. In: Wilkins RH (ed) Cerebral vasospasm. Raven, New York, pp 19–23
10. Spetzler RF, Wilson CB, Weinstein P et al. (1978) Normal perfusion pressure breakthrough theory. Clin Neurosurg 25:651–672

Development of Vasospasm After Subarachnoid Hemorrhage Under Nimodipine Therapy Before and After Aneurysm Surgery

P. Sanker, D. Terhaag, R.A. Frowein, and K. van Leyen[1]

Introduction

In recent years transcranial Doppler sonography for the measurement of cerebral blood flow velocity (BFV) and therapy with calcium antagonists have been employed both before and after aneurysm surgery. The aim of this study was to investigate the difference between preoperative and postoperative vasospasm under nimodipine therapy and the role that transcranial Doppler sonography follow-up examinations play in improving nimodipine therapy.

Patients and Methods

In 1987 and 1988 transcranial Doppler sonography was carried out on 94 patients with SAH and aneurysm surgery. The limits of normal BFV were defined as 30–80 cm/s; the pulsatility index (PI) ranged from 0.5 to 1.0 in the regions of the internal carotid artery (ICA), the middle cerebral artery (MCA), and the anterior cerebral artery (ACA). All patients initially received 1 mg/h nimodipine intravenously. Providing the blood pressure was stable, the dosage was increased to 2 or 3 mg/h successively. Parallel to decreasing BFV the dosage was then reduced continuously.

Results

In order to make the comparison of the quantitative extent of vasospasm in each of the patients easier and more objective, the BFV factor was determined. It is defined as the increase in BFV in relation to the normal BFV. A 2.5-fold increase in BFV can be observed in 63% of patients older than 60 years. Thus, the relative increase in BFV is even higher in the elderly than in other age groups. The PI measured at normal as well as maximum BFV rises with increasing age (Fig. 1). In the

[1] Neurochirurgische Universitätsklinik Köln, Joseph-Stelzmann-Str. 7, D–5000 Köln 41

Advances in Neurosurgery, Vol. 18
K.-A. Bushe, M. Brock, and M. Klinger (Eds.)

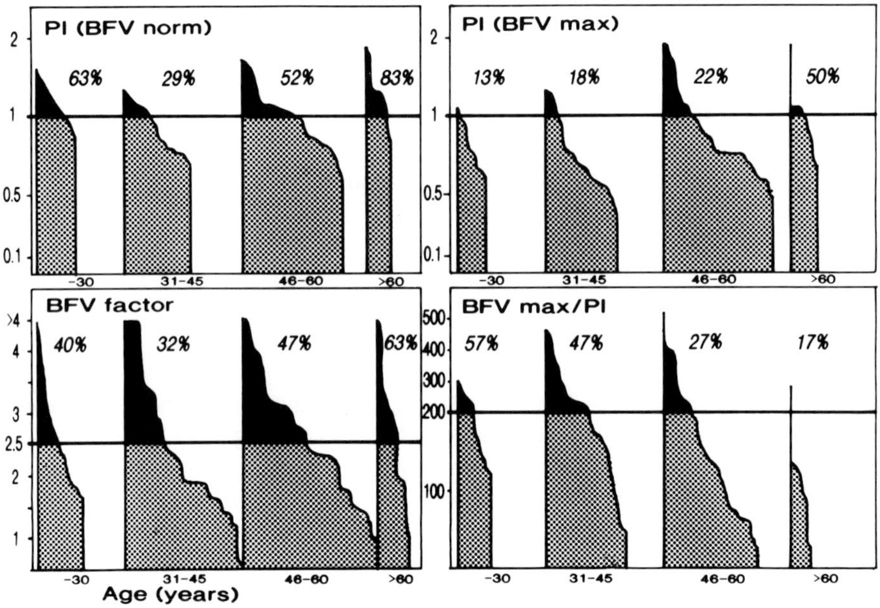

Fig. 1. Pulsatility index (*PI*) at normal and maximum cerebral blood flow velocity (*BFV*), BFV factor, and BFV maximum divided by PI in relationship to the patients' age

following figures, determination of the BFV maximum and its time dependency was based on the BFV factor.

Following SAH but before operation the BFV reached its maximum between the 6th and 9th days. Readings of more than 200 cm/s were seldom found. Normalization took place within 4 weeks providing no recurrent bleeding interfered (Fig. 2). In the postoperative course a rapid increase could be observed, again with a maximum on the 4th–6th days. The highest measurements reached 260 cm/s. They decreased again at a comparatively slow pace. Between the third and fourth postoperative week the BFV still exceeded 80 cm/s in 50% of the patients. Normalization sometimes needs 6 months or more (Fig. 2).

The behavior of the PI varied. In the early phase after SAH as well as after aneurysm surgery the PI was sometimes found to be higher and sometimes lower than normal. In the latter cases the minimum PI was measured on the 15th–17th days after SAH and on the 10th day after operation (Fig. 2). In patients who recovered quickly from the operation the BFV rose slowly, with a maximum on the 8th–10th days. With coma or disturbed consciousness for more than 3 days after the operation a rapid increase in BFV with a maximum on the 4th–6th days was observed. The absolute values of BFV were not higher, however. The PI

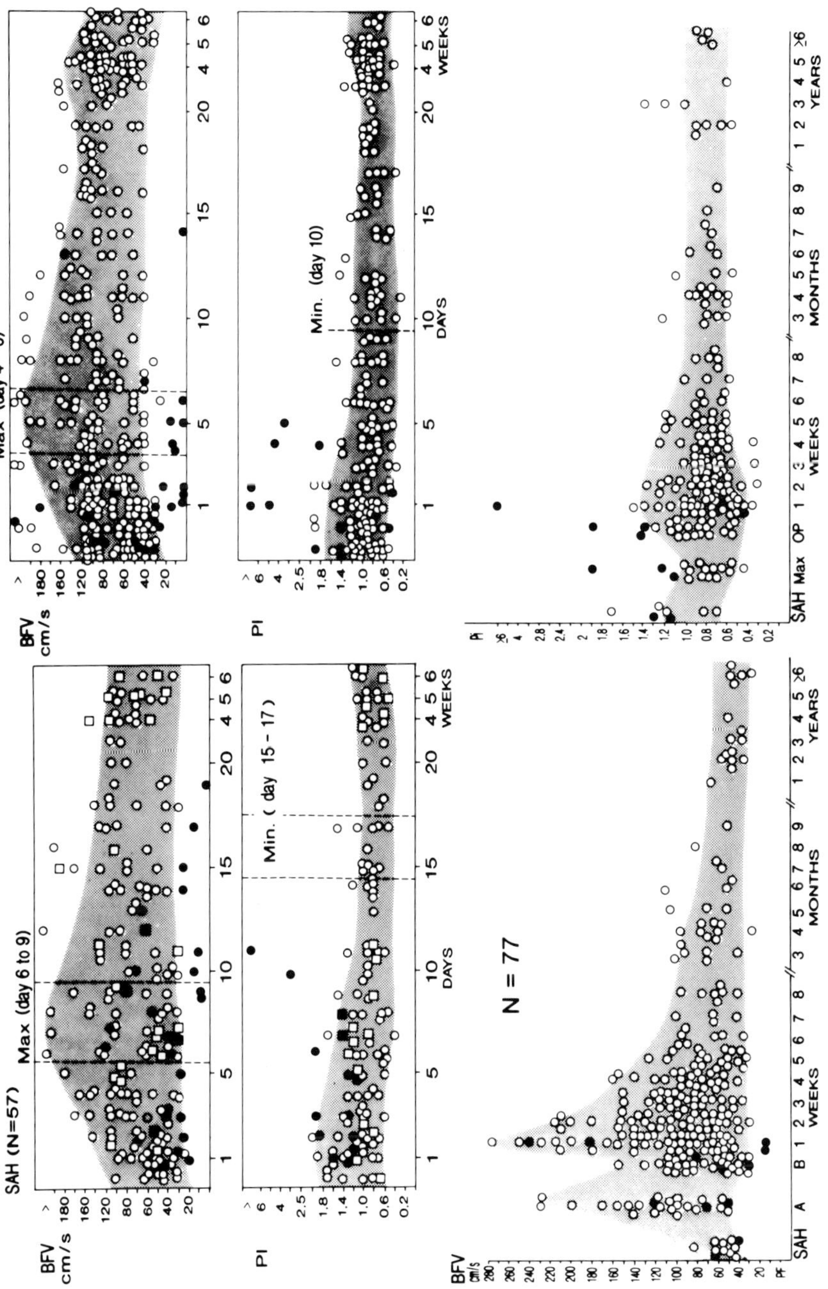

Fig. 2. Development of cerebral blood flow velocity (*BFV*) and pulsatility index (*PI*) before (*upper left*) and after (*upper right*) aneurysm surgery. *Below:* long-term follow-up of BFV and PI. *Open circles*, survivors; *closed circles*, nonsurvivors; *boxes*, last measurements before operation

minimum was reached on the 11th day in patients with a quick recovery; in those who persisted in a state of disturbed consciousness it took only 8 days (Fig. 3).

Two typical cases will illustrate the influence of nimodipine therapy before and after aneurysm surgery (Fig. 4):

Case 1: A 52-year-old female patient suffered an SAH. Angiography was undertaken immediately but showed no aneurysm. Because of problems concerning the blood pressure the fully alert patient received nimodipine orally. The BFV rose to 120 cm/s in the region of the right MCA within 3 days after SAH, and to 150 cm/s by the 5th day. On the 9th day the state of consciousness worsened and she complained of headaches. Nimodipine was raised to 2 mg/h intravenously. Due to low blood pressure values – but with receding BFV – the dosage was reduced to 1 mg/h. Two days later she got worse again; BFV measured 200 cm/s. Following intensified treatment with nimodipine the state of consciousness and BFV improved again to 100 cm/s on the 21st day after SAH. The repeat angiogram was operated on 4 weeks after SAH. Postoperatively the patient recovered quickly and BFV only rose to 140 cm/s, so that the nimodipine dosage could be reduced.

Case 2: In a 25-year-old patient recurrent SAH took place from an aneurysm of the ICA. The BFV in the MCA did not exceed 160 cm/s under nimodipine therapy (2 mg/h i.v.), but fell rapidly thereafter. The patient's condition improved. On the third postoperative day the dosage was lowered to 1 mg/h, maintaining normal BFV measurements. However, the BFV changed in the region of the ACA to 140 cm/s and 170 cm/s on the 4th and 7th days respectively. Severe headaches occurred again and the patient received 2 mg/h nimodipine until the 11th postoperative day. Nimodipine was not reduced again until after complete clinical restitution.

Discussion

Nimodipine therapy after SAH and aneurysm surgery cannot prevent vasospasm but does limit its extent. Before the operation the BFV reached its maximum between the 6th and 9th days, seldom persisting at over 200 cm/s. Postoperatively a rapid increase [4] led to the maximum already being reached on the 4th–6th days. BFV measurements can even far exceed preoperative values. Therefore, we recommend the i.v. administration of nimodipine starting with 1 mg/h and increasing to 2–3 mg/h from the date of SAH until the 10th postoperative day. Afterwards we change to oral administration [1]. According to our experiences a low oral dosage of 3 × 30 mg should be continued until 3 months after the operation since in some instances normalization of the BFV takes more than 6 months.

More important from the prognostic standpoint than the absolute value of BFV is the speed with which the BFV rises. Patients with a slow increase in BFV postoperatively (8th–10th days) recovered in a markedly shorter time than patients with a rapid increase in BFV on the 4th–6th days [2, 3]. Equally important is

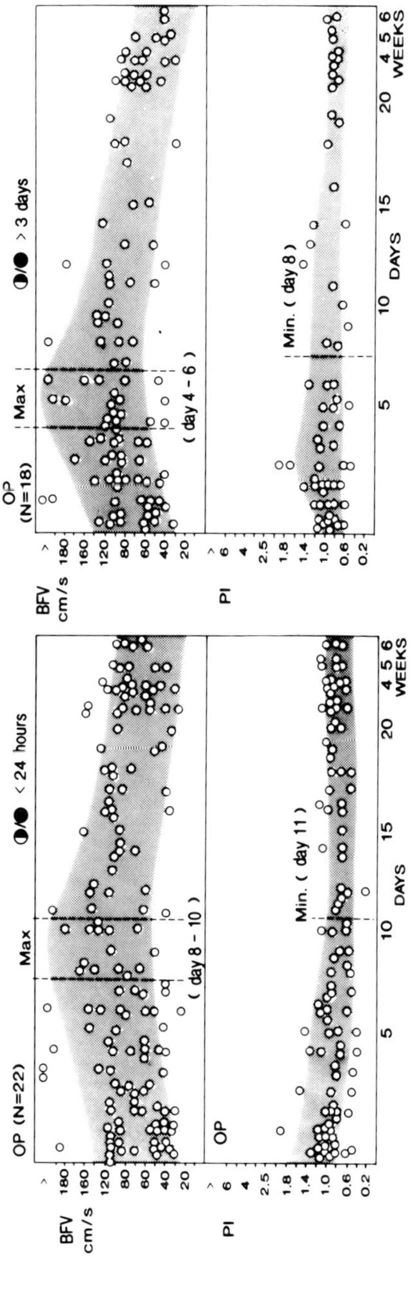

Fig. 3. Postoperative cerebral blood flow velocity (*BFV*) and pulsatility index (*PI*) in patients with recovery within 24 h (*left*) and in patients with disturbed consciousness for more than 3 days (*right*). *Semiclosed circles*, clouding of consciousness; *closed circles*, coma

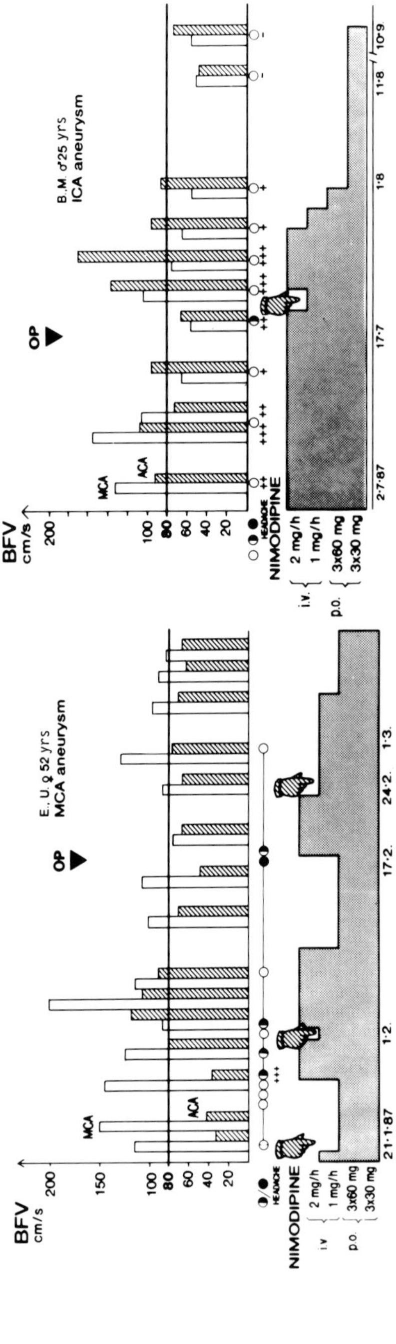

Fig. 4. Development of cerebral blood flow velocity (*BFV*) after SAH and aneurysm surgery. Influence of nimodipine therapy. *Left*: 52-year-old female patient, MCA aneurysm. *Right*: 25-year-old male patient, ICA aneurysm. *Open circles*, consciousness; *semiclosed circles*, clouding of consciousness; *closed circles*, coma

the BFV factor, i.e., the relative increase in BFV compared with the individual's normal BFV: a threefold increase in the BFV to 120 cm/s in elderly patients is as distinct an expression of vasospasm as an increase to 210 cm/s in younger patients with a normal BFV value of 70 cm/s.

Summary

After aneurysm surgery an increase in BFV is to be expected, with in some cases higher absolute measurements and an earlier maximum than before the operation. Therefore the continuous i.v. administration of nimodipine in a dosage of 2 mg/h is to be recommended until at least the 10th postoperative day. The risk of ischemic deficits is especially high in cases with a rapid increase in BFV and a maximum on the 4th–6th days and a high BFV factor.

References

1. Auer LM, Brandt L, Ebeling U et al. (1986) Nimodipine and early aneurysm operation in good condition SAH patients. Acta Neurochir (Wien) 82:7–13
2. Harders A (1986) Neurosurgical applications of transcranial Doppler sonography. Springer, Wien New York, pp 40–64
3. Harders A, Gilsbach J, Laborde G (1988) New aspects of vasospasm evaluated in 100 patients with aneurysm, subarachnoid hemorrhage and acute operation: a transcranial Doppler study. In: Walter W, Brandt M, Brock M, Klinger M (eds) Advances in neurosurgery, vol 16. Springer, Berlin Heidelberg New York, pp 67–72
4. Seiler R, Grolimund P, Zurbrügg H (1986) Die transkranielle Doppler-Sonographie. Schweiz Med Wochenschr 116:626–634

Prevention of Vasospasm with Nimodipine

J. Gilsbach and A. Harders[1]

Introduction

Vasospasm is a secondary complication of subarachnoid hemorrhage (SAH), occurring 4–14 days after the bleeding [18,42]. Angiographic investigations have revealed the incidence of "angiographic vasospasm" to be 30%–70% [3,14,17,25,35, 36,42]. This correlates well with the blood flow velocity changes in the basal arteries detected by transcranial Doppler sonography (TCD). A velocity increase of more than twice the baseline value as a sign of a marked vessel diameter reduction is found in TCD in about 70% of patients [1,18–20,37]. Angiographic vasospasm and "Doppler sonographic vasospasm" do not correlate with the clinical symptoms. Only 15%–30% [11,12,21,22] of patients with vasospasm after aneurysmal SAH develop a "clinical or symptomatic vasospasm" with transient or permanent delayed ischemic deficits (DIDs).

If prevention is attempted, prevention both of angiographic or Doppler sonographic vasospasm and of clinical and symptomatic vasospasm can be aimed for.

Prevention of Angiographic and Doppler Sonographic Vasospasm with Nimodipine

Placebo-controlled trials with control angiography during the critical period of vasospasm between the 4th and 14th days are rare. In a placebo-controlled study on the oral effect of nimodipine, Allen et al. [4] performed angiography only in patients with DIDs; hence the results are not conclusive. The same holds true for a similar study by Philippon et al. [32]. Results of control angiography in the British nimodipine study on oral nimodipine and the prevention of ischemia were not reported in detail [33]. Only in the discussion was it mentioned that spasm was not seen less often in nimodipine-treated patients than in placebo-treated patients. Only the placebo-controlled Canadian study [31] on oral nimodipine in poor condition patients offers some information on the incidence of vasospasm in angiograms

[1] Neurochirurgische Universitätsklinik, Hugstetter Str. 55, D–7800 Freiburg i.Br.

Advances in Neurosurgery, Vol. 18
K.-A. Bushe, M. Brock, and M. Klinger (Eds.)

routinely performed around day 8 after SAH. No significant difference in moderate to severe angiographic vasospasm was found between the two groups, with 64.3% in the nimodipine and 66.2% in the placebo group.

In the controlled intravenous study undertaken by Öhman and Heiskanen [29] only immediate postoperative angiography was performed; i.e., angiography was not performed during the critical period of vasospasm.

In controlled experimental studies no differences in the incidence and severity of vasospasm were seen in primates treated with or without nimodipine [9]. In one uncontrolled investigation with intravenous nimodipine and early surgery, Ljunggren et al. [28] found moderate to severe vasospasm in 27 out of 55 patients 6–14 days after SAH.

Transcranial Doppler sonographic findings from controlled series are not available. The incidence of Doppler sonographic vasospasm in uncontrolled series of patients treated with early surgery and intravenous nimodipine is in a range already known from control angiography after late surgery [18–20, 25, 38, 42]. Harders et al. [20], for example, found critical vasospasms (velocities 3 kHz or 120 cm/s) in 74% of their patients treated with early surgery and intravenous nimodipine. Seiler et al. [38] compared two consecutive series treated with and without nimodipine and operated on early and late. They found no difference in incidence, but a difference in the severity of the velocity increase: the patients treated with nimodipine had slower velocities in the basal arteries than those treated without. Whether this phenomenon is to be attributed to nimodipine, however, is not absolutely clear: An improvement of the collateral circulation by nimodipine would be followed by a slower velocity due to the reduction of the flow through the narrowed segment. The same slower velocity, however, could also be found if nimodipine widened the stenosed segment or prevented vasospasm of large arteries.

Neither the Doppler nor the angiographic information is sufficient to prove any effect of nimodipine on the occurrence and severity of vasospasm. These investigative methods do make it seem probable, however, that the incidence of vasospasm is not significantly influenced by nimodipine. And from the Doppler study it seems possible that nimodipine-treated patients have either a better collateral flow or a less marked vessel narrowing.

Prophylactic Treatment of Clinical or Symptomatic Vasospasm

A clinical analysis can also be undertaken to evaluate the beneficial effect of nimodipine on vasospasm: the incidence and severity of DIDs and the overall outcome must also reflect an improvement if nimodipine works.

A growing number of double-blind, placebo-controlled, prospective and randomized studies on the effect of nimodipine in the prevention of symptomatic vasospasm after SAH have been published [4, 29, 31–33]. In all (Fig. 1), a signi-

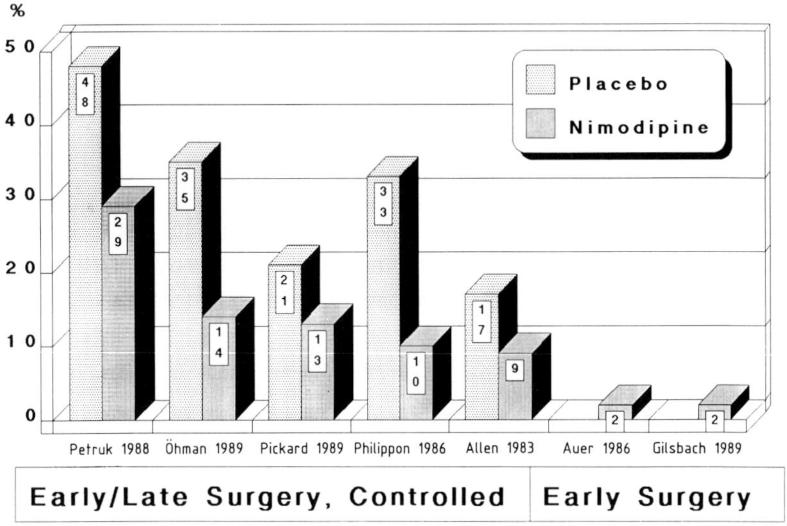

Fig. 1. DIDs after preventive administration of nimodipine in aneurysmal SAH

ficant reduction of DID with death and disability was observed in patients treated with nimodipine as compared with placebo-treated patients. Correspondingly, good to fair outcomes were observed more frequently in treated patients.

In all controlled trials on oral nimodipine the aneurysms were clipped late or were not treated surgically. Two controlled studies on the intravenous administration of nimodipine were published by Öhman and Heiskanen [29, 30] with early and later surgery. Once again, the patients treated with nimodipine had a lower incidence of delayed ischemia.

In these controlled studies, the incidence of DID with death and disability ranged between 9% and 29%. This incidence was clearly lower in patients treated with early surgery and intravenous nimodipine. In the unicenter studies from Ljunggren et al. [28], Auer [5], and Gilsbach and Harders [16] and in the multicenter trials published by Auer et al. [6] and Gilsbach et al. [15], the incidence of symptomatic vasospasm with severe disability and death ranged around 2%. This incidence was the lowest ever published and, interestingly, in nearly all cases vasospasms and complications were responsible for the DID and almost no patients suffered a permanent or lethal deficit from vasospasm alone.

Our Own Results

One hundred and eighty-nine consecutive cases operated on between April 1983 and March 1989 were reevaluated. A morbidity of 14% and a good outcome rate of 82% in the good condition group were nearly identical with the previously published series using the same protocol with early surgery and intravenous nimodipine [5, 6, 15, 28] (Fig. 2). The analysis of mortality and morbidity (Fig. 3) shows very clearly that vasospasm is a less important factor for an unfavorable outcome and, insofar as it is important at all, then only in combination with other complications (e.g., hypotension).

Discussion

The results of new surgical (e.g., early surgery) and pharmacological (e.g., nimodipine) methods to improve the outcome in aneurysmal SAH have often been compared with historical data. In the older literature incidences of DID with death and disability between 20% and 60% [23, 25, 26, 35, 36, 39–43] were reported. Compared with these figures, recently published results showing a 2% DID rate after early surgery and nimodipine are extremely good. In order to make a realistic comparison, however, only recent results of the classical treatment with early and

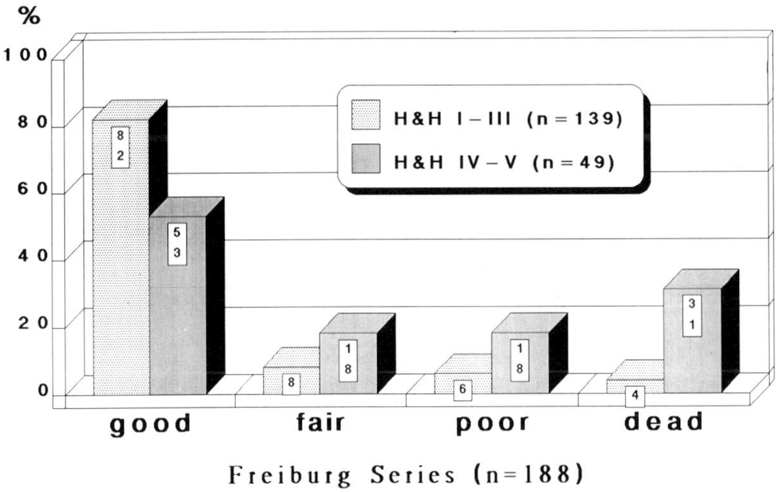

Fig. 2. Outcome after early aneurysm surgery and preventive intravenous nimodipine

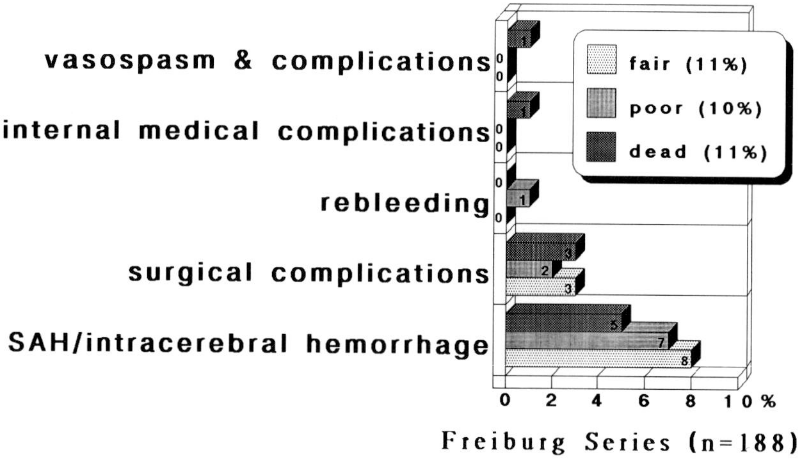

Fig. 3. Causes of mortality and morbidity after early aneurysm surgery and preventive intravenous nimodipine

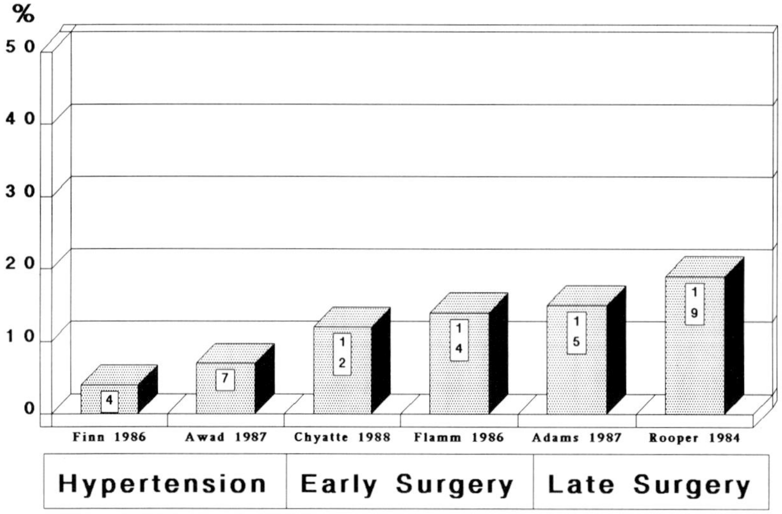

Fig. 4. DIDs after different treatment regimens without nimodipine

delayed surgery and induced hypertension and hypervolemia should be included. In those series wihout nimodipine the incidence of DID with death and disability ranges between 4% and 19% [2, 7, 8, 10, 13, 24, 26, 34] (Fig. 4). The best results were achieved with induced hypertensive and hypovolemic therapy started after the onset of ischemic symptoms [7,34]; early surgery alone was followed by an incidence of DID of 12%–14% [8,13].

The placebo-controlled trails on oral and intravenous nimodipine showed a significant reduction of ischemic events in treated patients (Fig. 1). The remaining number of patients who died or who were disabled due to vasospasm, however, was relatively high in the treated group, with 9%–29%. These results were not better than those after early surgery alone [13,26,27] or induced hypertension [7,34]. Only the results of the noncontrolled series with intravenous nimodipine and early surgery were better than all the other published trials [5,6,15,16,28]. This means that nimodipine has a preventive effect on DIDs but that the management regimen of the controlled studies does not produce better results compared with the most successful results of other treatment regimens without nimodipine.

Conclusion

Double-blind trials on nimodipine in delayed (and in the rare case, early) surgery prove an anti-ischemic effect of the drug. The overall results, however, are not convincing. Open uncontrolled prospective studies on intravenous nimodipine and early surgery demonstrate the lowest incidence of delayed ischemia with convincing overall results. Other factors that provoke or prevent delayed ischemia should not be underestimated (for example, blood pressure and surgical complications).

References

1. Aaslid R, Huber P, Nornes H (1984) Evaluation of cerebrovascular spasm with transcranial Doppler ultrasound. J Neurosurg 60:37–41
2. Adams HP Jr, Kassell NF, Torner JC, Haley EC Jr (1987) Predicting cerebral ischemia after aneurysmal subarachnoid hemorrhage: influences of clinical condition, CT results, and antifibrinolytic therapy. A report of the Cooperative Aneurysm Study. Neurology 37:1586–1591
3. Allcock JM, Drake CG (1965) Ruptured intracranial aneurysms – the role of arterial spasm. J Neurosurg 22:21–29
4. Allen GS, Hyo SA, Preziosi TJ et al. (1983) Cerebral arterial spasm – a controlled trial of nimodipine in patients with subarachnoid hemorrhage. N Engl J Med 308:619–624
5. Auer LM (1984) Acute operation and preventive nimodipine improve outcome in patients with ruptured cerebral aneurysms. Neurosurgery 15:57–66
6. Auer LM, Brandt L, Ebeling U et al. (1986) Nimodipine and early aneurysm operation for good condition SAH patients. Acta Neurochir (Wien) 82:7–13
7. Awad IA, Carter P, Spetzler RF, Medina M, Williams FW (1987) Clinical vasospasm after subarachnoid hemorrhage: response to hypervolemic hemodilution and arterial hypertension. Stroke 18:365–372
8. Chyatte D, Fode NC, Sundt TM (1988) Early versus late intracranial aneurysm surgery in subarachnoid hemorrhage. J Neurosurg 69:326–331

9. Espinosa F, Weir B, Overton T, Castor W, Grace M, Boisvert P (1984) A randomized placebo controlled trial of nimodipine after subarachnoid haemorrhage visualized by computerized tomographic scanning. Neurosurgery 60:1167–1175
10. Finn SS, Stephensen SA, Miller CA, Drobnich L, Hunt WE (1986) Observations on the perioperative management of aneurysmal subarachnoid hemorrhage. J Neurosurg 65:48–62
11. Fisher CM, Roberson GH, Ojemann RG (1977) Cerebral vasospasm with ruptured saccular aneurysm – the clinical manifestations. Neurosurgery 1:245–248
12. Fisher CM, Kistler JP, Davis JM (1980) Relation of cerebral vasospasm to subarachnoid hemorrhage visualized by computerized tomographic scanning. Neurosurgery 6:1–9
13. Flamm ES (1986) The timing of aneurysm surgery. Clin Neurosurg 33:147–158
14. Fletcher TM, Taveras JM, Pool JL (1959) Cerebral vasospasm in angiography for intracranial aneurysms. Incidence and significance in one hundred consecutive angiograms. Arch Neurol 1:38–47
15. Gilsbach JM, Ljunggren B, v Holst H et al. (1987) Outcome in 204 SAH patients subjected to early aneurysm surgery and intravenous nimodipine. Report on a multicenter double blind dose comparison study. In: Prevention and treatment of delayed ischemic dysfunction in patients with subarachnoid hemorrhage. An update. Barcelona, 7 Sept.
16. Gilsbach J, Harders A (1989) Morbidity and mortality after early aneurysm surgery – a prospective study with nimodipine prevention. Acta Neurochir (Wien) 96:1–7
17. Graf CJ, Nibbelink DW (1974) Cooperative study of intracranial aneurysms and subarachnoid hemorrhage. Report on a randomized treatment study. III. Intracranial surgery. Stroke 5:559–601
18. Harders AG, Gilsbach JM (1987) Time course of blood velocity changes related to vasospasm in the circle of Willis measured by transcranial Doppler ultrasound. J Neurosurg 66:718–728
19. Harders A, Gilsbach J, Laborde G (1988) New aspects of vasospasm evaluated in 100 patients with aneurysmal subarachnoid hemorrhage and acute operation. A transcranial Doppler study. Advances in Neurosurgery 16:67–72
20. Harders A, Gilsbach J, Hornyak M (1988) Incidence of vasospasm in transcranial Doppler sonography and its clinical significance: a prospective study in 100 consecutive patients who were given intravenous nimodipine and who underwent early aneurysm surgery. In: Wilkins RH (ed) Vasospasm. Raven, New York, pp 33–52
21. Heros RC, Zervas NT, Negoro M (1976) Cerebral vasospasm. Surg Neurol 5:354–362
22. Heros RC, Zervas NT, Varsus V (1983) Cerebral vasospasm after subarachnoid hemorrhage: an update. Ann Neurol 14:599–608
23. Kassell NF, Boarini DJ, Adams HP, Sahs AL, Graf CJ, Torner HC, Gerk MK (1981) Overall management of ruptured aneurysm: comparison of early and late operation. Neurosurgery 9:120–128
24. Kassell NF, Torner JC (1984) The international cooperative study on timing of aneurysm surgery – an update. Stroke 15:556–570
25. Kwak R, Niizuma H, Ohi T, Suzuki J (1979) Angiographic study of cerebral vasospasm following rupture of intracranial aneurysms: part 1. Time of the appearance. Surg Neurol 11:257–262
26. Ljunggren B, Brandt L, Kägström E, Sundbärg G (1981) Results of early operations for ruptured aneurysms. J Neurosurg 54:473–479
27. Ljunggren B, Brandt L, Sundbärg G, Säveland H, Cronquist S, Stridbeck H (1982) Early management of aneurysmal subarachnoid hemorrhage. Neurosurgery 11:412–418

28. Ljunggren B, Brandt L, Säveland H, Nilsson P-E, Cronquist S, Andersson KE, Vinge H (1984) Outcome in 60 consecutive patients treated with early aneurysm operation and intravenous nimodipine. J Neurosurg 61:864–873
29. Öhman J, Heiskanen O (1988) Effect of nimodipine on the outcome of patients after aneurysmal subarachnoid hemorrhage and surgery. J Neurosurg 69:683–686
30. Öhman J, Heiskanen O (1989) Timing of operation for ruptured supratentorial aneurysms: a prospective randomized study. J Neurosurg 70:55–60
31. Petruk KC, West M, Mohr G et al. (1988) Nimodipine treatment in poor grade aneurysm patients. Results of a multicentre, double-blind, placebo controlled trial. J Neurosurg 68:505–517
32. Philippon J, Grob R, Dagreou F, Guggiari M, Rivierez M, Viars P (1986) Prevention of vasospasm in subarachnoid hemorrhage. A controlled study with nimodipine. Acta Neurochir (Wien) 82:110–114
33. Pickard JD, Murray GD, Illingworth R et al. (1989) Effect of oral nimodipine on cerebral infarction and outcome after subarachnoid haemorrhage: British aneurysm nimodipine trial. Br Med J 298:636–642
34. Rooper AH, Zervas NT (1984) Outcome 1 year after SAH from cerebral aneurysm: management morbidity, mortality, and functional status in 112 consecutive good-risk patients. J Neurosurg 60:909–915
35. Saito I, Ueda Y, Sano K (1977) Significance of vasospasm in the treatment of ruptured intracranial aneurysms. J Neurosurg 47:412–429
36. Saito I, Shigeno T, Aritake K, Tanishima T, Sano K (1979) Vasospasm assessed by angiography and computerized tomography. J Neurosurg 51:446–475
37. Seiler RW, Grolimund P, Aaslid R, Huber P, Nornes H (1986) Cerebral vasospasm evaluated by transcranial ultrasound with clinical grade and CT-visualized subarachnoid hemorrhage. J Neurosurg 64:594–600
38. Seiler RW, Grolimund P, Zurbruegg H (1987) Evaluation of the calcium-antagonist nimodipine for the prevention of vasospasm after aneurysmal subarachnoid hemorrhage. A prospective transcranial Doppler ultrasound study. Acta Neurochir (Wien) 83:7–16
39. Sundt TM, Kobayashi S, Fode NC, Whisnant JP (1982) Results and complications of surgical management of 819 intracranial aneurysms in 722 cases. J Neurosurg 56:753–765
40. Taneda M (1982) The significance of early operation in the management of ruptured intracranial aneurysms – an analysis of 251 cases hospitalized within 24 hours after subarachnoid haemorrhage. Acta Neurochir (Wien) 63:201–208
41. Taneda M (1982) Effect of early operation for ruptured aneurysms on prevention of delayed ischemic symptoms. J Neurosurg 57:622–628
42. Weir B, Grace M, Hansen J, Rothdert C (1978) Time course of vasospasm in man. J Neurosurg 48:173–178
43. Yamamoto I, Hara M, Ogura K, Suzuki Y, Nakane T, Kageyama N (1983) Early operation for ruptured intracranial aneurysms: comparative study with computed tomography. Neurosurgery 12:169–174

The Clinical Value of Transcranial Doppler Sonography in Aneurysmal Subarachnoid Hemorrhage as an Indication for Cerebral Angiography and a Means of Controlling Nimodipine Treatment

C. Buchbender, P. Hammans, and S. Palkovic[1]

Introduction

Mortality and morbidity following aneurysmal bleeding depend on recurrent hemorrhage and the occurrence of cerebral vasospasm. The pathomechanism of cerebral artery narrowing following subarachnoid hemorrhage (SAH) is not exactly known; blood decomposition products are thought to cause vessel contraction with subsequent ischemia and more or less distinct neurological deficits. Later on, changes in vessel walls similar to those seen in proliferative vessel disease may be responsible for the persistent narrowing of cerebral arteries with maintenance of the neurological deficits [5].

Decisive progress in therapy led to findings showing that all agents which cause vessel contraction following SAH lead to an enhanced calcium flow into vessel muscle cells. Consequently the "brain artery specific" calcium entry blocker nimodipine was developed and widely used [7]. The first controlled trial of nimodipine in patients with SAH was published in 1983 by Allen et al. [1].

The diagnosis of vasospasm, which was made in 40% of the patients according to our own former findings, could originally only be established by means of angiography. However, for several years now, diagnosis has been easy using transcranial Doppler sonography (TCD).

Methods

Since January 1988 patients who were admitted following SAH have been examined by TCD using the TC2-64B device with built-in frequency analyzer and video printer (EME, Ueberlingen, West Germany). Daily examination of both middle cerebral arteries (MCA) was performed (Fig. 1), since former findings have shown that the MCA is always involved in cerebral vasospasm irrespective of the main localization of blood clots in the subarachnoid compartments. Furthermore, the MCA is the vessel which provides the highest reliability in the identification and

[1] Neurochirurgische Universitsklinik, Sigmund-Freud-Str. 25, D–5300 Bonn 1

Advances in Neurosurgery, Vol. 18
K.-A. Bushe, M. Brock, and M. Klinger (Eds.)

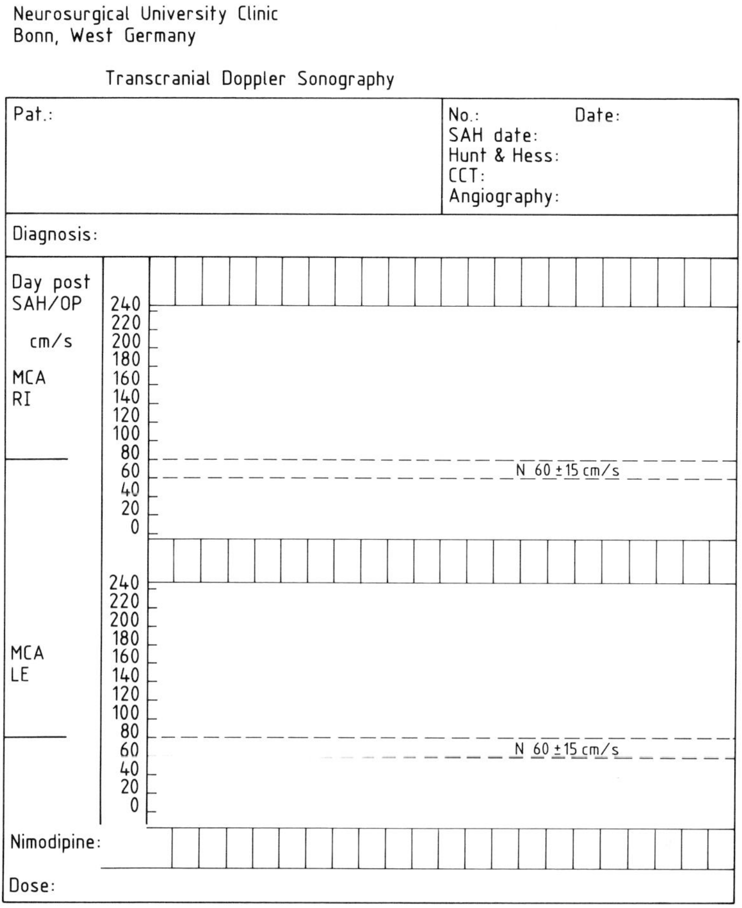
Neurosurgical University Clinic
Bonn, West Germany

Transcranial Doppler Sonography

Pat.:

No.: Date:
SAH date:
Hunt & Hess:
CCT:
Angiography:

Diagnosis:

Day post SAH/OP

cm/s

MCA
RI

240
220
200
180
160
140
120
100
80
60
40
20
0

N 60 ± 15 cm/s

MCA
LE

240
220
200
180
160
140
120
100
80
60
40
20
0

N 60 ± 15 cm/s

Nimodipine:

Dose:

Fig. 1. Form for daily control of TCD following SAH. Examination of MCA on both sides (or other vessels). Standard value of velocity in the MCA (60 ± 15 cm/s) is marked

interpretation of results using TCD [4, 5]. All patients underwent angiography when TCD failed to show vasospasm.

Nimodipine treatment was performed if TCD showed a velocity of 120 cm/s or more and contraindications were not present. The initial route of administration was intravenous for at least 10 days with a dose of 2 mg/h. The dose was increased within a short time to 3 mg/h if TCD showed a velocity higher than 250 cm/s, and the duration of intravenous administration was extended until the velocity reached a value less than 140 cm/s. Thereafter administration was continued orally with reduced amounts for 8 more days (Fig. 2).

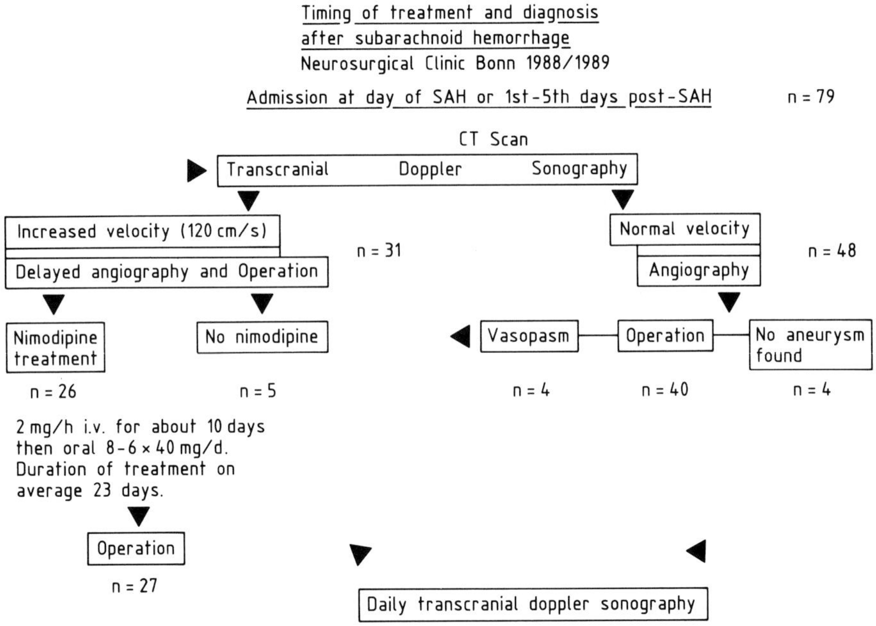

Fig. 2. Flow chart regarding timing of treatment and diagnosis following SAH in the Neurosurgical University Clinic, Bonn 1988/89. Illustration of duration and standard dosage of nimodipine treatment

Results

In 31 of 79 cases an enhanced velocity could be demonstrated. In 8 of those 31 cases a small acceleration of velocity (seldom higher than 240 cm/s) could be found at a discrete site of one vessel within 2 days post-SAH. This site was in accordance with the localization of the aneurysm, diagnosed later with angiography.

Between the 4th and 7th days post-SAH, moderate (velocity 140–200 cm/s) or severe (velocity higher than 200 cm/s) [8] vasospasm developed, which lasted 19 days on average. A good correlation could be demonstrated between the decrease in enhanced mean velocity and the improvement in the clinical state based on the Hunt and Hess grading system [6] (Fig. 3). Twenty-six of the 31 patients were treated with nimodipine for 23 days on average, the longest duration of therapy in any one case being 7 weeks. The mean velocity in those 26 nimodipine-treated patients was about 40 cm/s lower than the mean value for the remaining five patients with vasospasm who could not be treated.

The remaining 48 patients without vasospasm present on TCD underwent angiography, leading to the diagnosis of vasospasm in four cases; thus the rate of correspondence in respect of absent vasospasm was 92%. In one case with angiographically proven vasospasm a TCD examination was performed immediately, which confirmed the diagnosis, showing a velocity of 296 cm/s. After 2 h of local intra-arterial administration of nimodipine (0.2 mg/h) through the angiographic catheter, the velocity decreased to a value of 154 cm/s, accompanied by an improvement in the clinical state.

Five patients developed focal neurological deficits with computer tomographic signs of ischemia. One of the cases resulted from operative complications, while the remaining four patients showed vasospasm on TCD.

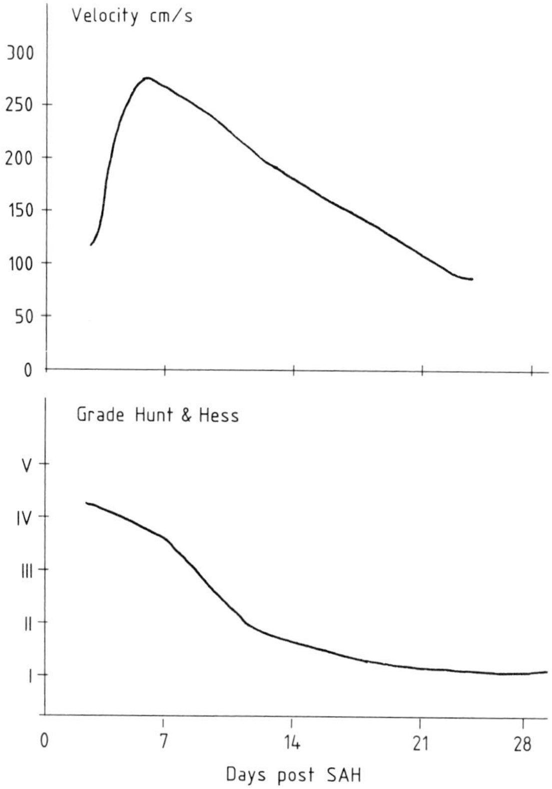

Fig. 3. Correlation between TCD velocity and neurological grading (Hunt and Hess grades I–V) following SAH in 79 patients

Discussion

The TCD finding of small accelerations in velocity at discrete sites of one vessel has also been described by several other authors and has led to the adoption of the term "early" vasospasm to indicate a local reaction to the rupture of the aneurysm [3, 8]. This is supported by the corresponding angiographic results concerning the aneurysm localization in our patients. This phenomenon of local velocity acceleration on TCD can provide a worthy diagnostic aid in establishing the diagnosis of SAH in cases of questionable bleeding.

In all cases of presumably vasospasm-conditioned cerebral ischemias, TCD showed increased velocities. Correlation betwen decrease in enhanced velocities and improvement in the clinical state has also been described elsewhere [8]. TCD-controlled nimodipine treatment thus seems to be suggestive especially regarding the individual differences in the duration and level of vasospasm; furthermore, TCD is noninvasive and can be repeated at any time. The method is helpful in identifying patients who run a high risk of developing neurological deficits due to vasospasm and who could benefit from nimodipine therapy. The accelerated dilatation of cerebral arteries following local administration of nimodipine, which we could prove by means of TCD, has been described previously by Böker et al. [2].

The 92% agreement between TCD and angiography in showing absent vasospasm proves TCD to be a valid criterion for indication of angiography. Highest validity can be obtained when a short interval is left between TCD and the angiographic examination, since the development of vasospasm may be short-term.

References

1. Allen GS et al. (1983) Cerebral arterial spasm – a controlled trial of nimodipine in patients with subarachnoid hemorrhage. N Engl J Med 308:619–624
2. Böker DK et al. (1985) Immediate postangiographic intraarterial treatment of cerebral vasospasm after subarachnoid hemorrhage with nimodipine. Report on 3 cases. Neurochirurgia 28:118–120
3. Fraser RAR (1980) Cerebral vasospasm: after 15 years in the laboratory. In: Wilkins RH (ed) Cerebral arterial spasm. Williams & Wilkins, Baltimore, pp 287–290
4. Harders A (1986) Vasospasm and aneurysm surgery. In: Harders A (ed) Neurosurgical applications of transcranial Doppler sonography. Springer, Wien New York, pp 40–64
5. Hossmann V (1985) Subarachnoidalblutung. In: Hossmann V, Grötz J, Schrör K (eds) Kalziumantagonisten und zerebrale Erkrankungen. Eine Standortbestimmung. Springer, Berlin Heidelberg New York, pp 41–60
6. Hunt WE, Hess RM (1968) Surgical risk as related to time of intervention in the repair of intracranial aneurysms. J Neurosurg 28:14–20

7. Schrör K (1985) Kalziumantagonisten – Physiologische Grundlagen, allgemeine und spezielle Pharmakologie und Prinzipien der klinischen Anwendung unter besonderer Berücksichtigung des Gefäßsystems. In: Hossmann V, Grötz J, Schrör K (eds) Kalziumantagonisten und zerebrale Erkrankungen. Eine Standortbestimmung. Springer, Berlin Heidelberg New York, pp 1–34
8. Seiler RW, Aaslid R (1986) Transcranial Doppler for evaluation of cerebral vasospasm. In: Aaslid R (ed) Transcranial Doppler sonography. Springer, Wien New York, pp 118–131

Clinical and Transcranial Doppler Sonography Evaluation During Treatment with Calcium Antagonists After Subarachnoid Hemorrhage

H. Schuster, T. Czech, A. Korn, and K. Ungersböck[1]

Between 1985 and 1988, the calcium antagonist nimodipine was given prophylactically, to prevent vasospasm, to nearly all patients admitted to the Department of Neurosurgery, University Hospital, Vienna, following aneurysmal rupture. Upon admission the patients were given nimodipine intravenously at 0.25–0.50 μg/kg body weight per minute. Subsequent dosage adjustment depended upon clinical symptoms of vasospasm. Dosage was limited by the occurrence of any fall in blood pressure that was poorly responsive to treatment and by an increased pulmonary shunt volume [6].

Patients and Methods

Our study included 72 patients admitted during 1987 and 1988 in whom a raised mean flow velocity (MFV) (> 80 cm/s), measured by transcranial Doppler sonography (TCD), had been recorded. There were 23 males and 49 females, and their ages ranged from 22 to 73 years. All 72 patients were operated upon. Nine patients had double aneurysms. Three of these underwent a second operation, while in the remaining six both aneurysms were clipped in the same operation. In 39 of the 72 patients (54%), the first operation was performed within 3 days after the subarachnoid hemorrhage (Fig. 1); in a further 24 patients (33%) it was performed within the following 10 days, and in eight patients (11%), later than 2 weeks after the hemorrhage. In one patient, no hemorrhage had occurred at all.

The following aneurysm localizations were found by angiography: ACoA 31, ICA 26, MCA 19, and one aneurysm each in the ACA, the pericallosal artery, the PCA, the basilar artery tip, and the vertebral artery.

Preoperative CT scans were evaluated according to the classification of Fisher [3]. Assessment of the preoperative clinical presentation using the scale of Hunt and Hess revealed the following distribution in the 72 patients with increased TCD flow velocity: grade 0 (no SAH) = 1 patient (1.3%), grade 1 = 17 patients (23.6%),

[1] Neurochirurgische Universitätsklinik Wien, Währinger Gürtel 18–20, A–1090 Wien

Advances in Neurosurgery, Vol. 18
K.-A. Bushe, M. Brock, and M. Klinger (Eds.)

grade 2 = 21 patients (29.1%), grade 3 = 25 patient (35.7%), grade 4 = 8 patients (11.1%).

All patients were checked regularly at least every 48 h by TCD of the MCA and the ACoA regions. Where the flow velocity showed a tendency toward rapid increase, the nimodipine dosage was increased until either clinical symptoms of vasospasm subsided or other parameters being monitored simultaneously (blood pressure, pulmonary shunt volume, intracranial pressure) prevented any further increase. Clinical results were assessed using the Glasgow Outcome Scale (GOS).

Results

Preoperative computer tomography (CT) revealed no blood in the subarachnoid space in nine patients (12.5%), only slight traces of blood in 16 patients (22%), and a massive hemorrhage in the basal cisterns in the remaining 47 patients (65%), in some cases extending into other subarachnoid spaces. Fourteen patients (19.5%) in all three groups suffered another intracerebral and/or intraventricular hemorrhage in addition to the subarachnoid hemorrhage.

The TCD values for the 72 patients ranged from slightly elevated MFV (80–120 cm/s) in 32% of patients, through moderately elevated MFV (120–160 cm/s) in

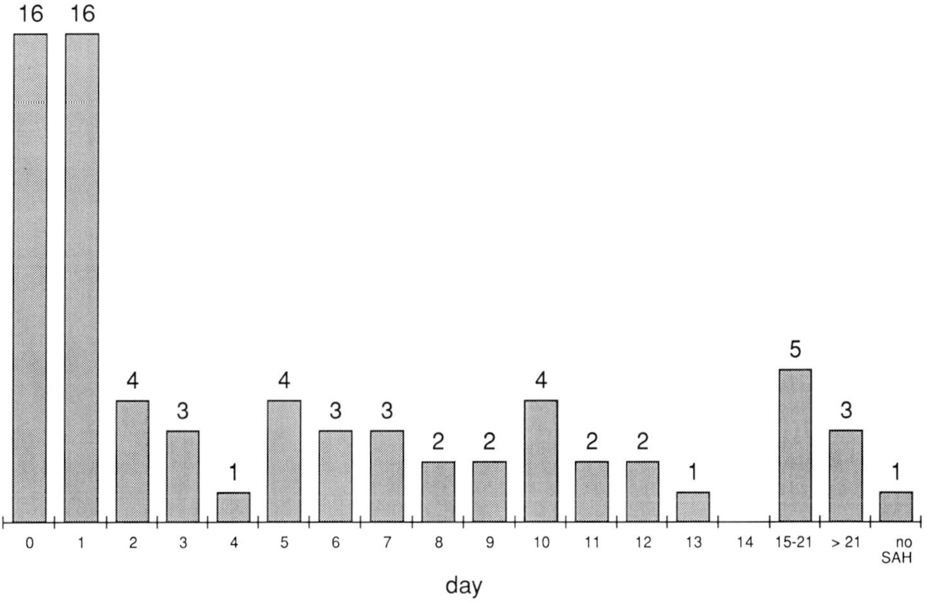

Fig. 1. Day of operation after SAH (72 patients)

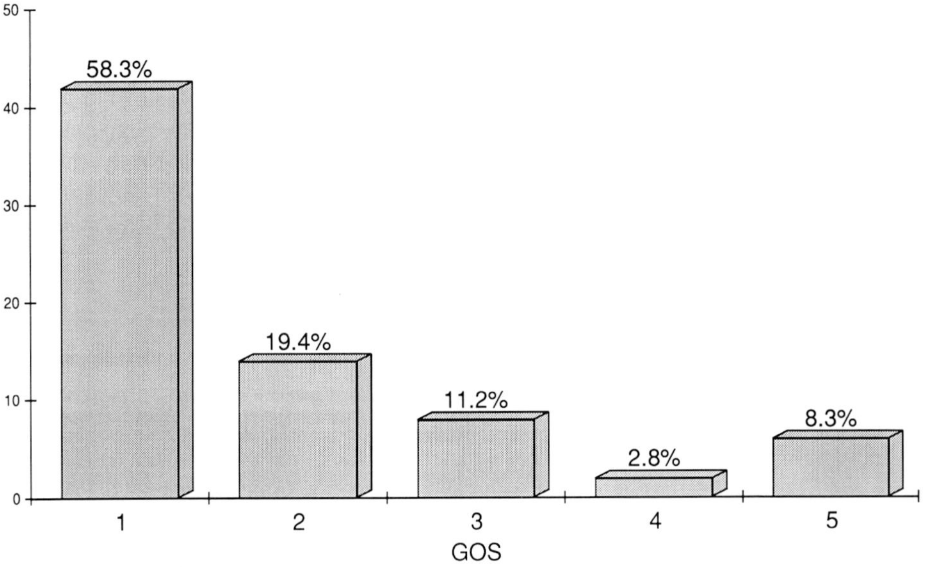

Fig. 2. Results according to the GOS (72 patients)

33.5%, to greatly elevated MFV (> 160 cm/s) in 34.5%. On the GOS, 58% of the patients were grade 1, and 19.5% grade 2. Thus a satisfactory result was achieved in 77.5% of patients; 22.5% reached only grades 3–5 on the GOS, and six of these patients died (8.3%, Fig. 2).

If we take into consideration the overall mortality and severe morbidity caused by vasospasm alone, the data are as follows: GOS grade 5 = 5.6%, GOS grade 4 = 1.4%, and GOS grade 3 = 8.3%. The evaluation of all the patients was carried out 5–18 months after the operation.

Comparison of the preoperative CT grading with the clinical outcome (GOS) made it apparent that patients with a massive accumulation of blood in the subarachnoid space (group 3) had a distinctly worse prognosis (Table 1) [3].

The correlation between the maximum measured mean flow velocity (MFV) and the clinical outcome according to the GOS is shown in Fig. 3. The degree of MFV increase had no direct relationship with the clinical outcome in our group of patients.

Of the 72 patients with increased flow velocity on TCD, 28 (39%) showed clinical symptoms of vasospasm as well. Of these patients, 21.5% had an MFV of 80–120 cm/s, 25% an MFV of 120–160 cm/s, and 53.5% an MFV of over 160 cm/s. These 28 patients had the following admission grades (Hunt and Hess): grade 1 = 7%, grade 2 = 28.5%, grade 3 = 46.5%, grade 4 = 18%. Of the 28 patients who had clinically manifest vasospasm despite treatment with the calcium antagonist,

Table 1. Results (GOS) according to CT grading (72 patients)

	GOS				
CT	I	II	III	IV	V
1	7	2	1		
2	12	2	1		
3	23	10	6	2	6

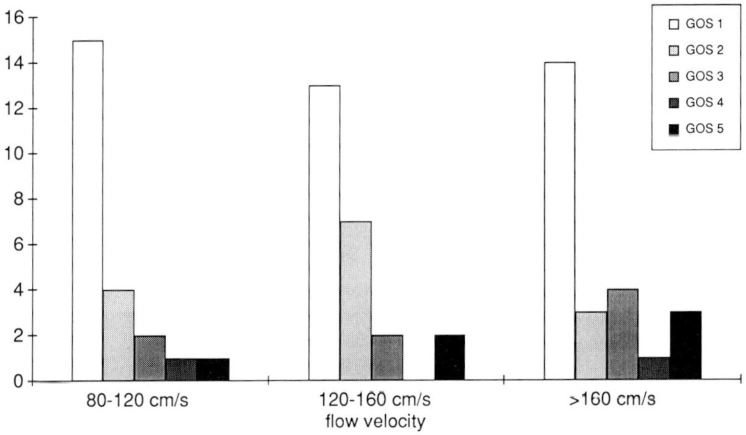

Fig. 3. Maximum MFV vs outcome (GOS) (72 patients)

50% had a good outcome (GOS 1–2). The mortality caused by vasospasm alone was 14.5% (GOS 5).

Discussion

In patients with TCD-proven vasospasm, clinical signs of vasospasm occurred in a not insignificant percentage of cases (39% of our group of patients) in spite of the prophylactic use of nimodipine. While those patients who had an MFV of over 160 cm/s (60.9%) were at higher risk of developing clinical signs, a slight or moderate increase in flow velocity also carried a certain clinical risk (total 27.6%). In particular, a very rapid increase in velocity (> 25 cm/s/day) often heralded clinical symptoms. This observation has been confirmed by other authors [4]. By appropriately monitored administration of nimodipine (dose increase with blood pressure monitoring, hypervolemia, administration of hypertensive drugs, monitoring of the pulmonary shunt and the intracranial pressure), the effects of clinical vasospasm can be partially controlled.

Neither a direct change in the TCD flow velocity due to a therapeutic increase in the nimodipine dosage nor a clinical improvement attributable with certainty to this increased dosage alone was observed in our patients. The effect on flow velocity has also not been reported uniformly in the literature [7, 10].

Nevertheless, many authors [1, 2, 7–9, 11] report a positive influence of prophylactic nimodipine administration on the outcome of patients with clinical vasospasm. In our patient group, where only 54% had an early operation within the first 3 days with nimodipine prophylaxis, the mortality solely attributable to vasospasm was 5.6%. These results, in patients unselected for their preoperative condition, show no significant difference from patient groups who had early operation and nimodipine, as described in the literature [1, 9], but show slightly better results as compared with an also unselected patient group without nimodipine [5].

Conclusions

Seventy-two patients with increased TCD flow velocity (mean MFV > 80 cm/s), all of whom had been operated on for an intracranial aneurysm and had received intravenous nimodipine prophylactically, were surveyed for the importance of different TCD values and CT upon their outcome. Slightly to greatly increased flow velocities were distributed evenly among the patients and did not correlate with the clinical results. Massive accumulations of blood in the basal cisterns as diagnosed by CT were prognostically unfavorable. In 39% of patients, clinical vasospasm occurred in spite of prophylactic use of nimodipine. The outcome for the 72 patients correlates well with that of comparable patient groups in the literature and is slightly more favorable than without nimodipine.

References

1. Allen GS, Ahn HS, Preziosi TJ et al. (1983) Cerebral arterial spasm – a controlled trial of nimodipine in patients with subarachnoid hemorrhage. N Engl J Med 308:619–624
2. Auer LM, Brandt L, Ebeling U et al. (1986) Nimodipine and early aneurysm operation in good condition SAH patients. Acta Neurochir (Wien) 82:7–13
3. Fisher CM, Kistler JP, Davis JM (1980) Relation of cerebral vasospasm to subarachnoid hemorrhage visualized by computerized tomographic scanning. Neurosurgery 6:1–8
4. Harders A, Gilsbach J (1987) Klinische Relevanz der transkraniellen Dopplersonographie bei der aneurysmatisch bedingten Subarachnoidalblutung. In: Widder B (ed) Transkranielle Doppler-Sonographie bei zerebrovaskulären Erkrankungen. Springer, Berlin Heidelberg New York, pp 72–78

5. Kassel NF, Torner JC (1986) The international cooperative study on timing of aneurysm surgery – an update. Stroke 15:566–570
6. Korn A, Aloy A, Ungersböck K, Czech T, Schuster H (1987) Importance of "extended" monitoring during therapy with calcium antagonists in neurosurgery. J Clin Mon 3:333–334
7. Pasqualin A, Battaglia R, Sidoti C, Chioffi F, Da Pian R (1988) Relationship between hemodynamic and anti-ischemic effects of nimodipine in SAH patients with ruptured intracranial aneurysms: the value of transcranial Doppler recording. 2nd International Conference on Transcranial Doppler Sonography, Salzburg, Austria
8. Perneczky A, Koos WT, Vorkapic P, Ungersböck K (1984) Die therapeutische Anwendung des Kalziumantagonisten Nimodipin bei intrakraniellem Vasospasmus nach Subarachnoidalblutung. In: Voth D, Glees P (eds) Der zerebrale Angiospasmus. de Gruyter, Wien New York, pp 493–500
9. Säveland H, Ljunggren B, Brandt L, Messeter K (1986) Delayed ischemic deterioration in patients with early aneurysm operation and intravenous nimodipine. Neurosurgery 18:146–150
10. Seiler RW, Grolimund P, Zurbruegg HR (1987) Evaluation of the calcium-antagonist nimodipine for the prevention of vasospasm after aneurysmal subarachnoid haemorrhage. Acta Neurochir (Wien) 85:7–16
11. Seiler RW, Reulen HJ, Huber PH, Grolimund P, Ebeling U, Steiger HJ (1988) Outcome of aneurysmal subarachnoid hemorrhage in a hospital population: a prospective study including early operation, intravenous nimodipine, and transcranial Doppler ultrasound. Neurosurgery 23:598–604

CBF Dynamics – Vasospasm After Subarachnoid Hemorrhage

J. Meixensberger, A. Brawanski, M. Rösener, M. Holzschuh, I. Danhauser-Leistner, U. Freise, and W. Ullrich[1]

Introduction

Besides rebleeding, vasospasm is one of the most devastating complications in the clinical course after subarachnoid hemorrhage (SAH). Although early aneurysm surgery minimizes the risk of rebleeding and improves the outcome of SAH patients in a good clinical condition, there is still the danger of secondary ischemic deficits. The development of delayed neurological deterioration is contributed to by changes in regional cerebral blood flow (rCBF). Under clinical conditions two noninvasive methods are available to study CBF dynamics: (a) noninvasive two-dimensional rCBF measurement using xenon 133, and (b) transcranial Doppler ultrasonography (TCD) determining blood flow velocity in the circle of Willis. The present study was undertaken to answer the following questions.

1. Does a correlation exist between rCBF and TCD in the time course of vasospasm after SAH?

2. Is there a disturbance of cerebrovascular reactivity to changes in $PaCO_2$ postoperatively after aneurysm clipping, and can this information be used as a prognostic factor predicting secondary ischemic deficits?

Patients and Methods

Our series comprised 21 patients (7 males and 14 females). Their ages ranged from 23 to 79 years, with an average of 44 years. Nine patients had ruptured aneurysms of the internal carotid artery (ICA); there were seven aneurysms of the anterior communicating artery, four of the middle cerebral artery (MCA), and one of the basilar bifurcation.

The clinical status of all patients was graded according to the scale of Hunt and Hess [4] preoperatively. Eleven patients were in grade I, four in grade II, four in grade III and two in grade IV. The extent of SAH on CAT scan was estimated

[1] Neurochirurgische Klinik und Poliklinik der Universität Würzburg, Josef-Schneider-Str. 11, D–8700 Würzburg

Advances in Neurosurgery, Vol. 18
K.-A. Bushe, M. Brock, and M. Klinger (Eds.)

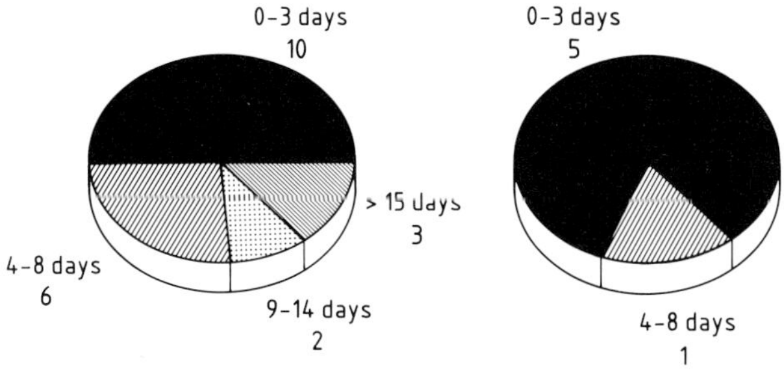

Fig. 1. Day of the operation after the last SAH (*left*) and the corresponding development of vasospasm (*right*)

according to the method of Davis et al. [2]. Most of the patients had a moderate to severe SAH (Davis 2/3). The day of the operation after the last SHA and the corresponding development of delayed neurological deficits are given in Fig. 1.

A mobile unit for the bedside measurements (Novo Cerebrograph 10a) with ten detectors, five placed over each area of the MCA, or a stationary unit (Novo Cerebrograph 32) with 32 detectors, 16 placed over each hemisphere, was utilized for the CBF measurements. Regional CBF was measured after intravenous administration of 10–15 mCi xenon 133. The rCBF was calculated as the initial slope index (ISI) according to Risberg et al. [8] using a bicompartmental model [7]. Immediately after the aneurysm operation we performed two CBF measurements with different $PaCO_2$ levels in order to determine CBF at rest and CBF reactivity to hyperventilation. Simultaneously we assessed the blood flow velocity in each ICA and MCA by TCD. Additionally we monitored continuously arterial blood pressure and determined arterial $PaCO_2$ blood gas content. In the follow-up after the operation we performed repeated measurements of CBF and blood flow velocity every 2–3 days.

Results

Correlation Between CBF and Blood Flow Velocity

Thirty-two measurements were analyzed with simultaneous CBF and TCD studies. Correlating the absolute ISI values of the MCA region to the data of blood flow velocity, we were unable to find any significant correlation for the operated and the contralateral hemisphere ($r = 0.31$). The average hemispheric CBF was 46 ± 10.5 ISI units for the operated hemisphere and 48.2 ± 12 ISI units for the contralateral hemisphere. The affected hemisphere showed higher flow velocities than the unaffected hemisphere. On the side of the ruptured aneurysm we determined critical

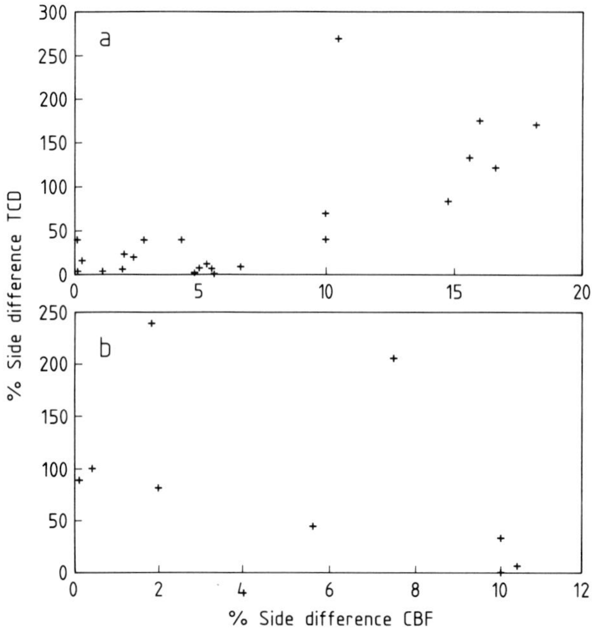

Fig. 2. Interhemispheric difference (%) in CBF and blood flow velocity (TCD) between side of ruptured aneurysm and unaffected hemisphere, **a** showing correlation between CBF and TCD ($r = 0.74$, $P < 0.001$) and **b** showing no significant correlation

flow velocities over 120 cm/s [3], with a maximum of mean flow velocity in the MCA of 230 cm/s in ten cases. On the unaffected hemisphere we never saw such critical flow velocities. Calculating the interhemispheric difference between the affected and unaffected hemisphere we found a correlation of CBF and blood flow velocity in 72% of our measurements ($r = 0.74$, $P < 0.001$) (Fig. 2a). On the other hand, there were measurements with significant interhemispheric differences for TCD in the MCA without any significant interhemispheric CBF difference in the corresponding region and vice versa (Fig. 2b). As a significant interhemispheric difference we assumed a 10% CBF difference and a 40% difference in blood flow velocity over corresponding areas.

Correlation Between CBF and TCD During Clinical Vasospasm

Figure 3 shows CBF and TCD values of five patients at the time of the maximum of the delayed neurological deficit and at the point when the neurological deficits had resolved. During the time course of clinical vasospasm significantly critical blood

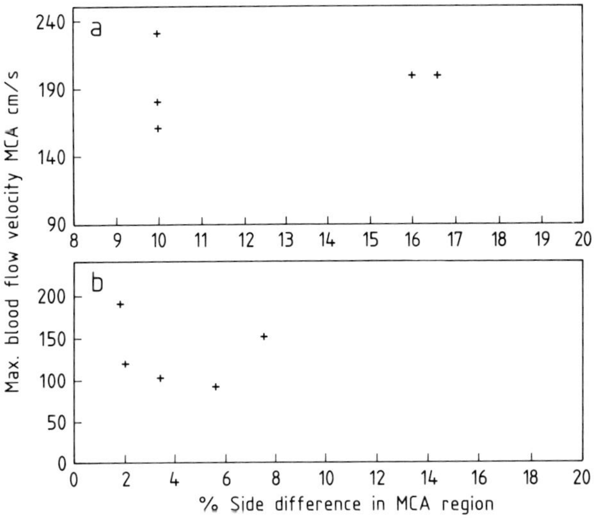

Fig. 3 a,b. Correlation between CBF and TCD in time course of vasospasm

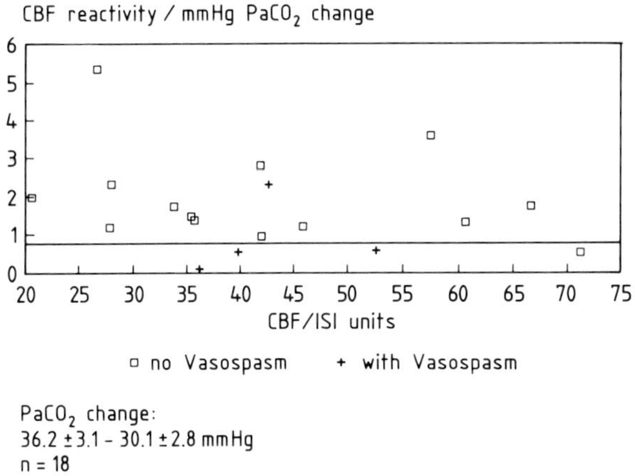

Fig. 4. Correlation of CBF and CBF-CO_2 reactivity

flow velocities (> 120 cm/s) in the MCA on the side of the ruptured aneurysm could be determined corresponding to a significant CBF reduction. At the time of clinical improvement the CBF values had normalized in all cases, but critical blood flow velocities were seen in two of five cases. In the time course there was a good correlation of clinical vasospasm to CBF, but not, in all cases, of vasospasm to TCD data.

CBF Reactivity to Changes in $PaCO_2$ After Aneurysm Clipping

Eighteen measurements immediately after the aneurysm operation were analyzed (Fig. 4). No correlation between CBF at rest and CBF-CO_2 reactivity could be found (r = 0.42). Four patients showed a decreased response to $PaCO_2$ changes and three of them developed vasospasm postoperatively. Only one of the patients with normal CBF-CO_2 reactivity later presented with delayed neurological deficits. As a normative value of CBF reactivity to $PaCO_2$ changes we used 0.75 ISI units per 1 mmHg $PaCO_2$ in accordance with Maximilian et al. [5].

Discussion

After SAH, cerebrovascular dynamics are important for the evaluation of the clinical status and the development of delayed neurological deficits. Two-dimensional 133 xenon rCBF measurement and the TCD are both methods to assess the cerebrovascular status. Although the blood flow velocity in the circle of Willis correlated well to CBF under normal physiological conditions [1], we did not find a sufficient correlation in pathological conditions in all cases. This may be due to the fact that blood flow velocity is determined in large vessels of the circle of Willis without obtaining any information about the peripheral vascular bed. Similar results were reported by Sekhar et al. using the stable xenon method for CBF determination [9]. Additionally, our data showed that in the time course of clinical vasospasm CBF can be normalized despite a still critical elevation of blood flow velocity.

The cerebrovascular capacity after SAH can be tested by changes in $PaCO_2$ and there are reports by other groups that the CBF-CO_2 response is impaired after SAH [6, 10]. An impairment of the CBF-CO_2 response may be a factor in the development of a delayed neurological deficit. Our data suggest a higher risk of clinical vasospasm under such conditions. However, further investigations are necessary to strengthen the prognostic value of this parameter.

References

1. Brass LM, Prohovnik I, Pavlakis S, Mohr JP (1987) Transcranial Doppler examination of middle cerebral artery velocity versus xenon rCBF: two measures of cerebral blood flow. Neurology 37 [Suppl 1]:85
2. Davis JM, Davis KR, Crowell RM (1980) Subarachnoid hemorrhage secondary to ruptured intracranial aneurysm: prognostic significance of cranial CT. AJR 134:711–715
3. Harders AG, Gilsbach JM (1987) Time course of blood velocity changes related to vasospasm in the circle of Willis measured by transcranial Doppler ultrasound. J Neurosurg 66:718–728

4. Hunt WE, Hess RM (1968) Surgical risk as related to time of intervention in the repair of intracranial aneurysms. J Neurosurg 28:14–20
5. Maximilian VA, Prohovnik I, Risberg J (1980) The cerebral hemodynamic response to mental activation in normo- and hypercapnia. Stroke 11:342–347
6. Messeter K, Brandt L, Ljunggren B, Svendgaard NA, Algotsson L, Romner B, Ryding E (1987) Prediction and prevention of delayed ischemic dysfunction after aneurysmal subarachnoid hemorrhage and early operation. Neurosurgery 20:548–553
7. Obrist WD, Thompson HK, Wang HS, Wilkinson WE (1975) Regional cerebral blood flow estimated by 133-xenon inhalation. Stroke 6:245–256
8. Risberg J, Ali Z, Wilson EM, Wills EL, Halsey JH Jr (1975) Regional cerebral blood flow by 133 xenon inhalation: preliminary evaluation of an initial slope index in patients with unstable flow compartments. Stroke 6:142–148
9. Sekhar LN, Wechsler LR, Yonas H (1988) The value of transcranial Doppler examination in the diagnosis of vasospasm following subarachnoid hemorrhage. Neurosurgery 22:813–821
10. Voldby B, Enevoldsen EM, Jensen FT (1985) Cerebrovascular reactivity in patients with ruptured intracranial aneurysms. J Neurosurg 62:59–67

Discrepancy Between the Results of TCD and the Clinical Status of Patients After SAH

E. Wranze-Bielefeld, W.A. Dauch, and B.L. Bauer[1]

Introduction

Transcranial Doppler sonography has become a widely accepted diagnostic method for the detection of intracranial vascular pathologies. It seems clear that vasospasm is often combined with neurological deficits. We use the terminology of Harders [3] in speaking about normal cerebral blood flow and subcritical and critical vasospasm detected by TCD (normal cerebral blood flow: velocity below 80 cm/s; slight vasospasm: velocity up to 120 cm/s; severe vasospasm: velocity above this value as detected in the MCA). In 1987 Seiler and Grolimund [6] described vasospasm found in patients who were clinically asymptomatic. They discussed this phenomenon and speculated that it is due to a lack of autoregulation and perhaps better collateralization in the affected area.

In 1988 at a meeting of the German EEG Society, Laumer [5] reported a discrepancy between the neurological status and accelerated blood flow velocity in 16% of his patients with SAH. As possible reasons he mentioned the existence of collateralization, hyperperfusion (i.e., accelerated velocity without vasospasm), compression of the blood vessels, and vasospasm too lateral to be detected by TCD. That there is a discrepancy in about a quarter of the cases caused us to investigate whether there might be systematic reasons for the discrepancy. Perhaps from this it would be possible to relativize the results of TCD in some cases.

Material and Method

We took into account all patients with SAH in the last 3 years, but not those patients with traumatic SAH because the results of the TCD are too heterogeneous in such cases, as Frowein [1] reported recently.

In this paper we describe 47 patients who suffered from aneurysms in the vessels shown in Fig. 1. The distribution of Hunt and Hess grades [4] is shown in Fig. 2,

[1] Neurochirurgische Abteilung, Philipps Universität Marburg, Baldinger Str. 1, D–3550 Marburg

Advances in Neurosurgery, Vol. 18
K.-A. Bushe, M. Brock, and M. Klinger (Eds.)

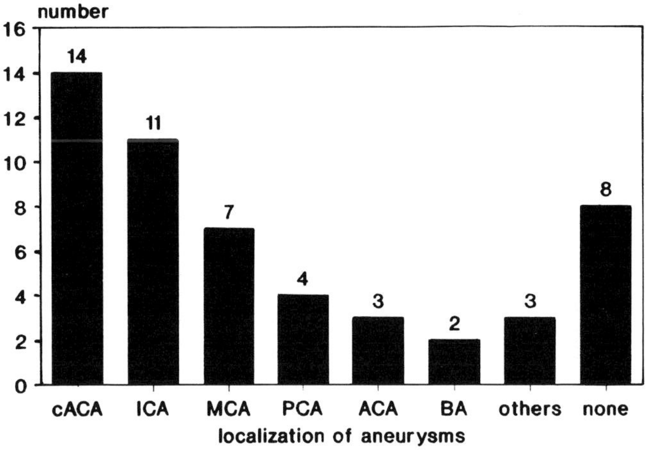

Fig. 1

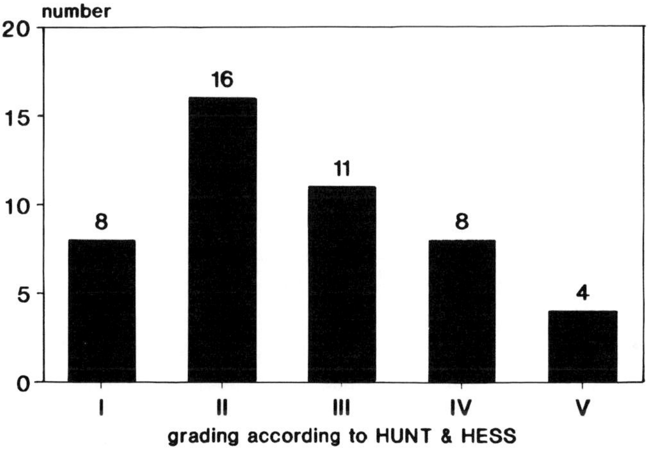

Fig. 2

and the age of the patients in Fig. 3. The sex distribution was 44.6% men and 55.4% women.

In 30% of the cases we must assume either that the patients had a second bleeding while in hospital or that the bleeding that brought the patient to our hospital was not the first. After we received the patients we did the TCD immediately. If no vasospasm was detected, we performed angiography and operated on the patients early. If a spasm was detected, we controlled the TCD until we found normal blood flow velocities. We then performed angiography and after this carried out a late operation.

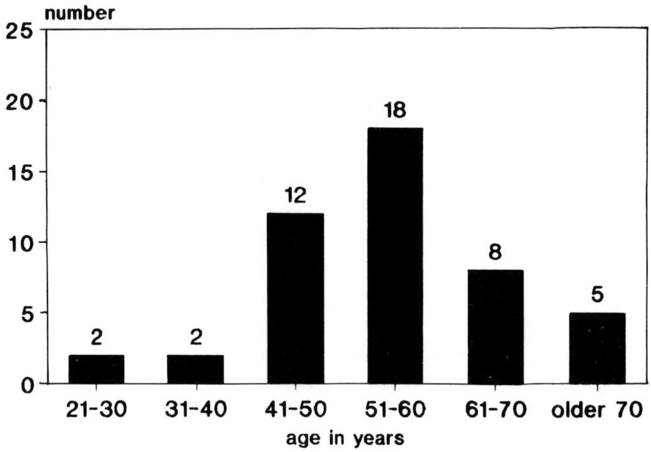

Fig. 3

Postoperatively we controlled patients nearly every day with TCD. All the patients were treated with Nimotop 2 mg/h until the 10th day postoperatively or until there was a clear reduction in the blood flow velocity.

Results

In seven patients (15%) we found no aneurysm although we could see the SAH on the CT scan. These patients clinically had an excellent status and we did not find vasospasm at any time. Ten patients (21%) died before we could perform the operation, and two (4%) died postoperatively. In 16 patients (34%) we were able to perform the early operation. In 13 patients (28%) we could not operate before the 10th day after bleeding, either because they had a very reduced clinical status (Hunt and Hess grade worse than III) or because they had severe vasospasm when they were admitted to our hospital.

In 15 of the 47 cases (32%) we found a discrepancy between the TCD and the clinical status. This discrepancy was found in two different groups:

1. Patients who did well clinically but had persistent severe vasospasm. This happened in three patients. Twice we found the spasm in the operated vessel (ICA) and once we found a persistent spasm on the operated side. In this case an aneurysm of the cACA had been operated upon.

2. Patients in whom we did not detect vasospasm or only slight acceleration in the blood flow velocity, but who had a very reduced clinical status. There were 12 such cases (21%).

When we examined the data we found the following interrelations: One important factor was the age of the patient. All patients older than 70 years showed such a discrepancy (5/5). Another important factor was the Hunt and Hess grade. Twelve patients had grade IV or V, and in 10 of these patients we did not find vasospasm (83%). The third factor we identified was bleedings in the posterior fossa: in neither of two such cases did we find a relation between TCD and clinical status. No other factor seemed to be important.

Discussion

Concerning group 1, we considered the possibility of a clip-induced stenosis. Because there were only three cases, we do not want to speculate further on this subject. Group 2, however, requires further thought. If one takes into account that there are few accepted facts about the origin of vasospasm one has to accept that the vasospasm is a reaction to the SAH. One may speculate that the acceleration in blood flow velocity is necessary for adequate blood perfusion in the swollen brain. Only when the flow velocity is too fast – one might call this an "accident" in the reaction – can neurological deficits occur. From this point of view one can define lack of blood flow velocity acceleration as a loss of autoregulation. It is easy to accept that the old brain has lost a good part of its former ability for autoregulation. The old and also the ill brain is not as able as the younger brain to react to the cue of SAH.

Regarding the bleedings in the posterior fossa, the data are not statistically independent. One patient was more than 70 years old and also bled from an aneurysm of the BA. Thus only one patient remains, not enough to be interpreted.

If we look at the patients with reduced clinical status we have to implement some limitation concerning their number. Two of them died before onset of vasospasm was possible. One patient was older than 70 years, so we should also eliminate his data from this group. We think that even in the remaining cases the brain was so damaged that it was not able to react with acceleration in the blood flow velocity.

In general it is said that vasospasm is responsible for the reduced clinical status. That is an explanation only in those cases in which spasm has been detected. Grote and Hassler [2] have shown some evidence that an interruption of the cerebral blood flow occurs at the onset of the bleeding and speculate that this is the reason for the bleeding's cessation. If one accepts this interpretation, the reduced clinical status appears in a different light. It cannot be reduced to vasospasm – the onset is much later; rather the reason is the slowed blood perfusion after the onset of the bleeding. One must conclude that the longer the cessation of the perfusion remains, the greater the damage. The worse the Hunt and Hess grade and the worse the ability of the brain to recover, the earlier neurological deficits might occur without vasospasm but rather because of the loss of or the reduction in the ability for autoregulation.

We conclude from our data, in combination with the results of Grote and Hassler, that:

1. The highly aged brain does not react with normal autoregulation which entails an acceleration in blood flow velocity in response to SAH.

2. A reduced clinical status (Hunt and Hess IV or V) has to be correlated in many cases (from our data in 50%) with the brain lesion caused by interruption of the cerebral blood flow. This might be the reason for poor autoregulation that does not allow vasospasm to occur.

Further research has to be done to find out whether those patients who recover from a poor clinical status are those who have an acceleration in blood flow velocity.

References

1. Frowein RA (1989) Die Prognose des schweren Schädel-Hirn-Traumas. Oral presentation: Therapie und Prognose schwerer Schädel-Hirn-Traumata. Marburg
2. Grote E, Hassler W (1988) The critical first minutes after subarachnoid hemorrhage. Neurosurg 22:654–661
3. Harders A (1986) Neurosurgical applications of transcranial Doppler sonography. Springer, Wien New York
4. Hunt EW, Hess RM (1968) Surgical risk related to time of intervention in the repair of intracranial aneurysms. J Neurosurg 28:14–19
5. Laumer R (1988) Diskrepanz zwischen neurologischem Status und der Flußgeschwindigkeit. Oral presentation: Deutsche EEG Gesellschaft
6. Seiler RW, Grolimund P (1987) Evaluation von Spasmen nach einer Subarachnoidalblutung mittels transkranieller Doppler-Sonographie. In: Widder B (ed) Transkranielle Doppler-Sonographie bei zerebrovaskulären Erkrankungen. Springer, Berlin Heidelberg New York

Aktuelle Rechtsfragen in der Neurochirurgie
in Forschung und Klinik

Current Legal Issues in Neurosurgical Research
and Treatment

Zur Verwendung menschlichen Gewebes, insbesondere von foetalem Gewebe in der Neurochirurgie

E. Deutsch[1]

Die Entnahme von Gewebe im allgemeinen

Das Persönlichkeitsrecht

Zur Entnahme menschlichen Gewebes, sei es Blut, Haut, Knochen oder andere Teile des menschlichen Körpers, ist die Einwilligung der betroffenen Person erforderlich. Wenn der Eingriff mit einem auch noch so geringen Risiko behaftet ist, z.B. bei einer Leber-Biopsie, muß diese Einwilligung in Form einer Einwilligung nach Aufklärung (informed consent) erteilt werden. Dabei ist es nicht notwendig, daß die Einwilligung wörtlich oder ausdrücklich erklärt wird. Vielmehr kann eine konkludente Einwilligung vorliegen, etwa infolge einer Zusammenarbeit zwischen Arzt und Patient, wenn der Patient den Empfehlungen seines Arztes folgt und wissentlich darin einwilligt, daß Gewebe oder andere Körperteile entnommen werden.

Umfang der Einwilligung und Eigentumsrechte

Die normale Verwendung. Indem der Patient die Entnahme von Gewebe gestattet, gibt er seine Einwilligung zur Durchführung normaler medizinischer Maßnahmen dieses Gewebe betreffend. Dies schließt die Genehmigung der Durchführung aller erforderlichen medizinischen Tests an dem entfernten Körperteil ein, auch wenn dieses dabei zerstört, aufgelöst oder in eine andere Form gebracht wird. Wenn das Gewebe einem anderen Patienten übertragen werden soll, kann diese Verwendung ebenfalls gestattet werden. Im allgemeinen kennt der Patient diese normale Verwendung oder jedenfalls die Grenzen, innerhalb derer eine normale Verwendung stattfinden kann. Wenn aber Körperteile zu einem Zweck entnommen werden, der dem Patienten nicht bekannt ist, sollte der Patient auch hierüber informiert werden. Das ist zum Beispiel der Fall, wenn entnommene Knochen durch künstliche Körperteile, etwa ein künstliches Hüftgelenk, ersetzt werden. Wenn die entnomme-

[1] Abteilung für Arzt- und Arzneimittelrecht des Juristischen Seminars, Universität Göttingen, Goßlerstr. 19, D–3400 Göttingen

Advances in Neurosurgery, Vol. 18
K.-A. Bushe, M. Brock, and M. Klinger (Eds.)

nen Knochen zermalen oder gehobelt werden, um dann in flüssiger Form für Unfallopfer verwendet zu werden, sollte der Patient über diese unerwartete Verwendung seiner Knochen aufgeklärt werden. Zumindest würde die nach Aufklärung erklärte Einwilligung des Patienten in diesen Fällen das Gewissen des Arztes entlasten.

Übeflüssiges Material. Manchmal wird mehr Körpergewebe, insbesondere Flüssigkeiten, entnommen, als unbedingt notwendig wäre. Das kann dann der Fall sein, wenn die eigentlich erforderliche Menge nicht bekannt ist, oder wenn es notwendig ist, auf unvorhergesehen erforderliche Tests oder Umstände vorbereitet zu sein. Gewöhnlich erklärt der Patient zwar sein Einverständnis zur Entnahme nicht nur des absoluten Minimums, sondern der normalen Menge von Gewebe und Körperflüssigkeiten. Häufig stellt sich jedoch die Frage, was mit dem überflüssigen Teil geschehen soll und darf. Manchmal wird der überflüssige Körperteil für weitere medizinische Zwecke verwendet, so im Falle von Körperflüssigkeiten, die Bestandteil von medizinischen Versuchen werden. Zum rechtlichen Aspekt der Verwendung überflüssigen Körpergewebes gibt es einen Meinungsstreit: Manche meinen, daß der Patient davon ausgehe, daß die überflüssigen Teile vernichtet werden. Andere sind der Ansicht, daß eine normale medizinische Verwendung, welche dem Patienten in keiner Weise Schaden zufügen kann, rechtlich möglich ist, da der Patient davon ausgehe, es handele sich um überflüssiges Material. Der Streit konnte bisher nicht entschieden werden. Möglicherweise kommt es insoweit darauf an, im Wege der Auslegung festzustellen, wovon der Patient hinsichtlich der Weiterverwendung des überlfüssigen Materials tatsächlich ausgeht.

Eigentumsrechte und Patientenschutz. Im Staat Kalifornien gab es einen Fall, der einen Patienten betraf, dessen Blut und Blutbestandteile außergewöhnliche Eigenschaften aufwiesen. Die ursprüngliche Diagnose für den Patienten lautete auf Haarzellen-Leukämie. Im Verlauf der Behandlung entnahmen seine Ärzte ihm Blut und andere Körpersubstanzen und entfernten seine Milz operativ. Der Patient behauptete anschließend, daß sein Arzt und andere ohne ihn aufzuklären oder seine Einwilligung zu erlangen seine Milz und andere Körpersubstanzen zu Forschungen benutzt hätten, welche den Arzt und andere in die Lage versetzten, Zell-Linien mit ungewöhnlichen Charakteristika festzustellen, welche wiederum in der Lage waren, wertvolle Pharmazeutika zu produzieren. Später wurde das entnommene Gewebe benutzt, um eine Zell-Linie zu entwickeln, für welche die Ärzte und die Universität ein Patent bekamen. Der Patient behauptete nunmehr, daß er, wenn er gewußt hätte, was geschehen sollte, nicht seine Einwilligung zur Milzentfernung zu diesen kommerziellen Zwecken gegeben hätte, darauf bestanden hätte, an der Kontrolle seines Blutes und seiner Körpersubstanzen beteiligt zu werden, den Forschern nicht erlaubt hätte, seine Körpersubstanzen einzig zu ihrem wirtschaftlichen Vorteil zu nutzen, und daß er sich an einem anderen Ort um eine Behandlung bemüht hätte. Er wollte an den so erlangten wirtschaftlichen Vorteilen beteiligt werden. Die Ärzte und die Universität erklärten, daß der Patient kein Eigentümerinteresse an seinen Körpersubstanzen habe, und daß die krankhafte Milz des Patienten keinerlei Wert habe; der Patient habe darüberhinaus auf sein erkranktes Organ verzichtet. Das Gericht entschied zugunsten des Klägers. Die Körpersubstanzen seien Eigentum des Patienten, erklärte das Gericht (Moore v. Regents of the University of California 249 Cal. Rptr. 494, California Court of Appeals, July 21st, 1988).

Foetales Gewebe

Einwilligung der Eltern oder der Mutter

Wenn der Foetus abortiert wird, gehört das foetale Gewebe nicht der Klinik oder dem Arzt. Vielmehr ist auch in den Richtlinien der NIH-Kommision betreffend die Forschung mit foetalem Gewebe festgelegt, daß für die Forschung mit foetalem Gewebe die Einwilligung entweder der Eltern oder, im Falle einer alleinstehenden Mutter, die Einwilligung der Mutter erforderlich ist. Wenn diese Einwilligung nicht erteilt wird, muß der Foetus vernichtet, bzw. mit ihm in der üblichen Weise verfahren werden. Wenn jedoch die Mutter oder beide Eltern in die Durchführung von medizinischen Versuchen mit dem foetalen Gewebe einwilligen, ist es rechtlich u.a. auch möglich, es anderen Patienten zu implantieren.

Der Kreis der berechtigten Personen

Da der Foetus, weil er nie zu einer lebenden Person geworden ist, kein Träger von eigenen Rechten ist, muß er auch nach seiner Abtötung als Körperteil der Mutter betrachtet werden. Deshalb hat die Mutter das Recht, den Foetus zu beseitigen und entsprechende Anweisungen zu erteilen. Wenn die Mutter verheiratet ist, haben nach Ansicht der herrschenden Lehre in der rechtswissenschaftlichen Literatur beide Elternteile dieses Recht. So wurde auch in einem Falle entschieden, wo in vitro befruchtete Eizellen vom Krankenhaus vernichtet wurden, ohne daß zuvor das Paar, welches das Sperma und die Eizellen zur Verfügung gestellt hatte, konsultiert worden war. Der District Court des Southern District of New York stellte dazu fest, daß das Krankenhaus das Persönlichkeitsrecht der Eltern verletzt habe und deshalb den Eltern zum Schadensersatz verpflichtet sei (Del Zio v. Presbyterian Hospital, United States District Court, Southern District of New York, 1978 (74 Civ. 3588).

Bedenken hat es allerdings wegen des Umstandes gegeben, daß es in der Mehrzahl aller Fälle die Mutter ist, welche ihre Einwilligung in den Schwangerschaftsabbruch erteilt. Warum, so wird gefragt, soll dann auch noch die Mutter berechtigt sein, über die abgetöte Leibesfrucht zu bestimmen? Bei der Beantwortung dieser Frage muß man jedoch berücksichtigen, daß die Mutter die einzige Person ist, die zumindest eine natürliche bzw. biologische Beziehung zum abortierten Foetus hat. Im Falle eines verheirateten Paares haben dann beide Elternteile eine solche rechtliche Position inne.

Grenzen der Erlangung foetalen Gewebes

Es gibt ethische Grenzen bei der Erlangung foetalen Gewebes. Es würde unter jedem denkbaren ethischen Gesichtspunkt unerträglich sein, eine Schwangerschaft, welche ohnehin abgebrochen werden soll, nur deshalb zu verlängern, um bestimmte Gewebe weiter heranwachsen zu lassen. Darüberhinaus gibt es Befürchtungen dahingehend, daß manche Frauen nur deshalb schwanger werden würden, um foetales Gewebe zu produzieren. Insbesondere wenn dies aus Profitstreben geschähe, wäre es ethisch verwerflich. Es würde den boni mores widersprechen, die Schwangerschaft und den heranwachsenden Foetus in dieser Weise zu entwürdigen [Laufs, Rechtliche Grenzen der Transplantationsmedizin, FS Narr (1988) 34, 44].

Implantation

Versuche zu Forschungszwecken

Für den Fall, daß die Einpflanzung foetalen oder anderen Gewebes noch keine anerkannte medizinische Behandlungsform darstellt, sondern noch im Rahmen medizinischer Forschung erfolgt, sind die Regeln über medizinische Versuche anwendbar. Dies zum einen deshalb, weil der Empfänger des Gewebes ein Mensch ist, und experimentelle medizinische Verfahren am Menschen auch als solche behandelt werden sollten. Zum anderen sind die Regeln über die Durchführung medizinischer Versuche aber vor allem deshalb anwendbar, weil das Gewebe entweder einem toten Foetus oder einem einwilligenden Menschen entnommen wurde. Die somit erforderliche Einwilligung nach Aufklärung muß sich deshalb auch darauf beziehen, daß es sich bei dem Heileingriff um einen medizinischen Versuch handelt. Im übrigen muß diese Maßnahme im Vergleich zu alternativen, herkömmlichen Eingriffen bessere Erfolge erwarten lassen. Schließlich sollte auch eine Beratung oder eine Überprüfung durch eine Ethikkommission/Human Subjects Protection Committee erfolgen.

Neuerdings hat es einige Spekulation darüber gegeben, ob Ethikkommissionen auch an Entscheidungen über die Durchführung anderer Versuche, die nicht die Implantation foetalen Gewebes zum Inhalt haben, beteiligt werden sollten. Auf den ersten Blick scheint es, als seien die normalen Regeln über medizinische Versuche am Menschen nicht anwendbar, da der Foetus gestorben ist. Unabhängig von einer solchen Betrachtungsweise ist jedoch die Einwilligung nach Aufklärung der Mutter bzw. des Elternpaares einzuholen. Präzise formuliert, geht es in diesem Zusammenhang nicht um die Einwilligung nach Aufklärung, wie sie üblicherweise bei der Durchführung klinischer Versuche erforderlich ist. Vielmehr handelt es sich hier um die Einwilligung einer Person, deren Persönlichkeitsrecht betroffen ist.

Einwilligung nach Aufklärung

Wie ich schon angedeutet habe, muß bei der Einpflanzung von Gewebe sowohl wenn sie zu Forschungszwecken, als auch wenn sie im Rahmen einer Routinebehandlung erfolgt, die Einwilligung nach Aufklärung des Empfängers vorliegen. Denn die Implantierung fremden Gewebes stellt sich bei theoretischer bzw. technischer Betrachtungsweise immer auch als eine Maßnahme dar, durch welche das Recht des Empfängers auf körperliche Unversehrtheit verletzt wird. Deshalb muß der Patient über den geplanten Eingriff, insbesondere darüber, daß fremdes Gewebe eingepflanzt werden soll, aufgeklärt werden. Er muß über die Vorteile und Risiken informiert werden und dann seine Einwilligung erteilen. Wie ich schon gesagt habe, gelten diese Überlegungen auch für den Spender des Gewebes. Wenn der spender ein abortierter Foetus ist, muß um die Einwilligung der Eltern bzw. der Mutter nachgesucht werden. In diesem Zusammenhang wurde kürzlich berichtet, daß, wenn foetales Gewebe einem anderen nasciturus injiziert wird, die Einwilligung der Mutter (des behandelten Foetus) auch die Implantation des Gewebes an dem lebenden Foetus deckt. Sie muß allerdings wissen, daß das Gewebe ihrem ungeborenen Kind injiziert werden soll.

Medizinische Indikation

Die Implantation fremden Gewebes kann nur auf der Basis einer soliden medizinischen Indikation erfolgen. Auch und gerade dann, wenn es sich dabei um einen Eingriff im Rahmen eines klinischen Versuchs handelt, wie es meist der Fall sein dürfte, muß eine Begründung vorliegen, die diese Behandlungsform rechtfertigt. Damit ist impliziert, daß die zu erwartenden Vorteile gegenüber den zu erwartenden Risiken überwiegen. Das Risiko/Vorteil-Verhältnis muß positiv sein. In diesem Zusammenhang jedoch muß der Jurist beiseite treten und das Podium dem Mediziner überlassen. Denn es kann nur Aufgabe des Juristen sein, die äußeren Grenzen dieses neuen medizinischen Verfahren aufzuzeigen, nicht jedoch, ein Urteil über die medizinische Indikation abzugeben.

Risiko und erforderlicher Standard in der Neurochirurgie – Überholtes, Notwendiges, Unerprobtes – (Neurochirurgisches Referat)

R. Wüllenweber[1]

Die Problematik dieses Themas beinhaltet, daß am Ende wahrscheinlich mehr offene Fragen als Antworten übrigbleiben. Es geht bei diesem Thema um die Frage, ob unter Abwägung zwischen Gefährdung des Patienten und Nutzen für den Patienten – das Risiko für den Patienten – unter den aktuellen Standards in Diagnostik und Therapie mehr verstanden werden kann als ein „Minimalkonsens in einer defensiven Neurochirurgie", mit anderen Worten: Bedeutet Berücksichtigung der aktuellen Standards mehr als eine „Einigung auf den kleinsten gemeinsamen Nenner"?

G. Carstensen hat in einer Veranstaltung der Nordrheinischen Akademie für ärztliche Fort- und Weiterbildung Ende Januar 1989 sinngemäß den Standard so definiert, daß er den jeweiligen Stand der naturwissenschaftlichen Erkenntnis und ärztlichen Erfahrung repräsentiert, der zur Erreichung des ärztlichen Behandlungszieles erforderlich ist und sich in der Erprobung bewährt hat. Er betont, daß der Standard der täglichen Erfahrung entnommen werden müsse, daß er einen normativen Charakter mit steuernden Eigenschaften habe, daß aber jeder Versuch, den Standard frühzeitig festzulegen oder zu dogmatisieren, den Fortschritt hemmen würde, daß es vielmehr zu den Berufspflichten des Arztes gehöre, den Standard ständig zu prüfen und weiter zu entwickeln. In einem naturwissenschaftlich fundierten Fach wie der Medizin könnten nur diejenigen den Standard festsetzen, die nach der Sachkunde hierzu befähigt seien. Nur sie könnten entscheiden, wann ein Heilverfahren als wissenschaftlich anerkannt, überholt oder sorgfaltswidrig zu gelten habe. Mit anderen Worten: Die Rechtsprechung könne zwar die Grenze ziehen, nicht aber über den Inhalt des Standards befinden.

Wenden wir uns den derzeit gültigen Standards in Diagnostik und Therapie zu, so wird schnell klar, daß „der Teufel auch hier im Detail steckt".

In der *Diagnostik* ist wohl unumstritten, daß eine nichtinvasive Methode der invasiven vorzuziehen ist, sofern der Informationswert beider Methoden übereinstimmt. Am Beispiel Computertomographie versus Myelographie in der Diagnostik von Erkrankungen der Wirbelsäule und des Rückenmarks liegt das überwiegende Risiko der Myelographie nach wie vor in der Benutzung jodhaltiger Kontrastmittel. In der Praxis zeigt sich mehr und mehr, daß auch ein Risiko in der Computertomogra-

[1] Haager Weg 69, D–5300 Bonn 1

Advances in Neurosurgery, Vol. 18
K.-A. Bushe, M. Brock, and M. Klinger (Eds.)

phie besteht, wenn man berücksichtigt, daß für Radiologen, die mit unzureichenden neurologisch-neurochirurgischen Kenntnissen nicht nur Diagnosen stellen, sondern auch Therapievorschläge machen, die Amortisation ihrer teuren Apparate bei steigendem Konkurrenzdruck einen hohen Stellenwert hat. Berücksichtigt man die allgemeine „Strahlenhysterie", die nicht erst nach Tschernobyl angeheizt wird, so wäre die risikoärmste diagnostische Methode die Kernspintomographie, wobei offen bleibt, ob starke Magnetfelder Einfluß auf den Organismus haben können, aber auch, ob unser auf Solidarität beruhendes Sozialsystem die enormen Kosten auf Dauer verkraften kann. Die Frage, die sich neu stellt, ist, ob wir bei der Festlegung eines Standards bei der Beachtung naturwissenschaftlicher Erkenntnisse und klinischer Erfahrungswerte ökonomische Faktoren außer acht lassen dürfen.

Bei der *Therapie* möchte ich versuchen, das Problem des lumbalen Bandscheibenvorfalles zu erläutern. Es stellt sich die Frage, ob neuere Therapieverfahren wie die Chemonukleolyse oder das favorisierte Verfahren der Absaugung von Bandscheibengewebe zum heutigen Standard gehören. Die bei unserem Kongreß in Hannover noch hochgelobte Chemonukleolyse hat in den letzten Jahren an Attraktivität erheblich eingebüßt. Bei einer sauberen Indikationsstellung, bei der nur ein geringer Prozentsatz von Patienten für das Verfahren überhaupt infrage kommt, folgte in unserem Krankengut in ca. 50% einer Chemonukleolyse die Operation. In wenigen Fällen ist die Chemonukleolyse auch heute noch indiziert. Als Standard kann ich sie ebensowenig anerkennen wie die anderen Methoden, über die auf diesem Kongreß diskutiert wurde. In zwei Jahren wird sich die Frage des Standards der „semioperativen" Behandlung der Bandscheibenvorfälle erneut stellen.

Ein weiteres Problem betrifft die intraoperative Röntgenkontrolle bei lumbalen Bandscheibenvorfällen. Zahlreiche Gutachtenaufträge, die mir von Gutacherkommissionen und Schlichtungsstellen zugingen, befassen sich mit der Frage, ob eine intraoperative Röntgenkontrolle der Bandscheibenhöhe angezeigt ist, d.h. dem Standard entspricht. Meine Meinung dazu ist klar: Wenn ein erfahrener Operateur eine Bandscheibe, die er intraoperativ – dem klinischen Befund entsprechend – als prolabiert beurteilt, ausräumt, so entspricht das dem geltenden Standard. Verlangt man aber, daß in diesem Stadium durch Röntgenkontrolle die Etagenhöhe bestätigt wird, so muß man in Kauf nehmen, daß eine Strahlenbelastung erfolgt, die eine Gonadenschädigung bei lumbaler Lokalisation des Prozesses nicht ausschließt. Die radiologische Kontrolle kann deshalb nur mit einem Bildwandler erfolgen, der dem neuesten technischen Standard entspricht und damit die geringste Strahlenbelastung garantiert. Ein solches Gerät steht in vielen neurochirurgischen Kliniken sicher nicht zur Verfügung, man muß aber berücksichtigen, daß auch in den kleinen neurochirurgischen Abteilungen die Bandscheibenoperation den größten Anteil der Eingriffe überhaupt darstellt. Die Anforderung an den geschilderten technischen Standard kann heute nur mit großen Geldmitteln erfüllt werden, die in den wenigsten Kliniken zur Verfügung stehen dürften.

Wenden wir uns einem weiteren Beispiel der Therapie zu, so ist heute die Carotis-Unterbindung zur Behandlung eines sackförmigen Aneurysmas des Circulus Willisii sicher überholt. Die meisten von uns dürften zustimmen, wenn ich behaupte, daß die Versorgung derartiger Aneurysmen mit Hilfe des Operationsmikroskops zum Standard gehört. Es kann aber kein Zweifel bestehen, daß ein erfahrener Operateur auch heute noch mit gutem Erfolg Aneurysmen operieren kann, ohne auf das Mikroskop zurückzugreifen, d.h. im Einzelfall kann es mehrere Standardmethoden geben, wobei der von Carstensen ebenfalls erwähnte Standpunkt betont werden soll, daß eine bewährte Methode mit guten Ergebnissen nicht deshalb zu verlassen ist, weil sich ein anderes therapeutisches Prinzip durchsetzt.

Das gleiche gilt auch für die *medikamentöse Behandlung:* Man muß zur Kenntnis nehmen, daß heute fast jeder kleinste Fieberschub mit Antiobioticis behandelt wird, obgleich Acetylsalicylsäure oder Paracetamol den gleichen Effekt haben können. Entscheidend ist in einem solchen Fall die persönliche klinische Erfahrung: Ist der Temperaturanstieg das erste Zeichen einer Sepsis, so ist die alleinige Behandlung mit Wadenwickeln ohne Zweifel ein Behandlungsfehler, der Mitursache eines tödlichen Ausganges sein kann. Ist eine Thromboseprophylaxe mit Heparin Standard, auch wenn steigende Zahlen postoperativer Nachblutungen nicht zu übersehen sind? Über den psychotherapeutisch-suggestiven Effekt von homöopathischen Zubereitungen brauchen wir nicht zu diskutieren, weil die Präparate nicht schaden. Anders ist es allerdings, wenn beim Anwender ideologische Gründe maßgeblich sind, die dazu führen, daß dem Patienten vielleicht dringend indizierte Medikamente vorenthalten werden.

Die Frage der Früh- oder Spätoperation nach einer Aneurysmablutung berührt das wahrscheinlich schwierigste Kapitel, den Standard in der *Indikationsstellung.* Von vornherein dogmatisch festzulegen, daß nur Patienten im Stadium Hunt und Hess I oder II primär operiert werden dürfen, ist sicher falsch, wenn man die zugegeben kleine Zahl von Patienten berücksichtigt, bei denen es zu einer intracerebralen Massenblutung gekommen ist, und die im tiefen Koma nur durch eine perakute Operation eine kleine Chance der Lebenserhaltung haben.

Ich habe von meinen ersten chirurgischen Lehrern gelernt, daß man keine Patienten aufgeben dürfe, bevor nicht die allergeringste operative Chance ausgeschöpft worden sei. Dieser Standpunkt gilt sicher nicht für die Operationsindikation maligner Gliom-Rezidive. Als Standard muß die Indikation im Einzelfall aufgrund der Lokalisation, aufgrund des Allgemeinzustandes des Patienten und der gesamten Lebenssituation einschließlich der Lebenserwartung beurteilt werden.

Es kann kein anderes Kriterium geben bei geriatrischen Patienten, bei denen die Beurteilung der zu erwartenden Lebensqualität und der entsprechenden Lebensumstände im Vordergrund steht. Standard kann nicht bedeuten, das Leben von pflegeabhängigen, dementen Patienten auf jeden Fall operativ zu verlägern. Wir haben im vergangen Jahr bei einer bewußtlosen 84jährigen Patientin ein epidurales Haematom mit dem Erfolg operiert, daß die Patientin wieder in der Lage war, in ihre eigene Wohnung zurückzukehren und sich ohne fremde Hilfe selbst zu versorgen. Das ist sicher die Ausnahme, die aber zeigt, daß es keine Standards gibt, die uns hindern, im Einzelfall etwas zu riskieren ohne leidlich sichere Beurteilung der Prognose. Weniger dramatisch ist die Operationsindikation bei alten Patienten mit einer Stenose des Spinalkanals, sei es cervikal oder lumbal. Als Regel kann man eigentlich nur die häufig unwägbare Abwägung der Behandlungschancen bezeichnen. Das beinhaltet die Frustration, wenn es zu keinem Behandlungserfolg kommt, aber auch die Dankbarkeit der Patienten, die nach einer solchen Operation wieder laufen können, wie mir noch vor zwei Wochen eine 78jährige Patientin erzählte, die nach einer Laminektomie über drei Etagen wieder in der Lage ist, Konzerte und Theater zu besuchen. Das einzige Prinzip kann in diesen Fällen nur das „nihil nocere" sein, zugegeben ein Prinzip, das sich nicht nur an medizinischen Kriterien, sondern an ethischen Standards im Einzelfall verankern läßt. Um die Frage der Standards in der Indikationsstellung abzuschließen, so möchte ich feststellen, daß nur derjenige fachkompetent ist, der über die ärztliche Erfahrung verfügt, die ihm erlaubt, aus einer Vielzahl von ähnlich gelagerten Fällen, die aus dem Fundus der Einnerung quasi als Referenzmethode aufsteigen, im aktuellen Fall zu entscheiden.

Nun zu den *Standards in der Operationstechnik.* An erster Stelle sollte nach wie vor das gewebeschonende Operieren stehen, ein Prinzip, das in allen operativen Disziplinen unverändert gültig ist. Die Fortschritte der modernen Anaesthesie erlauben heute ohne Schaden für den Patienten längere Operationszeiten als früher. Das sollte aber nicht dazu verführen, ein ebenfalls altes chirurgisches Prinzip aufzugeben: das zügige, zielgerichtete Operieren. Die Katastrophe der Infektion mit HIV-positiven Blutkonserven muß eine Warnung an alle diejenigen sein, die glauben, daß es auf den Verbrauch von einigen Konserven mehr oder weniger heute nicht mehr ankommt. Ich bin der Ansicht, daß zum Beispiel mikroneurochirurgische Operationen maligner Gliome nicht als Standard angesehen werden sollten.

In engem Zusammenhang mit der Operationstechnik, aber selbstverständlich auch der Routine des einzelnen Operateurs, steht der Standard in Stress-Situationen. Solche Situationen werden bei Flugpiloten simuliert. Wenn ich sehe, welche große Begeisterung Computerspiele bei meinen Mitarbeitern auslösen, so frage ich mich, ob man nicht auch neurochirurgische Operationssituationen simulieren und mittels des Computers trainieren kann. Welche Rolle der Stress bei Operationen spielt – ein Stress, der teilweise sogar quantifizierbar ist – haben die Arbeiten von Poimann u. Mitarb. aus der Würzburger Klinik über die vegetativen Reaktionen bei Bandscheibenoperationen gezeigt. Möglichkeiten, den Stress zu reduzieren, hat der einzelne Operateur nur in begrenztem Umfang. Mir ist in Erinnerung geblieben, was mein Freund Tibor Mérei sagte, als wir vor 15 Jahren die ersten Bypass-Operationen gemeinsam machten: „gut ausschlafen und vor der Operation keine Tasse Kaffe und keine Zigarette". Das ist ein zwar simpler aber wirksamer Standard zur Reduzierung von Stress-Situationen.

Ein abschließendes Wort zu den Standards in der Operationstechnik: Mich bedrückt es, wenn ich in großen Aneurysma-Statistiken lese, in einem wie hohen Prozentsatz es intraoperativ zur Aneurysma Ruptur gekommen ist bzw. zur Anwendung temporärer Clips. Ich möchte in diesem Zusammenhang an die eindrucksvollen Untersuchungen von Stolke u. Mitarb. erinnern, die auf dem Bonner Kongreß vorgetragen wurden zum Thema „Wie gut sind die guten Ergebnisse nach Aneurysma-Operationen?"

Standards in der Gutachterpraxis sind ein Problem, das ohne Zweifel durch Laienfunktionäre angeheizt wird, von Organisationen, denen das „Wohl des Patienten" angeblich am Herzen liegt. Ich habe im März 1989 in einer Fernsehberichterstattung von einem Gewerkschaftsfunktionäre gehört, daß erst „wenn diese Generation von Gutachtern ausgestorben ist, es möglich wird, den Patienten gerecht zu werden". In dieser Einstellung spiegelt sich sowohl das psychologisch verständliche Kausalitätsbedürfnis des einzelnen Patienten als auch die Anspruchshaltung des modernen Menschen an den Sozialstaat. Das Schicksal, das den einzelnen Menschen mit einer Erkrankung schlägt, kann nur ein Betriebsfehler sein, der dem Arzt oder dem Staat angelastet wird. Umso dringlicher ist es, daß wir uns in der Beurteilung von Zusammenhangsfragen absolut einig sind. Dafür zwei Beispiele: Für die Frage des ursächlichen Zusammenhanges zwischen Trauma und lumbalem Bandscheibenvorfall ist der profunde Arbeit von Frowein u. Mitarb., die sie seinerzeit in Berlin vorgetragen haben, auch heute noch als Standard anzusehen.

Was die Frage des Zusammenhanges zwischen Hirntumoren und Trauma anbelangt, so sollten wir uns darüber im klaren sein, daß diese Entscheidung auf einer Konvention zwischen Neurochirurgen und Neuropathologen beruht, die allein auf das Zusammentreffen von Trauma, Zeit, Ort und Tumorenentstehung abzielt, ohne kausal begründet zu sein.

Bollingers Spätapoplexie, die R. Frowein im vorigen Jahr bei dem Fortbildungskurs in Nauheim zur Diskussion gestellt hat, wird heute durch neuroradiologische Befunde bei traumatischen Carotis-Dissektionen in Verbindung mit der transkraniellen Dopplersonographie in neuem Licht gesehen, ein Beispiel dafür, daß mit neuen Untersuchungsmethoden neue Erkenntnisse in der Zusammenhangsfrage möglich sind.

Das *Problem der Qualitätskontrolle* ist auf vielen Gebieten gelöst und damit zum Standard erhoben. Untersuchungsergebnisse klinischer Labors werden regelmäßig im Rahmen der Qualitätskontrolle beurteilt und sind als Standard demnach überprüfbar. Die Qualitätskontrolle neurochirurgischer operativer Tätigkeit wird in der Kommission für Qualitätskontrolle nachgeprüft. Dabei zeigt sich allerdings das Manko, daß sich gerade die kleinen Abteilungen nicht an der Qualitätsstudie beteiligen.

Zur *Sorgfaltsplicht* des Neurochirurgen zählt neben der selbstverständlichen Sorgfalt in Diagnostik und Therapie auch die *Dokumentationspflicht,* wobei in naher Zukunft die Dokumentation mit Hilfe des Computers zum Standard wird.

Dieser Standard wird trotz verkürzter Arbeitszeit der Sorgfaltspflicht Genüge tun und mehr Zeit für die Betreuung der Patienten ermöglichen. Voraussetzung ist nicht nur die Anschaffung der Geräte, sondern auch das Engagement, sich mit der Computertechnik zu befassen, was bei vielen jungen Kollegen erfreulicherweise vorhanden ist.

Sind *operative Innovationen* unter dem Handicap einer massiven Reduzierung der Zahl der Tierversuche noch möglich? Die mikroneurochirurgische Technik, die zweifellos im Tierversuch trainiert werden muß, hat sich in Verbindung mit einer Renaissance der deskriptiven Anatomie entwickelt, die wir Forschern wie Johannes Lang und Rabischong, aber auch jungen Kollegen, z.B. aus der Berner Klinik, verdanken. Dabei entwickeln sich neue Standards, die durchaus als innovativ bezeichnet und vervollkommnet werden können.

Der Kreis schließt sich. Ich hoffe, durch einige Beispiele gezeigt zu haben, daß die Skepsis, die sich im „kleinsten gemeinsamen" ausdrückt, vielleicht übertrieben ist. Ein für jeden Arzt unverzichtbares Kapitel muß ich ausklammern, da die Besprechung den Rahmen dieses Vortrages sprengen würde. Es handelt sich um die Beachtung von Standards in der ärztlichen Ethik. Wer die Entwicklung in den letzten Jahren aufmerksam verfolgt hat, wird feststellen, daß sich ein Spannungsverhältnis entwickelt hat zwischen der vorwiegend aus den USA kommenden utilitaristischen Betrachtungsweise und der Verantwortungsethik, wie sie beispielsweise Jonas vertritt. Diese Entwicklung hindert micht nicht, an dem persönlichen Bekenntnis zur individuellen Verantwortung für jeden einzelnen Patienten, aber auch unser Fach – die Neurochirurgie – festzuhalten.

Risiko und erforderlicher Standard in der Neurochirurgie – Überholtes, Notwendiges, Unerprobtes – (Juristisches Referat)

H.-L. Schreiber[1]

Alle ärztliche Behandlung erfolgt mehr oder weniger unter Risiko, d.h. der Gefahr des Mißerfolges oder sogar einer Schädigung des Patienten. Man hat zugespitzt, aber sachlich nicht falsch gesagt, ärztliches Eingreifen tausche das Risiko des natürlichen Verlaufes der Krankheit gegen das erfahrungsgemäß auch statistisch in der Regel geringere Risiko der medizinischen Behandlung.

Ohne ihren Charakter als Wissenschaft damit umfassend bestimmen zu wollen, kann doch gesagt werden, daß die Medizin in ihrem Kern eine praktische Naturwissenschaft ist, sie entwickelt auf naturwissenschaftlicher Basis Technologien zur Bekämpfung bzw. Linderung von Krankheiten.

Die Neurochirurgie betreibt expansive Technologie, damit erzielt sie außerordentliche Erfolge, geht aber ständig erhebliche Risiken ein, handelt je nach Methode und Situation in unterschiedlichem Maß „auf Risiko".

Ebenso wie das Risiko des sogenannten natürlichen, d.h. unbehandelten Verlaufs der Krankheit trägt der Kranke das Risiko eines vom Arzt nicht verschuldeten Mißerfolges der Behandlung. Der Arzt schuldet seine Leistung, aber nicht den Erfolg der Behandlung. Der behandlungsspezifische Fehlschlag ist Risiko des Patienten.

Der Arzt haftet nur, soweit er Mängel in der Qualität seiner Tätigkeit zu verantworten hat, die für eine Verschlechterung des Gesundheitszustandes ursächlich geworden sind. Der bloße Mißerfolg macht weder zivil- noch strafrechtlich haftbar, sondern erst die Abweichung von dem, was ein Arzt in der konkreten Situation leisten kann und muß.

Haftung begründet eine Verletzung der in der Behandlung erforderlichen Sorgfalt, der sie auslösende Behandlungsfehler wird umschrieben als die Verletzung der für einen ordentlichen und gewissenhaften, durchschnittlich befähigten Arzt in einer konkreten Situation gebotenen Sorgfalt.

Was denn die Sorgfalt in der jeweiligen Situation fordert, das findet der Arzt im geschriebenen Recht nicht, es wäre auch ganz unmöglich, das alles im Gesetz, und sei es noch so ausführlich, festzulegen.

[1] Staatssekretär im Niedersächsischen Ministerium für Wissenschaft und Kunst, Prinzenstr. 14, 3000 Hannover 1

Advances in Neurosurgery, Vol. 18
K.-A. Bushe, M. Brock, and M. Klinger (Eds.)

Der Standard der Behandlung, der zweite Begriff in unserem Thema, ist es, der hier Orientierung geben soll. Er bestimmt sich, so die allgemein in Rechtsprechung und Literatur verwendeten Formeln, nach dem dem Arzt verfügbaren Stand der medizinischen Wissenschaft und Praxis zur Zeit der Behandlung. Das Wort „Standard" wird in ganz unterschiedlichem Sinne gebraucht. In Reiseprospekten bezeichnet „Standard" die unterste Kategorie im Unterschied zu Komfort oder gar Luxus. So darf der Begriff in unserem Sinne nicht verstanden werden, auch nicht nur, wie Herr Wüllenweber meines Erachtens zutreffend gesagt hat, als kleinster gemeinsamer Nenner einer defensiven Medizin.

Erforderlicher Standard meint vielmehr das, was ein – wie es in der Rechtsprechung heißt – vernünftiger, durchschnittlich befähigter, gewissenhafter und besonnener Angehöriger des jeweiligen Verkehrskreises, also hier ein Arzt, speziell ein Neurochirurg, an Kenntnissen, Können und Aufmerksamkeit erbringen kann und muß. Vom Betroffenen, dem Patienten, aus gesehen ist es das an Leistung und Sorgfalt, was er normalerweise in einer ordentlichen Klinik erwarten darf.

Das sind nun bisher alles recht abstrakte, allgemeine, keine präzisen, leicht handhabbaren Kategorien, die einem unmittelbar sagen könnten, welche Behandlungsmethode angewandt werden muß, um nicht die Grenze zur Haftung zu überschreiten. Aber für die Bestimmung des erforderlichen Verhaltens und des zulässigen Risikos vermag der „Standard" doch einiges zu leisten. Ich werde das gleich näher darzulegen versuchen. Aber zunächst noch einige Erläuterungen zur näheren Bestimmung des „Standard".

Er ist weniger als eine strikte Rechtsregel, die generell gilt. Standards sind beweglich, sie ändern sich und gelten nur für die jeweilige individuelle Situation einer Behandlung. Daher erscheint der „Standard der ärztlichen Wissenschaft und Praxis" als Mittel zur Beurteilung der Richtigkeit ärztlichen Verhaltens richtig.

Ähnlich gilt das für den Sicherheitsbereich im Bauordnungs- und Umweltrecht für den geforderten Standard anerkannter Regeln von Sicherheit und Technik. Standard ist nicht nur eine Beschreibung tatsächlich praktizierten Verhaltens, sondern enthält normative Elemente im Sinne von anerkanntem, in der Praxis für richtig und erforderlich angesehenem Verhalten. Standard meint nicht bloß übliche, sondern die erforderliche Sorgfalt, nicht etwa einen verbreiteten Schlendrian oder übliche Mißstände. Andererseits enthält „Standard" auch ein faktisches Element: das, was eine Disziplin wie die Neurochirurgie normalerweise tut, geht in ihn ein. In der amerikanischen Literatur wird zutreffend gesagt, Standard enthalte die Verweisung auf ein in der Lebenswirklichkeit befolgtes und akzeptiertes Normalverhalten.

Was bedeutet das für die Haftung und das Haftung vermeidende Verhalten des Arztes?

1. Der erforderliche Standard operativer Behandlung als Parameter der Haftung mißt sich an den wissenschaftlich/praktischen Möglichkeiten einer Behandlung zur jeweiligen Zeit.
   Ärztliche Tätigkeit ist ein offenes, sich ständig wandelndes, in Veränderung und Auseinandersetzung begriffenes Feld, das häufig von rivalisierenden Methoden bestimmt ist. Es kann keinen Kodex, keine abgeschlossene Zusammenstellung etwa der „richtigen" Operationsmethoden geben, keinen DIN-Katalog der neurochirurgischen Operationen.
2. In letzter Linie sind es zwar die Gerichte, die im Streitfall über den Standard und das zulässige Risiko zu entscheiden haben. Unsere Rechtsordnung weist ihnen diese Aufgabe der Streitentscheidung, wie auf vielen anderen Gebieten, so auch auf dem Gebiet des ärztlichen Verhaltens, zu.
   Die Standards sind auch juristische, sie sind richterliche Urteilsmaßstäbe. Sachlich sind es aber wesentlich ärztliche Maßstäbe. Das Recht kann von sich aus nicht sagen, was ärztlich möglich und nötig ist. Es verweist primär auf medizinische Kriterien zurück.
   Das Recht ist dabei für die Beurteilung in einer Art „Grenzkontrolle" zuständig, was zum Schutze von Leben und Gesundheit an ärztlich Möglichem eingehalten werden muß.
   Das gilt etwa für organisatorische Vorkehrungen im Krankenhaus, für die Regeln der Arbeitsverteilung, für die Notwendigkeit der Weiterbildung, für die Regeln der Arbeit im Team, für die Notwendigkeit einer Geräteausstattung etc. Im eigentlichen medizinischen Kernbereich der Diagnose, der Indikation und der Therapie sind die Gerichte weitgehend auf die Sachverständigen aus der Medizin angewiesen. Auf sie kommt es entscheidend an. So kann der Jurist zu den vielfältigen Fragen der Op-Technik, die Herr Wüllenweber hier angesprochen hat, eigentlich nichts sagen, etwa zur Überlegenheit oder Unterlegenheit der Chemonukleolyse oder zur intraoperativen Röntgenkontrolle bei lumbalen Bandscheibenvorfällen.
3. Orientierung am Standard hat erhebliche Gefahren: Standardbildung erfolgt durch eine an der Vergangenheit bisheriger Praxis orientierte Erwartung und enthält daher ein statisches, rückwärts gewandtes Element. Ausrichtung am Standard kann – begünstigt durch das Vordringen rechtlicher Kontrollen in der Medizin – zur Orientierung am üblichen, am wenigsten Gefahr einer Haftung begründenden Operationsverfahren führen, das nicht das für den Patienten aussichtsreiche sein kann. „Standardisierung" bei der Weiterentwicklung neurochirurgischer Operationsmethoden wäre gefährlich. Zutreffend scheint mir die Feststellung Steffens, Vorsitzender des Arzthaftungssenats beim BGH, daß sich ärztliches Vorgehen prinzipiell nicht auf einen abgeschlossenen Regelko-

dex stützten könne, sondern für den jeweiligen Behandlungsfall die Regel erst in der Behandlung finden müsse. Dafür müsse dem Arzt ein ausreichender Beurteilungs- und Entscheidungsspielraum bleiben, den auch die Haftung nicht verkürzen dürfte. Anderes würde zur Überdiagnose, zur auf eingefahrene Methoden fixierten, statistisch orientierten Therapie führen und letztlich zum Nachteil des Patienten ausschlagen. Wo es divergierende, vertretbare verschiedene Methoden gibt, dort kann es keinen Standard geben.

Versuchen wir, das bisher Gesagte noch wenigsten etwas näher zu konkretisieren.

1. Bei mehreren Behandlungsmöglichkeiten soll dem Prinzip des „sichersten Weges" der Vorzug gegeben werden. Das erscheint im Hinblick auf den Zweck der Therapie prinzipiell richtig: Bei gleich erfolgversprechenden Wegen ist derjenige zu wählen, der das geringste Risiko für den Patienten enthält. Das gilt auch für die Diagnosemethoden. Prinzipiell ist die nicht invasive Methode zu wählen, wenn sie hinreichende Erkenntnis verspricht. Schwierig ist die von Herrn Wüllenweber aufgeworfene Frage, ob auch ökonomische Faktoren bei der Diagnose zu beachten sind: Grundsätzlich ist die sicherste und risikoärmste Methode zu wählen, die Myelographie hätte danach zurückzutreten, weil sie durch die Benutzung jodhaltiger Kontrastmittel riskanter erscheint. Geringfügig belastende, aber sonst diagnostisch gleichwertige Verfahren dürfen wohl gewählt werden, wenn sie gut, aber etwas kostengünstiger sind. Prinzipiell muß wohl nicht stets das teuerste, aufwendigste, am wenigsten belastende Verfahren gewählt werden, wenn ein ökonomisch günstigeres zur Verfügung steht, das gleich gute Ergebnisse bringt und keine nennenswerten Risiken enthält.
2. Der bisherige „Standard" sollte nicht hindern, ein neues, auch mit Risiken behaftetes Verfahren zu erproben, wenn die Vorteile im Hinblick auf den Therapieerfolg so groß sind, daß die Risiken zumutbar erscheinen.
3. Ein auf einem Kongreß vorgestelltes, erst an relativ wenigen Fällen praktiziertes Verfahren wird noch nicht zum Standard. Man wird es niemandem zum Vorwurf machen, wenn er es noch nicht übernimmt. Es gibt allerdings auch überholte Verfahren. Präzise den Punkt oder den Tag zu nennen, an dem ein neues, unter Risiko erprobtes Verfahren in den gebotenen Standard umschlägt, bin ich nicht in der Lage. Auch bisherige Verfahren dürfen weiter verwendet werden, wenn sie Erfolg haben und keine unzumutbaren Risiken aufweisen, solange ein neues, sicher überlegenes Verfahren nicht zur Verfügung steht.
4. Zweifelhaft kann sein, ob eine Methode mit größerem Risiko zulässig ist, wenn sie eine durchgreifendere Heilungschance bietet. Bei erheblichen Gefahren durch die Krankheit scheint mir das zulässig. Bei geringen Gesundheitsgefahren werden riskante Methoden zu ihrer Besserung nicht zulässig sein.
5. Sicher ist richtig, daß Standards es nicht hindern, in einem Einzelfall mit Gründen etwas abweichend vom Standard zu riskieren, wenn kein anderes gesichertes Ver-

fahren zur Verfügung steht, das geringere Risiken aufweist.
Wenn Wüllenweber plastisch von der „unabwägbaren Abwägung" der Behandlungschancen spricht, so bezeichnet das treffend eine Situation, in der kein Standard gilt, der es verbieten würde, ein Risiko einzugehen.

Versuchen wir zusammenzufassen.

1. Der „Standard der gebotenen Sorgfalt" ist ein notwendiges, aber kein absolutes Instrument, die erforderliche Behandlung zu bestimmen. Er bezeichnet bestimmte Regelerfordernisse, die mangels besserer Alternativen gelten.
2. Der „Standard" legt keineswegs eine bestimmte Methode ein und für allemal fest. Er ist vielmahr veränderlich, er läßt Raum für andere zu erprobende Methoden, deren Erfolgsaussicht nicht außer Verhältnis zu den begleitenden Risiken steht.
   Die Gefahr der Standardisierung besteht in der rückwärts gerichteten Festschreibung bisheriger Methoden.
3. Wesentliches Instrument zur Feststellung des „Standard" als Haftungsmaßstab sind die medizinischen Sachverständigen. Ihnen obliegt die schwierige Vermittlung zwischen medizinischer Methode und juristischer Haftung nach dem Standard der gebotenen Sorgfalt.

Literatur

Deutsch E (1983) Arztrecht und Arzneimittelrecht
Laufs A (1988) Arztrecht, 4. Aufl.
Schreiber H-L (1984) Rechtliche Maßstäbe des medizinischen Standards. Langenbecks Arch Chir 364:294
Schreiber H-L (1985) Arzthaftung, Perspektiven der Rechtswissenschaft. Beitr Gerichtl Med XLIII:2
Steffen E (1986) Neue Entwicklungslinien der BGH-Rechtsprechung zum Arzthaftungsrecht, 2. Aufl.

Neuere Entwicklung in der Rechtsprechung zur Aufklärung

H.-H. Franzki[1]

Wer vor einer ärztlichen Zuhörerschaft über die neuere höchstrichterliche Rechtsprechung zur Aufklärungspflicht sprechen soll, muß mit einer Enttäuschung beginnen. Der Bundesgerichtshof hat seine Anforderung an den Umfang der Aufklärungspflicht trotz aller Bedenken, die hiergegen von seiten der Ärzteschaft erhoben worden sind, nicht eingeschränkt. Er hat sie eher noch verschärft. Als vor Jahren eine Verletzung des Facialisnervs bei einer Tympanoplastik trotz eines Risikos von 1:2000 als typische Komplikation bezeichnet wurde, über die aufzuklären sei [1], warfen Mediziner die Frage auf, ob bei einer so geringen Komplikationsdichte hier noch von einer Typizität gesprochen werden könne. Den Bundesgerichtshof hat dies nicht beeindruckt. Er hat in einer späteren Entscheidung [2] eine Darmperforation bei einer Rektoskopie als aufklärungspflichtiges Risiko bezeichnet, obwohl der Sachverständige hier die Komplikationsdichte mit allenfalls 1:10000 bezeichnet hatte. Er steht auf dem Standpunkt, auch bei solcher statistischen Seltenheit treffe den Patienten das Risiko, wenn es sich verwirkliche, zu 100%. Vom Arzt sei deshalb zu verlange, über ein solches eingriffsspezifisches Risiko selbst dann aufzuklären, wenn kein Mediziner auch nur entfernt daran denke, deswegen den Eingriff in Frage zu stellen; denn selbst ein verständiger Patient könne gute persönliche Gründe haben, auch in Fällen eindeutiger medizinischer Indikation den Eingriff abzulehnen und das Risiko nicht einzugehen.

Es mag Ärzten schwerfallen, für diese Rechtsprechung Verständnis aufzubringen. Sie ist auch innerhalb der Justiz nicht unumstritten. Der Senat des OLG Celle, in dem ich den Vorsitz führe, hat in einem Urteil vom 15.06.1981 B3 ausgeführt, so sehr die Rechtsprechung dazu verpflichtet sei, dem Patienten die Ausübung seines verfassungsrechtlich gewährleisteten Selbstbestimmungsrechts zu sichern, so sehr müsse sie sich andererseits doch davor hüten, die Anforderungen an die Aufklärungspflicht der Ärzte zu übertreiben. Sie würde damit den Ärzten praktisch unerfüllbare Pflichten auferlegen, ohne dem wahren Wohl des Patienten zu dienen. Es gehöre auch zu den hohen Berufspflichten des Arztes, nicht nur zu heilen, sondern auch Trost zu spenden, den Genesungswillen des Kranken, diesen wichtigen Heilungsfaktor, nicht zu lähmen, sondern zu stärken und den Patienten nicht in

[1] Präsident des Oberlandesgerichts, Leberstr. 47, D-3100 Celle

Advances in Neurosurgery, Vol. 18
K.-A. Bushe, M. Brock, and M. Klinger (Eds.)

unnötige Ängste zu versetzen. Der Arzt müsse deshalb auch Rücksicht nehmen auf die psychosomatische Ausnahmesituation, in der sich der schwerkranke Patient gerade wegen des Klinikaufenthaltes regelmäßig befinde, und dürfe diesen Patienten nicht über sein Krankheitsleid hinaus seelisch unnötig belasten und dadurch den Kranken womöglich von der Einwilligung in eine dringend gebotene Behandlung abschrecken. Die Aufklärung dürfe deshalb nicht uferlos werden.

Dies entspricht auch heute noch meiner Auffassung. Ich darf jedoch nicht verschweigen, daß der Bundesgerichtshof [4] das Urteil zwar bestätigt, gegen unsere Ausführungen zur Aufklärungspflicht jedoch Bedenken erhoben hat. Es ist deshalb meine Pflicht, Ihnen im folgenden weiterhin die Auffassung unseres höchsten Gerichts zu diesen Fragen mitzuteilen.

Doch bevor ich hier zu Einzelheiten komme, mag eine kurze allgemeine Bemerkung vorangestellt sein.

Angesichts der alten Frage „salus aut voluntas aegroti suprema lex?" hat es der Arzt gerade als seine hohe Aufgabe anzusehen, den Patienten von der Wichtigkeit und Notwendigkeit der anempfohlenen Behandlung zu überzeugen und dadurch salus und voluntas zur Kongruenz zu bringen, wobei auch die Eröffnung der Diagnose geboten sein wird, wenn anders dem Patienten der Ernst der Lage und die Behandlungsbefürftigkeit des Leidens nicht klarzumachen sind. Gelingt dem Arzt diese Überzeugungsarbeit, bei der mit Zustimmung des Patienten auch Angehörige eingeschaltet werden können, nicht, so sind ihm die Hände gebunden, mag ihm die Entscheidung des Patienten auch noch so unverständlich sein. Das gilt auch, wenn der Patient aus weltanschaulichen Gründen, z.B. als Zeuge Jehovas, eine u.U. lebensrettende Bluttransfusion verweigert.

Der Arzt sollte dieses Gespräch, das essentieller Teil seiner Zuwendung an den Patienten ist, nicht als eine ihm von der Rechtsprechung aufgedrängte, lästige und eigentlich überflüssige Formalie betrachten und auch nicht vordergründig unter dem Gesichtspunkt haftungsrechtlicher Konsequenzen führen. Richtig verstanden ist es eine vertrauenbildende Maßnahme, durch die er den Patienten auch vor späteren Enttäuschungen bewahren und Verantwortung mit ihm teilen kann.

Trotz des eingangs gemachten Hinweises auf die weiterhin strengen Anforderungen des Bundesgerichtshofs an die Aufklärungspflicht ist zur Beruhigung der Ärzteschaft zu sagen, daß der Kulminationspunkt der Verurteilungen wegen Aufklärungspflichtverletzungen längst überschritten ist. Das mag einerseits daran liegen, daß tatsächlich in größerem Umfang aufgeklärt wird, andererseits aber auch daran, daß die Rechtsprechung dem Arzt die Beweisführung erleichtert hat und von dem Patienten verlangt, daß er bei seinem Vorwurf der Aufklärungspflichtverletzung darlegt, wie er sich denn bei richtiger Aufklärung verhalten hätte und ob er in diesem Fall in einen echten Entscheidungskonflikt geraten wäre.

Der Bundesgerichtshof hält an seiner Rechtsprechung fest, daß der Patient nur „im großen und ganzen" aufzuklären sei, mit ihm also keine ins Detail gehenden medizinischen Gespräche geführt werden müssen und dem Patienten u.U. auch eine Fragepflicht obliegt, wenn er Näheres wissen will. Risiken, die als allgemein bekannt vorauszusetzen sind oder hinter der Größe des allgemeinen Operationsrisikos an Bedeutung zurücktreten, sind nicht aufklärungspflichtig. Wer sich als Patient einer größeren Operation mit entsprechendem Krankenlager unterzieht, braucht deshalb in der Regel über das allgemeine Thrombose- und Embolierisiko ebensowenig aufgeklärt zu werden wie über die allgemeine Gefahr einer Nachblutung, einer Infektion oder eines späteren Narbenbruchs. Dagegen geht der Bundesgerichtshof außerordentlich weit in der Mitteilungspflicht in bezug auf eine Mortalitätsrate, Funktionsbeeinträchtigung wichtiger Organe, nachhaltige Störungen des Bewegungsapparates, dauernde Belastungen für die künftige Lebensführung (z.B. künstlicher Darmausgang), Dauerschmerzen, Zwang zu langdauernder Schonung und erforderlich werdende weitere Operation.

Zur näheren Veranschaulichung seien beispielhaft folgende Fälle aufgeführt, die allen Disziplinen der Medizin entnommen sind und in denen eine Aufklärungspflicht bejaht wurde:

- Strahlenschäden
- Rekurrensparese bei Schilddrüsenoperation
- Facialis- und Trigeminusverletzungen bei Eingriffen im Ohr- und Kieferbereich
- Akzessoriuslähmung bei Lymphknotenexstirpation
- Halbseitenlähmung bei der Angiographie
- Armplexuslähmung oder Rückgratschädigung durch Röntgen- oder Kobaltbestrahlung
- Hodenatrophie nach Leistenbruchoperation
- Harnleiterverletzung bei der abdominalen Hysterektomie
- Hüftkopfnekrose nach etwaigem Fehlschlagen einer Abduktions-Osteotomie
- Sehnerv- oder Gehörschädigung durch aggressive Medikamente
- Atem- und Herzstillstand durch Stellatumblockade.

Etwas ausführlicher seien hier einige Fälle aus der Neurochirurgie dargestellt:

Vor Bandscheibenoperationen wird in der Regel der Hinweis auf das Risiko einer Querschnittslähmung gefordert. In einem Fall, in dem der Patient nach Verwirklichung dieses Risikos dem Arzt vorwarf, er habe hierauf nicht deutlich genug hingewiesen, hat der Bundesgerichtshof jedoch ausgeführt [5], sofern das schwerwiegende Operationsrisiko im Gespräch überhaupt ernsthaft angesprochen worden sei, könne nicht beanstandet werden, daß der Arzt den Patienten zu beruhigen und ihm unter zutreffender Darstellung der geringen Wahrscheinlichkeit des Schadenseintritts Furcht und Hemmungen vor der Operation zu nehmen versucht habe. Das sei sogar seine ärztliche Aufgabe. Bei der Beurteilung, ob während eines solchen Gesprächs der Arzt ein angesprochenes Risiko unzulässig verkleinert oder ob er zulässigerweise den Patienten therapeutisch beeinflußt habe, sei Vorsicht geboten. Die verantwortungsvolle Führung eines solchen Aufklärungsgesprächs im Einzelfall sei, sofern nur insgesamt dem Patienten ein zutreffendes Bild von dem Eingriff und dessen Folgen vermittelt wurde, Sache des Arztes, dem insoweit keine rechtlichen Vorschriften gemacht werden könnten. Der Arzt sei auch nicht verpflichtet, von sich aus dem Patienten mehr oder weniger genaue Prozentzahlen über die Möglichkeit der Verwirklichung eines Risikos mitzuteilen. Vielmehr sei es Sache des „im großen und ganzen" zutreffend informierten Patienten, von sich aus den Arzt zu befragen, wenn er weitere Einzelheiten wissen wolle.

Andererseits hat der Bundesgerichtshof [6] eine Aufklärungspflichtverletzung wegen unzulässiger Verharmlosung des Operationsrisikos in einem Fall bejaht, in dem der Patient vor einer Trigeminusoperation ausdrücklich um genaue Auskunft über das Risiko der vorgeschlagenen Behandlungsmethode gebeten hatte, der Arzt hierauf aber nur vom Verlust einiger Härchen beim Ausrasieren der Operationswunde am Kopf gesprochen hatte.

Ferner hat der Bundesgerichtshof [7] eine Aufklärungspflicht in bezug auf Nervenschädigungen bejaht, die bei der Bandscheibenoperation infolge der Knie-Ellenbogen-Lage (Häschenstellung) eintreten können. Er ist hier den Ausführungen des Sachverständigen gefolgt, der erklärt hatte, daß es sich insoweit zwar, was Dauerschäden anbelange, um eine seltene Komplikation handele, die aber jedenfalls nicht außerhalb des Bereichs der Wahrscheinlichkeit liege.

Das OLG Hamm [8] hat eine Aufklärungspflichtverletzung angenommen, weil der Patient zwar über das Risiko einer Wirbelsäulenoperation, nicht aber über den Grad der Dringlichkeit des Eingriffs aufgeklärt worden war. Es handelte sich um eine Entlastungslaminektomie, Probeexzision und Duraplastik bei einem intramedullären Tumor. Dem Patienten war die nach Auffassung des Sachverständigen falsche Darstellung gegeben worden, ohne den Eingriff werde es schon in ganz kurzer Zeit zu einem Querschnittssyndrom mit allen Konsquenzen kommen. Die Sachverständigen waren dagegen der Auffassung, hierzu hätte es einer jahrelangen Entwicklung bedurft.

In einem anderen Fall hat das OLG Stuttgart [9] entschieden, vor einer Operation zur Entlastung des durch eine Fehlstellung des Epistropheuszahns eingeengten Halswirbels sei der Patient über das Risiko der Tetraplegie aufzuklären. Dagegen hat das OG Bremen [10] eine Aufklärungspflicht vor einer dringend gebotenen Halswirbeloperation in bezug auf das sehr seltene Risiko einer Stimmbandlähmung verneint. Schließlich ist noch eine Entscheidung des OLG Köln [11] zu erwähnen, nach der vor der Operation eines Aneurysma in der linken Gehirnhälfte der Patient über das Risiko einer Halbseitenlähmung und einer Aphasie aufzuklären sei. Dies mag an Beispielen aus der höchstrichterlichen Praxis genügen.

Gelegentlich wird vom Patienten der Vorwurf erhoben, der Behandlungserfolg hätte ein besserer sein können, wenn sich der Arzt zu einer Verlegung in ein größeres Haus, insbesondere eine Universitätsklinik, entschlossen hätte. Der Bundesgerichtshof hat hier jedoch eine Aufklärungspflicht darüber, daß es personell oder apparativ besser ausgestattete Häuser gebe, verneint, sofern in dem Krankenhaus, in dem der Patient behandelt worden ist, jedenfalls fachärztlicher Standard gewährleistet war [12].

Zum Schluß sei noch einiges über die Aufklärungsmodalitäten bemerkt.

Das Aufklärungsgespräch kann seine Aufgabe nur erfüllen, wenn es von der richtigen Person, in der rechten Form, zur rechten Zeit und unter geeigneten Begleitumständen geführt wird.

Die Regel sollte sein, daß derjenige Arzt das Aufklärungsgespräch führt, der auch den Eingriff vornimmt. Es ist jedoch zulässig und im Klinikbetrieb mitunter unumgänglich, daß diese Aufgabe auf einen anderen Arzt, z.B. den diensthabenden Stationsarzt übertragen wird. Niemals darf sich der Arzt im Krankenhaus jedoch darauf verlassen, daß schon der einweisende (Haus-) Arzt das notwendige Aufklärungsgespräch geführt hat. Ferner darf niemals die Aufklärung dem nachgeordneten nichtärztlichen Personal überlassen werden, das auf weiterführende Fragen des Patienten keine zuverlässigen Auskünfte geben könnte.

Die Gesprächsführung muß auf den Bildungsgrad, die Aufnahmefähigkeit und Belastbarkeit des Patienten abgestellt sein. Unangebracht und geradezu unethisch ist eine Brutal- oder Horroraufklärung, die den Patienten so verängstigt, daß er einer dringend gebotenen Behandlung nicht zustimmt. Der Bundesgerichtshof [13] hat zwar vor vielen Jahren zur Frage der therapeutischen Rücksicht ausgeführt, die Aufklärung dürfe nur dann unterbleiben, wenn sie zu einer ernsten und nichtbehebbaren Gesundheitsschädigung des Patienten führen würde. Rein psychische Belastungen rechtfertigten dagegen ein Verschweigen der Diagnose und der Risiken nicht. Sie müßten vielmehr in Kauf genommen werden. Erstaunlicherweise spielt diese Frage bei der Verteidigung von Ärzten fast nie eine Rolle. Ich möchte nicht so weit gehen wie der Bundesgerichtshof, dem Arzt aber dann dringend empfehlen, daß er es in den Krankenpapieren vermerkt, wenn er aus therapeutischer Rücksicht bewußt von einer vollständigen Aufklärung absieht. Hier kann übrigens das Gespräch mit den nächsten Angehörigen ratsam sein, um die Belastbarkeit des Patienten oder seinen mutmaßlichen Willen zu erforschen.

Der Patient kann auf die Aufklärung verzichten, wenn er nicht beunruhigt werden möchte oder vertrauensvoll die Entscheidung seinem Arzt überlassen will. Auch hier ist ein Vermerk in den Krankenpapieren, besser noch eine vom Patienten unterzeichnete Erklärung zu empfehlen.

Die Aufklärung soll zur rechten Zeit stattfinden, in der Regel also so zeitig vor dem Eingriff (spätestens am Vorbend einer geplanten Operation), daß der Patient seine Entscheidung in Ruhe bedenken, mit Angehörigen besprechen und notfalls auf Wunsch auch noch (zumindest telefonisch) einen anderen Arzt oder seinen Hausarzt konsultieren kann.

Auch Ort und Gelegenheit sind für das Aufklärungsgespräch von Bedeutung. Es sollte nicht im Zuge der Visite, möglichst aber auch nicht im Mehrbettzimmer in Gegenwart von Bettnachbarn und Besuchern geführt werden, weil der Patient hier Hemmungen haben wird, seine Ängste zu offenbaren und Fragen zu stellen.

Um sich dem Patienten wirklich verständlich zu machen, ist eine Vermeidung medizinischer Fachausdrücke geboten. Da auch Ausländern das Selbstbestimmungsrecht zusteht, ist bei einem der deutschen Sprache nicht mächtigen Patienten ein Dolmetscher oder eine sonst zur Übersetzung geeignete Person zuzuziehen, sofern die Zeit es erlaubt.

Minderjährige eheliche Kinder werden rechtlich durch beide Elternteile vertreten. Sind sich diese nicht einig, muß das Vormundschaftsgericht entscheiden. Kann diese Entscheidung aus zeitlichen Gründen nicht herbeigeführt werden, handelt der Arzt so, wie es dem Wohle des kindlichen Patienten am besten entspricht.

Zur Frage, wie der Arzt sich Kenntnis vom übereinstimmenden Elternwillen verschafft, hat der Bundesgerichtshof [14] eine in der Praxis nicht leicht zu handhabende Dreistufentheorie aufgestellt. Bei alltäglichen Erkrankungen des minderjährigen Patienten kann er davon ausgehen, daß der das Kind begleitende Elternteil ermächtigt ist, für den anderen Teil mitzuhandeln. Bei ärztlichen Eingriffen schwerer Art mit nicht unbedeutenden Risiken muß sich der Arzt durch Rückfrage beim erschienenen Elternteil vergewissern, ob diese Ermächtigung vorliegt, kann aber in der Regel auf wahrheitsgemäße Auskunft vertrauen. Bei besonders weitreichenden Entscheidungen über die Behandlung mit erheblichen Risiken – und darum kann es gerade in der Neurochirurgie gehen – kann der Arzt nicht darauf vertrauen, daß der eine Elternteil freie Hand hat, für den anderen mitzuhandeln. Hier muß er sich die Gewißheit von solcher Ermächtigung verschaffen.

Für Aufklärung und Einwilligung ist keine Schriftform vorgeschrieben. Weit verbreitet ist im klinischen Betrieb die Verwendung von Formularen. Sie sind jedoch ziemlich wertlos, wenn sie nur die vorgedruckte Erklärung enthalten, daß mit dem Patienten der (nicht näher beschriebene) Eingriff besprochen worden ist, er über „alle Risiken" aufgeklärt ist und in alles eingewilligt hat. Mehr als die Tatsache eines Gesprächs ist mit einem solchen Schriftstück nicht zu beweisen. Beweiskräftig sind solche Dokumente nur, wenn der Eingriff – am besten vom Arzt handschriftlich – bezeichnet ist, die erwähnten Risiken in Stichworten aufgezählt sind, ebenso eine in Betracht kommende Handlungsalternative erwähnt ist und die darauf folgende Einwilligungserklärung vom Patienten unterzeichnet und vom Arzt sogleich gegengezeichnet ist.

Merkblätter und Broschüren können das Aufklärungsgespräch nur vorbereiten, niemals aber ersetzen. Der Bundesgerichtshof [15] steht dem Formularwesen offenbar zunehmend skeptisch gegenüber und will auch dem Arzt eine faire und reale Chance der Beweisführung geben, der das vertrauensvolle Gespräch mit dem Patienten unter vier Augen vorzieht und von der Einholung einer Unterschrift ganz absieht. In diesem Fall ist es jedoch dringend geboten, sogleich nach dem Aufklärungsgespräch einen Vermerk in den Krankenpapieren zu machen, in dem der vorgesehene Eingriff, die dem Patienten mitgeteilten Risiken, eine etwaige Behandlungsalternative und die Tatsache der Einwilligung oder deren Verweigerung festzuhalten sind und der vom Arzt mit Datum zu unterzeichnen ist. Ein solcher zur rechten und unverdächtigen Zeit gemachter Vermerk ist eine große Hilfe bei der Beweisführung und wird im allgemeinen vor Gericht Glauben finden.

Rechtsprechungsnachweise

1. BGH NJW 1980, 1905
2. BGH NJW 1984, 1395
3. OLG Celle VersR 1981, 1184
4. BGH VersR 1982, 1142
5. BGH NJW 1984, 2629
6. BGH VersR 1964, 614
7. BGH NJW 1985, 2192
8. OLG Hamm AHRS KZa 5350/11
9. OLG Stuttgart AHRS KZa 4350/5
10. OLG Bremen AHRS KZa 4350/6
11. OLG Köln AHRS KZa 4350/10
12. BGH NJW 1988, 763
13. BGH NJW 1959, 814; NJW 1972, 335; NJW 1984, 1397
14. BHG NJW 1988, 2946
15. BGH NJW 1985, 1399

Akürzungen

AHRS KZa	Arzthaftpflicht-Rechtsprechung-Kennzahl
BGH	Bundesgerichtshof
NJW	Neue Juristische Wochenschrift (Jahrgang und Seite)
OLG	Oberlandesgericht
VersR	Versicherungsrecht (Jahrgang und Seite)

Die ärztliche Aufklärung – Eine Gratwanderung zwischen juristischem Muß und ärztlichem Tun?

Th. Wallenfang, E. Ewig und K. Dei-Anang[1]

Die ärztliche Aufklärung – mit den Worten Karl Jaspers – zeichnet sich aus durch die Verantwortung für die Richtigkeit der Aussage und für die Wirkung auf den Kranken [7]. Die Problematik dieses sowohl von Juristen als auch von Ärzten anerkannten Anspruchs wird besonders deutlich, wenn die Krankheit selbst dem sog. „verständigen Patienten" als abstrakte juristische Konstruktion die innere Verarbeitung des Aufklärungsgesprächs versperrt.

Dies gilt typischerweise für neurochirurgische Patienten, die äußerlich unauffällig, jedoch im Gespräch eine eingeschränkte Merk- und Konzentrationsfähigkeit, häufig gepaart mit einer Stimmungslabilität, erkennen lassen. Zwei Fälle aus dem neurochirurgischen Alltag wollen die Problematik verdeutlichen. Der Patient, wie im ersten Fall (Abb. 1) mit einem bösartigen frontalen Hirntumor, wurde sich zwar seiner Krankheit bewußt, fraglich blieb jedoch, ob er Bedeutung und Tragweite der erfolgreichen Operation erfaßte, die einen Aufschub bewirkte, aber nichts an der kurzen Lebenserwartung änderte.

Nicht weniger problematisch sind Krankheitsbilder, bei denen Patienten mit rasenden Kopfschmerzen, die sie aus heiterem Himmel überfallen, von jeder operativen Maßnahme eine Erlösung erwarten. Wie im zweiten Fall (Abb. 2) stellt sich dem Arzt immer wieder die Frage, kann der Patient, durch die Subarachnoidalblutung unerwartet in einen extremen Zustand versetzt, die völlige Aufklärung verstehen bzw. verkraften.

Während die Rechtsprechung dem Wahrheitsgehalt den größeren Stellenwert einzuräumen scheint, neigt der Arzt dazu, die Aufklärung dem Krankheitsbild stärker anzupassen.

Begibt sich der Arzt in Fällen wie den oben geschilderten deshalb notweniger Weise auf eine Gratwanderung?

Nach Auffassung der Rechtsprechung muß das Aufklärungsgespräch den Patienten eine allgemeine Vorstellung von der Schwere des Eingriffes und seinen spezifischen Gefahren vermitteln. Der Arzt darf weder beschönigen noch die Folgen schlimmer darstellen als sie sind. Er muß die typischen Folgen sowohl der Ope-

[1] Neurochirurgische Klinik der Johannes Gutenberg-Universität, Langenbeckstr. 1, D–6500 Mainz 1

Advances in Neurosurgery, Vol. 18
K.-A. Bushe, M. Brock, and M. Klinger (Eds.)

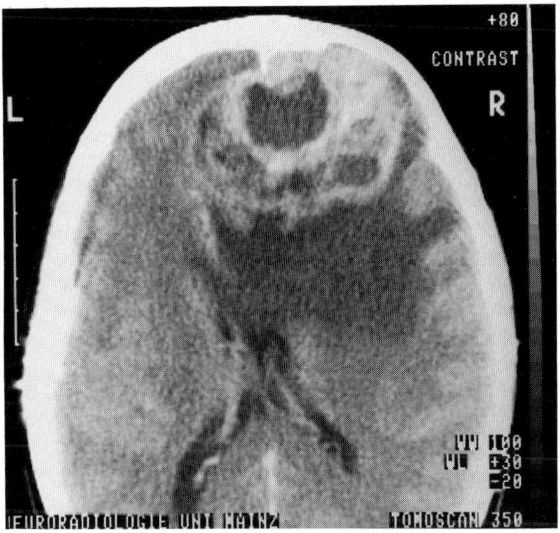

Fig. 1. Rechts frontaler Tumor mit perifokalem Hirnödem. Nach Exstirpation des Tumors – histologisch Glioblastoma multiforme – und Radiatio Hinweise für ein Rezidiv schon nach 5 Monaten. Die Aufklärungsgespräche, die bei dem hirnorganischen Zustandsbild des Patienten wiederholt durchgeführt wurden, bezogen sich nur auf die Operation

ration als auch die der Nichtbehandlung und die möglichen, wahrscheinlich nicht ganz seltenen Gefahren dem Patienten, dessen Verständnisfähigkeit außerdem zu überprüfen ist, mitteilen und das alles so schonend wie möglich [1–3, 5, 6]. Der Arzt bewegt sich dabei zwischen einem Zuwenig und einem Übermaß an Aufklärung. Beides führt zur Unwirksamkeit der Einwilligung mit der Folge, daß er sowohl zivil- als auch strafrechtlich zur Verantwortung gezogen wird [1].

Ausgangspunkt dieser Kriterien ist das durch Artikel 2 Abs. 2 im Grundgesetz verfassungsrechtlich garantierte Selbstbestimmungsrecht des Patienten. Er soll durch die Aufklärung in die Lage versetzt werden, das Für und Wider des Eingriffes gegeneinander abzuwägen. Seine Entscheidungsfreiheit wird aber nach juristischem Verständnis nur gewährleistet, wenn er alle Informationen – einschließlich der ungünstigsten Prognose – erhält, die für den Abwägungsprozeß von ausschlaggebender Bedeutung sind [1, 5].

Es erscheint äußerst fraglich, ob in der juristischen ex-post-Betrachtung nicht von einem Patienten ausgegangen wird, der auch bei Schwersterkrankungen, die immer seelisches Leid einschließen, noch in der Lage ist, das Für und Wider des Eingriffes nüchtern zu analysieren. Einem solchen Gesprächspartner begegnet man jedoch höchst selten in der klinischen Praxis. Das trifft in besonderem Maße für neurochirurgische Patienten zu, denen ein operativer Eingriff an Schädel und Hirn

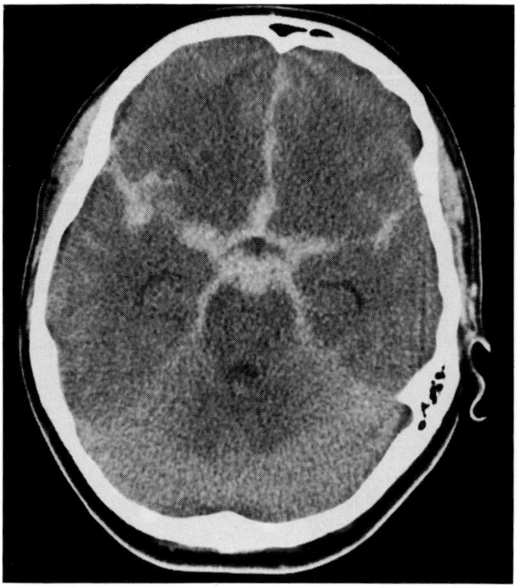

Fig. 2. Ausgedehnte Subarachnoidalblutung durch ein Trifurkationsaneurysma links. Patientin war bis auf extreme Kopfschmerzen mit Meningismus neurologisch unauffällig. Das Aufklärungsgespräch erfolgte kurz vor der Operation mit der schmerzgequälten Patientin

bevorsteht. Denn solche Eingriffe wecken beim Patienten lebensbedrohliche Empfindungen und die Furcht – in dem Bewußtsein, daß vom Gehirn alle körperlichen und geistigen Funktionen gesteuert werden – eine Wesensveränderung zu erfahren. Typisch dafür ist die nicht seltene zaghafte Frage: „Werde ich verrückt, Herr Doktor?"

Die Angst wird dadurch verstärkt, daß dem Patienten anders als bei sonstigen chirurgischen Eingriffen nur eine vage Vorstellung von der Art der operativen Maßnahmen zu vermitteln ist. Vor dem Hintergrund dieser Not und den erlebten Symptomen, wie Konzentrationsschwäche, psychische Labilität und Lähmungen, erfährt der Patient die ärztliche Aufklärung.

Die psychische Belastung, der der Patient ausgesetzt ist, wird von juristischer Seite unterschätzt. Sie kann unseres Erachtens aber die Willens- und Entscheidungsfähigkeit des Patienten soweit beeinträchtigen, daß eine wirksame Einwilligung und damit die Wahrnehmung des Selbstbestimmungsrechts [6] überhaupt in Frage gestellt ist.

Aus diesem Grund hat die Aufklärung in den eingangs geschilderten Fällen den Anforderungen der Rechtsprechung nicht genügt.

Dem Tumorpatienten wurde die in jedem Fall schlechte Prognose nicht mitgeteilt, die ihm die Ablehnung der Operation freigestellt hätte.

Bei der in ihrer Vigilanz nicht eingeschränkten Aneurysmapatientin verblieben Zweifel, ob sie unter dem Eindruck der Kopfschmerzen die Risiken der Operation erfassen konnte. Selbstverwirklichung und Freiheit der Entscheidung waren bei ihr von vorne herein eingeschränkt, da die Aufklärung sich auf die Dringlichkeit des Eingriffes beschränkte, was nach strenger juristischer Betrachtung einer Bevormundung durch den Arzt gleichkommt. Der in der Rechtsprechung anerkannte Grundsatz – je dringender der Eingriff, desto geringer die Anforderung an den Umfang der Aufklärung – half im konkreten Fall nicht weiter. Denn der Eingriff erfolgte nicht unter solchem Zeitdruck, daß die Reduktion der Information auf die Operation an sich und ihre Dringlichkeit nach Kriterium der Rechtsprechung zu rechtfertigen gewesen wäre. Andererseits bedeutete aber schon allein die Mitteilung einer keinen längeren Aufschub duldenden Operation einschließlich der Darstellung ihrer besonderen Risiken eine außerordentliche psychische Belastung.

Im Prinzip hat der BGH zwar anerkannt, daß im Ausnahmefall die Pflicht zur Aufklärung eingeschränkt sein kann. Das soll jedoch nur gelten, wenn die Aufklärung zu einer „ernsten und nicht behebbaren Gesundheitsschädigung" (BGH JZ 29, 176, 185) führen würde [2]. In der Entscheidungspraxis des BGH ist diese Lücke für den Arzt bis heute noch nicht zum Tragen gekommen. Im Zweifel räumt er dem Selbstbestimmungsrecht den Vorrang ein. Das geht soweit, daß im Interesse des Selbstbestimmungsrechts eine Verschlechterung des körperlichen Befundes und selbst eine Resignation für die verbleibende Lebenszeit (BGH JZ, 183, 302, 304) in Kauf genommen werden soll [5,6].

Diese Einstellung gegenüber dem Kranken mutet uns zu rechtspositivistisch an. Sie verunsichert den wenig erfahrenen Mediziner und verstärkt seine Neigung zur schematisierten Totalaufklärung und erhöht damit die Gefahr einer defensiven Medizin.

Vor allem aber widerspricht sie in ihrer Rigorosität der ärztlichen Fürsorgepflicht des Arztes, körperliches *und* seelisches Leid vom Kranken abzuwenden, und sie widerspricht dem von der Weltgesundheitsorganisation (WHO) definierten Begriff von Gesundheit. Denn die WHO beschreibt Gesundheit als einen Zustand völligen körperlichen, seelischen und sozialen Wohlbefindens und nicht nur das Freisein von Krankheit [4].

Es geht nicht darum, die ärztliche Aufklärungspflicht weitgehend einzuschränken, sondern gemessen an dem Gesundheitsbegriff der WHO dem seelischen Wohl des Patienten neben seiner körperlichen Integrität den ihm entsprechenden Stellenwert einzuräumen.

Nur wer den Anforderungen des BGH uneingeschränkt zu folgen vermag, begibt sich also nicht auf eine Gratwanderung.

Einen geeigneten Lösungsansatz, diesem für den Arzt unzumutbaren Dilemma zu entrinnen, sehen wir darin, dem Arzt im Einzelfall einen Entscheidungsspielraum zuzubilligen, der mit weitgehenden Dokumentationspflichten verbunden ist.

Literatur

1. Bodenburg R (1981) Entzerrung der ärztlichen Aufklärungspflicht: Grundaufklärung und Einschätzungsprärogative. NJW 12:601–605
2. Deutsch E (1980) Das therapeutische Privileg des Arztes: Nichtaufklärung zugunsten des Patienten. NJW 24:1305–1307
3. Deutsch E (1982) Neue Aufklärungsprobleme im Arztrecht. NJW 46:2585–2588
4. Deutsch E (1983) Arztrecht und Arzneimittelrecht. Springer, Berlin Heidelberg New York, S 331–332
5. Eberbach WH (1986) Die ärztliche Aufklärung unheilbar Kranker. MedR 4:180–186
6. Giesen D (1987) Zwischen Patientenwohl und Patientenwille. JZ 6:282–290
7. Jaspers K (1983) Die Idee des Arztes. In: Wahrheit und Bewährung – Philosophien für den Alltag. Pier (SP 268)

Straf- und zivilrechtliche Aspekte neurochirurgischer Tätigkeit

G. Lausberg[1]

Die zunehmende Verrechtlichung aller Lebensbereiche hat wie selbstverständlich auch zu einer Verrechtlichung der ärztlichen Tätigkeit geführt.

Unter der alarmierenden Überschrift „Ein gefährlicher Beruf" kritisierte im Jahre 1911 der Strafrechtler Kohler in ungewöhnlich scharfer Form „vorzugsweise mit dem bekannten Schlagwort von der Weltfremdheit der Richter" ein Urteil des Reichsgerichts, mit dem der Schuldspruch des Landgerichts Breslau gegen einen Gynäkologen wegen fahrlässiger Tötung bestätigt wurde. Dennoch bestand damals sicherlich kein Anlaß zur Beunruhigung der Ärzte im allgemeinen.

Ein dreiviertel Jahrhundert später zeigt jedoch eine kritische empirische Bestandsaufnahme, daß die befürchtete „Gefährdung des ärztlichen Berufs" durch das Strafrecht keine bloße Fata Morgana mehr, sondern eine – für Ärzte bittere – Realität ist, die nicht im Interesse einer optimalen, vom gegenseitigen Vertrauen zwischen Arzt und Patienten getragenen Krankenversorgung liegen kann und den Juristen nachdenklich stimmen sollte (soweit ein Zitat von Ulsenheimer 1987).

Ulsenheimer erörtert auch die Gründe für den Anstieg straf- und zivilrechtlicher Verfahren gegen Ärzte.

Es ist zum einen ein reduziertes Vertrauensverhältnis zwischen Patient und Arzt, das mehr einer geschäftsmäßigen Beziehung gewichen ist. Es ist aber auch das insbesondere von den Medien postulierte „Recht auf Gesundheit" und schließlich die Schuldsuche bei fehlendem Behandlungserfolg.

Zweitens ist heutigentags der „kritische" – sprich mißtrauische – Patient häufiger anzutreffen als dies früher der Fall war. Die Ursachen dürften wiederum in den Berichten der Medien mit sehr ausführlichen Erörterungen von Kunstfehlerprozessen, schließlich aber auch der Vorwurf einer „gnandenlosen Apparatemedizin", die den Patienten zum vermeintlich Kritischen werden läßt. Gerade beim Vorwurf der Apparatemedizin ist die Grenze zwischen Humanität und sogenannter Inhumanität eine Gratwanderung, denn die Anwendung der Apparate bleibt so lange human, wie der Patient noch zu retten ist, und wird nach Auffassung der Kritiker in dem Moment inhuman, wenn für den Patienten keine Rettungschancen

[1] Neurochirurgische Klinik, Knappschaftskrankenhaus Bochum-Langendreer, Universitätsklinik in der Schornau, D–4630 Bochum-Langendreer

Advances in Neurosurgery, Vol. 18
K.-A. Bushe, M. Brock, and M. Klinger (Eds.)

Diejenigen, die täglich in der Akuttherapie auf Intensivstationen tätig sind, wissen, daß der Übergang zwischen Hoffnung und Hoffnungslosigkeit für das weitere Leben des Patienten ein schmaler Grat ist.

Die Fortschritte in der Medizin bewirken auch bei den Patienten eine größere Erwartungshaltung. Es sind hier oft die speziellen Kenntnisse und Erfahrungen Einzelner, die gewissermaßen als medizinischer Standard aufgefaßt werden. Fries, Präsident der Deutschen Gesellschaft für Orthopädie, hat anläßlich des Jahreskongresses 1988 seiner Gesellschaft davor gewarnt, von Ärzten Unfehlbarkeit zu erwarten. Er führte aus, daß die Symptome einer Krankheit oft verschieden interpretierbar sind und die Möglichkeit der Fehlinterpretation somit in der Natur der Sache liege. Sofern nicht Fahrlässigkeit oder schuldhaftes Verhalten beim Arzt vorlägen, seien die Folgen einer ärztlichen Fehlinterpretation somit für den Patienten „schicksalhaft". Nach Fries werde dies aber „heute nicht mehr gern akzeptiert, weil es zur Regel geworden sei, für einen erlittenen Schaden einen verantwortlichen Dritten haftbar zu machen". „Dies führe zu immer mehr Prozessen gegen Ärzte und die Gerichtspraxis zeige, daß die an den Arzt gestellten Ansprüche tatsächlich im Bereich der Unfehlbarkeit lägen."

Nach Ulsenheimer werden viele Strafanzeigen erstattet, da das staatsanwaltliche Ermittlungsverfahren ebenso bequem wie kosten- und risikolos für den Zivilprozeß die erforderlichen Unterlagen und Beweise beschaffen könne.

Ein weiterer allerdings vorbeugend abwendbarer Grund für die Zunahme juristischer Forderungen gegen Ärzte dürfte im Ärger über Behandlungsbedingungen, hier insbesondere aber das oft fehlende ausführliche ärztliche Gespräch besonders bei aufgetretenen Komplikationen sein.

Schließlich ist noch Kollegenschelte oder Unachtsamkeit, oft vielleicht sogar Absicht im Gespräch mit dem Patienten über das Ergebnis der Behandlung des vorbehandelnden Arztes zu nennen.

Als letztes muß noch auf das Gewinnstreben mancher Patienten aus einem Behandlungsergebnis verwiesen werden. Hier ist der Fall eines Kindes zu nennen, bei dem eine erfolgreiche Herzoperation durchgeführt wurde. Die schriftliche Einwilligung zu dieser Operation lag aber nur von einem Elternteil vor, weil der operierende Arzt irrtümlich davon ausgegangen war, daß frühere Gespräche mit beiden Eltern über den Umfang der Operation für die Aufklärung zur Durchführung der Operation ausreichend waren. Nach erfolgreicher Operation klagte das Kind vertreten durch die Eltern auf Rechtswidrigkeit des Eingriffes, zu dem nur die Mutter die Einwilligung erteilt hatte. Das Bundesgericht führt in seiner Urteilsbegründung aus: „Erst recht durfte der Beklagte (oder operierende Arzt) nicht ohne weiteres annehmen, die Mutter des Klägers werde alsbald ihrem Ehemann den Inhalt des Gesprächs vom 5. Juni 1984 richtig und vollständig wiedergeben und dieser werde schon von sich aus tätig werden, wenn er anders als seine Ehefrau dem Eingriff nicht zustimmen wolle." Der Bundesgerichtshof führt weiter aus: „Die angefochtene Entscheidung beruht auf dem dargelegten Rechtsfehler. Sie erweist sich auch nicht deswegen schon jetzt im Ergebnis als zutreffend, weil nach den derzeitigen Erkenntnissen die Operation des Klägers am 8. Juni 1984 erfolgreich verlaufen ist. Das Berufungsgericht hat von seinem Standpunkt aus folgerichtig bisher offen gelassen, ob dem Kläger durch den rechtswidrigen Eingriff ein Schaden entstanden oder ob, was vielleicht für das Feststellungsbegehren ausreichen könnte, in Zukunft noch ein operationsbedingter Schaden zu erwarten ist. Der Kläger hat dazu entgegen der Ansicht der Revision ausreichend unter Beweisantritt vorgetragen, wie sich insbesondere aus dem Tatbestand des landgerichtlichen Urteils ergibt, mag es den ihn vertretenden Eltern auch vorrangig um eine Sanktion wegen des nach ihrer Ansicht eigenmächtigen Vorgehens des Beklagten (operierenden Arztes) gehen."

Schließlich hat der Bundesgerichtshof festgestellt: „Jeder Arzt, vor allem auch der Beklagte als Chefarzt der Abteilung einer Universitätsklinik, muß wissen, daß es der Einwilligung *beider* Elternteile eines noch nicht selbst einwilligungsfähig minderjährigen Kindes zu einem ärztlichen Eingriff bedarf." (BGH VI ZR 288/87)

Worin liegt nun die objektive medizinische Basis eines Straf- oder Zivilverfahrens gegen einen behandelnden Arzt?

Hat das alte Motto „Wo Sorgfalt ist – ist keine Schuld" noch Gültigkeit. Dieses Motto hat bezogen auf die unmittelbare ärztliche Behandlung noch Gültigkeit, schließt aber heutigentags juristische Notwendigkeiten z.B. der Dokumentation und Aufklärung ein. Nach eigener Gutachtererfahrung ist es in der Regel nicht Schlamperei, die den betreffenden Ärzten vorgeworfen werden muß, oft aber zu geringes Wissen und zu geringe Erfahrung – nicht mit einer Operationsmethode – aber mit der Handhabung möglicherweise auftretender Komplikationen. In vielen Fällen sind zu geringe juristische Kenntnisse, ohne die heutigentags ein Arzt seine Tätigkeit nicht mehr ausüben kann, die Basis für die Einleitung derartiger Verfahren. Hier ist die Garantenstellung des Arztes zu nennen, der bei der Übernahme einer Behandlung diese in vollem Umfang unter – wie es juristisch heißt – in der konkreten Situation der Behandlung gebotenen Sorgfalt durchzuführen hat, und wenn er an die Grenzen seines Wissens oder seiner Erfahrung kommt, Ärzte benachbarter Fachgebiete oder erfahrenere Ärzte des eigenen Fachgebietes zuziehen muß.

Denn die juristische Beurteilung eines „Operationserfolges" – die juristische Bezeichnung auch für einen Operationsmißerfolg – erfolgt ausschließlich unter juristischen Aspekten, wie an dem zitierten Urteil des Bundesgerichtshofes zu der Operationseinwilligung nur eines Elternteils erkennbar wird.

Juristische Relevanzen können sich ergeben in

1. Diagnostik
2. Therapie
3. Verkürzung der mittleren Verweildauer aus Kostengründen
4. Organisationsmängeln
5. Aufklärungsversäumnissen
6. Dokumentationsmängeln

In der Diagnostik spielt in unserem Fachgebiet insbesondere die nicht oder zu spät durchgeführte Computertomographie im Zusammenhang mit Schädelhirntraumen eine große Rolle. Hier sind insbesondere kleinere Krankenhäuser betroffen, die diese Untersuchungsmethode an ihrem Hause nicht zur Verfügung haben und oft aus Kostengründen aber auch aus Gründen des Transportweges diese diagnostische Maßnahme nicht routinemäßig einsetzen können.

Diese Problematik wird in folgendem Fall erkennbar:

59jähriger Mann, Schädelhirntrauma am 02.11.1983 mit primärer Bewußtlosigkeit, am folgenden Tage delirant, Transport in eine ca. 100 km entfernte Neurochirurgische Klinik zur Computertomographie. Es findet sich nur eine Kontusion. Auf Wunsch der Angehörigen Rückverlegung des Patienten in die überweisende Klinik. Sechs Tage später zweite CT-Kontrolle, unveränderter Befund. Patient unverändert delirant. Zwei Wochen später nach Angaben der Ehefrau Verschlechterung des Zustandes, der bei einer drei Tage später durchgeführten neurologischen Untersuchung nicht objektiviert werden konnte. Ein weiterer Tag später akute Einklemmungssymptomatik. Überweisung in Neurochirurgische Klinik. Dort Diagnostik und Operation eines großen sekundären Subduralhaematoms. Tödlicher Ausgang einige Monate später.

In diesem Falle erkannte der Gutachter – ein Neurologe – einen Diagnosefehler an und das Krankenhaus wurde zum Schadenersatz verurteilt.

In einem weiteren Fall wurde ein 55jähriger Patient in einem kleinen Krankenhaus nach einem Schädelhirntrauma, das er als alkoholisierter Mopedfahrer erlitten hatte, aufgenommen. Der Patient kam zu Fuß in die Ambulanz. Als Vorkrankheiten hatte er einen alkoholbedingten Leberschaden, ein Glaukom und einen alten Herzinfarkt. Der Blutalkoholwert betrug 2,09 Promille. Nach Versorgung einer Kopfplatzwunde und einer Röntgenaufnahme des Schädels, die einen Frakturverdacht ergab, wurde der Patient zwei Stunden nach Klinikaufnahme schläfrig, was als Alkoholwirkung fehlinterpretiert wurde. Eine weitere Stunde später wurde der Patient bewußtlos. Wiederum eine Stunde später wurde die Verdachtsdiagnose einer Hirnblutung gestellt. Aus organisatorischen Gründen war es nicht möglich, den Patienten zu intubieren, so daß sich der Transport in eine Spezialklinik noch um weitere 1 1/2 Stunden verzögert hat. Trotz Operation tödlicher Ausgang vier Tage später.

Wenn im vorliegenden Fall eine Computertomographie in der Klinik möglich gewesen wäre, wäre die Diagnose mit Sicherheit zu einem früheren Zeitpunkt gestellt worden.

Eine große Häufung von zivilgerichtlichen Verfahren zeigt sich in den letzten Jahren im Zusammenhang mit Bandscheibenoperationen, wobei der berechtigte oder unberechtigte Vorwurf einer falschen Höhenlokalisation im Vordergrund steht. Als Reaktion darauf wurde in einigen Kliniken eine Art Überdiagnostik durchgeführt, indem trotz eindeutigen Operationsbefundes noch eine Beweissicherung der operierten Bandscheibenhöhe durch ein Röntgenbild mit im ausgeräumten

Zwischenwirbelraum liegendem Instrument erfolgt. Manche Kliniken versuchen, die juristischen Folgen eines Höhenfehlers durch Aufklärung über die Operation in möglicher falscher Höhe zu begegnen. Dies kann nach meiner Auffassung jedoch nicht zu einer Exkulpierung führen. Mit weiterer Zuspitzung der Kostensituation im Gesundheitswesen ist es durchaus möglich, daß zukünftig Regreßansprüche wegen der krankenhausökonomisch geforderten Liegezeitverkürzung resultieren können, denn ein Arzt kann sich straf- und zivilhaftungsrechtlich *nicht* durch die Berufung auf das Wirtschaftlichkeitsprinzip entlasten. Hier wird die insbesondere für kleinere Krankenhäuser fatale Schere zwischen gebotener Sorgfalt und entstehenden Kosten offenbar. Schreiber sagt dazu: „Maßgeblich muß für die Behandlung und die Haftung die erforderliche Sorgfalt bleiben."

In einzelnen Fällen sind auch Regreßansprüche wegen zu früher Verlegung in andere Kliniken erfolgt, die nicht nur aus Kostengründen, sondern auch aus Bettenmangel auf der Intensivstation vorgenommen wurden.

Beim Organisationsverschulden kann die sogenannte „Anfängeroperation" dann juristisch Konsequenzen haben, wenn nicht wie der Bundesgerichtshof fordert „der Standard eines erfahrenen Chirurgen" bei der Operation gewährleistet sei, d.h. die Operation nicht nur von einem Anfänger durchgeführt wird, sondern auch der Assitent nicht die Erfahrung des Standards eines erfahrenen Chirurgen aufweist.

Das Organisationsverschulden wird oft aber aus einer zu geringen Personalausstattung sowohl im ärztlichen, oft auch im pflegerischen Bereich resultieren können. Dazu die weitere Erläuterung des bereits angesprochenen Falles:

Als der bewußtlose Patient um 17.30 Uhr auf die Intensivstation in der kleinen Chirurgischen Abteilung verlegt wurde, war zunächst kein Arzt da, der den Patienten untersuchen konnte, da beide diensthabenden Chirurgen operierten, die diensthabende Anästhesie-Chefärztin mit einem Notarztwagen unterwegs war und der diensthabende Anästhesist neben der chirurgischen noch eine gynäkologische Operation narkosemäßig zu betreuen hatte. Nach Beendigung der chirurgischen Operation stellten die Chirurgen um 18.15 Uhr die Verdachtsdiagnose einer Hirnblutung, waren aber nicht in der Lage, den Patienten zu intubieren. Der von der Narkose der gynäkologischen Operation zugezogene Anästhesist versuchte die Intubation ohne Relaxation, da er keine Narkoseschwester zur Verfügung hatte, die im Operationssaal stand. Die Intubation gelang nicht, und konnte schließlich erst 19.30 Uhr nach Ende der gynäkologischen Operation durchgeführt werden. Der Patient wurde intubiert und danach sofort der Spezialklinik überwiesen. Bei besserer Ausstattung von Operations- und Anästhesieabteilung wäre hier der sicherlich verlaufsbeeinträchtigende Zeitverlust von etwa 1 1/2 Stunden vermieden worden.

Ein ganz wesentlicher Aspekt für die Basis juristischer Auseinandersetzungen ist die mangelnde Aufklärung. Hier ist es wichtig, sich zu verdeutlichen, daß für den Patienten der sicherste Weg zur Realisierung eines Ersatzanspruches über die unzulängliche Aufklärung führt.

Das OLG Frankfurt hat in einem Urteil aus dem Jahre 1988 postuliert: „Läßt sich feststellen, daß der Arzt seiner dem Patienten geschuldeten Verpflichtung zur Aufklärung im konkreten Fall nicht oder nicht vollständig nachgekommen ist, so fehlt es automatisch an der notwendigen Einwilligung des Patienten. Ohne eine solche Einwilligung wird aber jeder Eingriff an sich rechtswidrig." (Bestätigt durch BGH-Beschluß AZ: VI ZR 98/88)

In solchen Fällen ist besonders auf die Beweislastumkehr zu verweisen. Der Arzt muß beweisen, daß er ausreichend aufgeklärt hat und für den Fall einer unzureichenden Aufklärung muß der Arzt den Beweis führen, daß der Patient auch unter voller Aufklärung dem Eingriff zugestimmt hätte.

Schließlich ist ein letzter ebenfalls wesentlicher Aspekt für die Basis juristischer Auseinandersetzung im Zusammenhang mit einer ärztlichen Behandlung zu nennen: die fehlende oder mangelhafte Dokumentation. Nachdem die Verpflichtung zur Dokumentation in die ärztliche Standesordnung aufgenommen wurde, ergeben sich hier oft gravierende Ansatzpunkte für einen Falschbehandlungsvorwurf. Dies gilt insbesondere im operativen Bereich, wenn etwa aus den Krankenblattunterlagen keine Hinweise auf eine Begründung des operativen Eingriffs bzw. auf die Operationsindikation niedergelegt sind. Hier ist es in jedem Falle dringend zu empfehlen, im Operationsbericht eine eingehende Begründung für die Operationsindikation voranzustellen oder diese auf einem gesonderten Blatt zu vermerken.

Zusammengefaßt besteht einerseits keine Veranlassung zum Fatalismus, obgleich der Arztberuf heutigentags mehr als zu jeder früheren Zeit ein juristisch gefahrengeneigter Beruf ist. Neben Sorgfalt und medizinischem Wissen ist ein juristisches Basiswissen die ärztliche Behandlung in ihrem gesamt Spektrum betreffend unabdingbar. Trotzdem werden sich juristische Folgerungen vielleicht nicht total vermeiden lassen. Ihnen wird aber in den meisten Fällen erfolgreich begegnet werden können, wenn Aufklärung und Dokumentation vollständig sind und eine – erstaunlicherweise oft fehlende! – sorgfältige Erläuterung des eigenen ärztlichen Handelns vorgelegt wird.

Literatur

Fries G (1989) Kongreßbericht in Medical Tribune 24.01.1989

Kohler (1911) (zitiert nach Ulsenheimer 1987)

Schreiber HL (1988) Sorgfaltspflicht und Wirtschaftlichkeit. Referat vor dem „Arbeitskreis Ärzte und Juristen" 25.11.1988, Frankfurt

Ulsenheimer K (1987) Ein gefährlicher Beruf: Strafverfahren gegen Ärzte. Der Frauenarzt 28:19–33

Erfahrungen einer „Gutachterkommission für Ärztliche Behandlungsfehler" mit Behandlungsfehlern im neurochirurgischen Fachgebiet

H.M Mehdorn, B. Hoffmann und W. Grote[1]

Einleitung

Angesichts der gerade in den letzten Jahren in der Öffentlichkeit vermehrt erhobenen Vorwürfe wegen ärztlicher Behandlungsfehler erscheint es sinnvoll, einige Erfahrungen der seit 13 Jahren bestehenden „Gutachterkommission für Ärztliche Behandlungsfehler bei der Ärztekammer Nordrhein" darzustellen. Ziel der vorliegenden Untersuchung war es insbesondere, die gegen Neurochirurgen bzw. auf neurochirurgischem Fachgebiet erhobenen Vorwürfe wegen Behandlungsfehlern retrospektiv zu analysieren, um möglichst Konsequenzen für den klinischen Alltag zu ziehen.

Die „Gutachterkommission für Ärztliche Behandlungsfehler bei der Ärztekammer Nordrhein", im folgenden abgekürzt „GAK", wurde 1975/76 durch die Ärztekammer Nordrhein gegründet. Das in den Statuten festgelegte Ziel war, „durch objektive Begutachtung ärztlichen Handelns den durch einen Behandlungsfehler Geschädigten die Durchsetzung begründeter Ansprüche und dem Arzt die Zurückweisung unbegründeter Ansprüche zu erleichtern" [1]. Während es in den ersten Jahren nach Gründung der Gutachterkommission möglich war, lange zurückliegende Behandlungen hinsichtlich eines Behandlungsfehlers überprüfen zu lassen, wurde 1981 die Frist von 5 Jahren nach Behandlung festgelegt, innerhalb derer ein Antrag gestellt werden sollte.

Methodik

Zwei unterschiedliche Ansätze wurden gewählt, um die Thematik zu untersuchen. Zum einen wurden die in der elektronischen Datenverarbeitungsanlage der GAK unter dem Stichwort „Neurochirurgie" (bezogen auf die behandelnde Klinik anonym) gespeicherten Aktenunterlagen aus den Jahren 1975/76 bis 1988 hinsichtlich der Art der angeschuldigten Behandlung sowie hinsichtlich der Entscheidung der GAK analysiert. Hierbei wurden zwar nur die bereits entschiedenen Anträge aus-

[1] Neurochirurgische Klinik, Unversitätsklinikum, Essen, Hufelandstr. 55, D–4300 Essen

Advances in Neurosurgery, Vol. 18
K.-A. Bushe, M. Brock, and M. Klinger (Eds.)

gewertet; da aber fast alle Anträge innerhalb von 2 Jahren nach Antragseingang entschieden werden, ist der Anteil der nicht berücksichtigten Anträge gering.

Zum anderen wurde angestrebt, die zahlenmäßige Bedeutung der Vorwürfe wegen Behandlungsfehlern in Bezug auf die insgesamt im Ärztekammerbereich Nordrhein tätigen neurochirurgischen Kliniken sehen zu können. Zu diesem Zweck erfolgte eine Umfrage bei den 11 Kliniken hinsichtlich der Zahl der Betten und ärztlichen Mitarbeiter, sowie der in den letzten 5 Jahren durchschnittlich ambulant und stationär behandelten Patienten; außerdem wurden, um typische Operationen auszuwählen, die Zahlen der Bandscheiben- und Hirntumoroperation abgefragt. Diese Zahlen wurden in Relation gesetzt zu den von den Kliniken mitgeteilten Zahlen der Anträge an die GAK bzw. anhängige Gerichtsverfahren bezüglich der Patienten, die in den Jahren 1984 bis 1988 behandelt wurden. Acht Kliniken sandten den ausführlichen Fragebogen zurück, der teilweise durch zusätzliche Stellungnahmen und Problemberichte ergänzt war.

Ergebnisse

In den Jahren 1976 bis 1988 gingen bei der GAK 7380 Anträge mit Vorwürfen wegen ärztlicher Behandlungsfehler ein. Hierunter fanden sich 122 Anträge mit Vorwürfen wegen neurochirurgischer Behandlungsfehler, entsprechend 1,65% aller Anträge. Die Abb. 1 zeigt die zeitliche Entwicklung der Antragseingänge: deutlich ist in den letzten 7 Jahren eine kontinuierliche Zunahme der Antragseingänge zu erkennen. Außerdem war in den Jahren 1981 bis 1986 eine deutliche Steigerung des Anteils der Vorwürfe zu verzeichnen, die durch die GAK anerkannt wurden, so daß für 32,5% der im Jahre 1986 eingegangenen und abgeschlossenen Anträge Behandlungsfehler anerkannt wurden. Erfreulicherweise liegt aber der Anteil der als berechtigt anerkannten Behandlungsfehlervorwürfe auf neurochirurgischem Fachgebiet mit 15,6% deutlich niedriger als der entsprechende Prozentsatz für das Gesamtkollektiv (22,9%).

Darüber hinaus zeigt sich, daß, bezogen auf das Jahr des Antragseinganges, die Anzahl der gegen Neurochirurgen erhobenen Vorwürfe nicht dem allgemein steigenden Trend folgt, vielmehr in den letzten Jahren weniger Behandlungsfehler vorgeworfen wurden.

Betrachtet man die zeitliche Verzögerung zwischen durchgeführter Behandlung und Eingang des Antrags auf Überprüfung eines Behandlungsfehlervorwurfes, so zeigt sich, daß 58% der Anträge innerhalb der ersten 2 Jahre nach durchgeführter Behandlung eingingen. Bezogen auf die Art der Behandlung läßt sich allerdings weiter zeigen, daß die „am Rücken" operierten Patienten später ihre Vorwürfe bringen als die an Gehirntumoren, Gefäßprozessen und anderen Krankheiten behandelten Patienten.

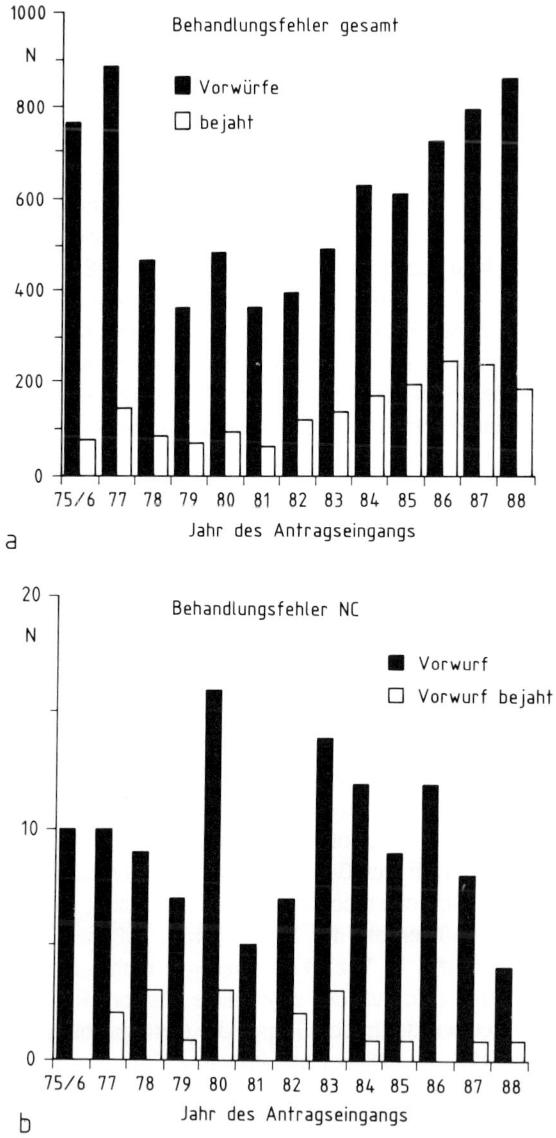

Abb. 1a,b. Anträge zum Vorwurf wegen ärztlicher Behandlungsfehler

Die im einzelnen vorgeworfenen Fehler bei neurochirurgischer Behandlung sind in der Tabelle 1 aufgeführt. Der überwiegende Teil der 122 angeschuldigten Fehler bezieht sich auf spinale Prozesse, insbesondere auf Bandscheibenoperation. Neunzehnmal wurde wegen einer postoperativen Verschlechterung der neurologischen Symptomatik der Vorwurf fehlerhafter Behandlung erhoben, wobei als schwerwiegendste Komplikation das Auftreten eines hochgradigen Kaudasyndroms angesehen werden mußte. In 7 Fällen wurde durch die GAK eine fehlerhafte Behandlung gesehen. In 6 dieser Fälle war nach Auftreten der neurologischen Symptomatik keine weiterführende Diagnostik zur Klärung ihrer Ursache erfolgt, so daß operative Therapie erst verzögert, in einem Fall erst am 13. Tag nach Auftreten des Kaudasyndroms, durchgeführt wurde. Diese Fehlergruppe ist die schwerwiegendste unter allen angeschuldigten Fehlern. Siebzehnmal beantragten Patienten wegen postoperativer Restbeschwerden, Fortbestehen der präoperativen Beschwerden oder typischen Postdiskektomiesyndroms eine Überprüfung der bei ihnen durchgeführten Behandlung. In keinem Fall konnte durch die GAK ein Behandlungsfehler gesehen werden. Ebenso ließen sich bei den wegen Rezidiven operierten Patienten keine Hinweise auf fehlerhafte Behandlung finden. Bei 3 von 4 Patienten hingegen, die angaben, zunächst an einer falschen Höhe ihres nachgewiesenen Bandscheibenvorfalls operiert worden zu sein, wurde dieser Vorwurf anerkannt, während bei dem vierten die „klinisch richtige Höhe" operiert wurde. Fachgerecht versorgte Duraverletzungen wurden nicht als fehlerhafte Behandlung angesehen, ebensowenig die als „Lagerungsschäden" im Rahmen der Knie-Ellbogen-Lagerung der Patienten in seltenen Fällen auftretenden Durchblutungsstörungen des Rückenmarkes aufgrund einer Kompression der A. spinalis anterior durch Osteophyten, die früher meist als Plexusläsionen interpretiert worden waren. Insgesamt ergibt sich hinsichtlich der neurochirurgischen Therapie spinaler Probleme, daß 14 von 76 angeschuldigten Fehlern bestätigt wurden, entsprechend 18,4% anerkannte Behandlungsfehler.

Hingegen wurden von den Vorwürfen ärztlicher Behandlungsfehler hinsichtlich anderer Erkrankungen lediglich 5 von 46, entsprechend 10,2% anerkannt. Als Fehler anerkannt wurde im Rahmen der Therapie der Hirntumoren, daß bei einem nachgewiesenen Tumor im Bereich des 4. Ventrikels eine von rechts frontal durchgeführte Biopsie des Mittelhirns erfolgte, der eigentliche Tumor (Ependymom) hingegen erst später angegangen wurde. In der Gruppe der Gefäßprobleme wurde als fehlerhafte Behandlung anerkannt, daß bei schrittweiser Embolisation eines über die A. carotis externa versorgten Gesichtsangioms nicht ausreichend Angiographie-Kontrollen zwischen den einzelnen Embolisationsschritten vorgenommen worden waren und anschließend eine Hemiparese resultierte. Von den 7 Patienten in der Gruppe der peripheren Nerven waren 3× Restbeschwerden nach Ulnarisschädigung, einmal ein Carpaltunnel mit angeschuldigt inkompletter Spaltung des Ligaments sowie ein Tarsaltunnelsyndrom, das erfolglos operiert wurde. Eine Ischiadicusschädigung bei zu fest angezogenem Tourniquet wurde als Folge einer fehlerhaften Behandlung angesehen. Auffallend war, daß 4 von 6 Patienten

Tabelle 1. Vorwürfe fehlerhafter Behandlung

Angeschuldigte Fehler	Fehlerhafte Behandlung bestätigt		
Rücken			
Bandscheiben-Operation			
Verschlechterung der Neurologie	19	7,	davon 6 zu spät diagnostiziert
Restbeschwerden	18	0	
Rezidiv	5	0	
„falsche Höhe"	5	3	
Duraverletzung	4	0	
„Lagerungsschaden"	4	0	
Spondylodiszitis	4	0	
sonstige Entzündung	4	0	
Bauch-/Gefäßverletzung	2	1,	weil nicht diagnostiziert
Ösophagus/N. recurrens	2	0	
Redon zurückgelassen	1	1	
spinale Tumoren	2	0	
Wirbelfrakturen	1	1	
Diagnostik	5	1	Gesamt 14/76 = 18,4%
Andere NC-Operationen			
Hirntumoren	11	1	
Gefäße	6	1	
periphere Nerven	7	1	
Trigeminusneuralgie	6	0	
Katheter (belassen/Sepsis)	3	2	
Stereotaxie	3	0	
sonstiges	10	0	Gesamt 5/46 = 10,2%

mit Trigeminusneuralgie, die sich an die Gutachterkommission wandten, im Auftreten einer Keratitis den Ausdruck eines Behandlungsfehlers sahen, während die beiden anderen weiterhin Schmerzen, bzw. einen Dauerschmerz angaben.

Um nun die Häufigkeit der Behandlungsfehlervorwürfe in Relation zu den erbrachten ärztlichen Leistungen zu sehen, erfolgte die Auswertung der beschriebenen Fragebögen. Bei Einschluß aller 11 Kliniken (wobei die Operationszahlen von 3 Kliniken lediglich teils auf telefonisch nachgefragten, teils schriftlich mitgeteilten Schätzwerten beruhen) kann angenommen werden, daß im Bereich der Ärztekammer Nordrhein durch Neurochirurgen pro Jahr ca. 4500 Bandscheiben- und 2400 Hirntumoroperationen vorgenommen werden. Die 76 im Laufe der letzten 13 Jahre gesammelten Behandlungsfehlervorwürfe entsprechen somit einer Häufigkeit von Behandlungsfehlervorwürfen bei Bandscheiben- und anderen Rückenoperationen von 0,13%, wobei 0,024% der durchgeführten Behandlungen durch die GAK als fehlerhaft eingestuft wurden. Hinsichtlich der Hirntumoroperationen entsprechen die 11 Anträge an die GAK 0,035% der durchgeführten Behandlungen, die bejahten Fehler entsprechen 0,0032% Häufigkeit, bezogen auf 2400 Operationen pro Jahr. Berechnet man die entsprechenden Zahlen für die Behandlung der Trigeminusneuralgie und die operative Therapie bei peripheren Nerven so kommt man auf 0,13% bzw. 0,15% Anträge/Behandlung an die GAK, wobei bei den peripheren Nerven 0,03% der Anträge/Behandlung akzeptiert wurden. Somit liegt der Anteil der von seiten der Patienten als fehlerhaft eingeschätzten Behandlungen für die Bandscheibenoperationen, die Behandlungen wegen Trigeminusneuralgie und Operationen an peripheren Nerven im gleichen Bereich; deutlich abgehoben hiervon sind hingegen die Anträge in der Gruppe der Hirntumoren, insbesondere wenn man den Anteil der bereits als fehlerhaft anerkannten Behandlungen betrachtet, der um den Faktor 10 unter dem für die Bandscheibenoperation liegt.

Betrachtet man nun abschließend die von den 8 Kliniken angegebenen Zahlen, so zeigt sich, daß bei 3100 jährlich durchgeführten Bandscheibenoperationen und 2100 jährlich durchgeführten Hirntumor-Operationen in den Jahren 1984 bis 1988 21 Anträge an die Gutachter-Kommission gestellt wurden, wobei lediglich ein Fehler anerkannt wurde. Bei 5 noch schwebenden Verfahren wurden 15 Vorwürfe abgelehnt. Sieben der 21 Verfahren wurden von den Patienten bzw. ihren Rechtsanwälten vor Gericht weiterverfolgt, sämtliche wurden als schwebend angegeben. Bei 2100 jährlich durchgeführten Hirntumoroperationen wurden 4 Anträge an die GAK gestellt, sämtliche wurden als Fehler abgelehnt, 2 Patienten verfolgten ihre Interessen weiter vor Gericht.

Diskussion und Zusammenfassung

Die Tätigkeit der im Bereich der Ärztekammer Nordrhein tätigen neurochirurgischen Kliniken ist im Untersuchungszeitraum einer Vielzahl von Vorwürfen wegen fehlerhafter ärztlicher Behandlung ausgesetzt worden. Erfreulicherweise lag der Anteil der Vorwürfe, bei denen objektiv ein Behandlungsfehler anerkannt werden mußte, unter dem entsprechenden Anteil der insgesamt gestellten 7380 Anträge. Als wesentlichster Fehler erwies sich, wenn keine sofortigen diagnostischen und therapeutischen Konsequenzen aus einem in direktem zeitlichen Zusammenhang nach einer Bandscheibenoperation aufgetretenen Defizit bis hin zum Kaudasyndrom gezogen wurden, sondern entsprechende Maßnahmen erst mit Verzögerung einsetzten und daher ein deutliches neurologisches Defizit verblieb. Positiv war auch hier, daß in den letzten 5 Jahren kein derartiges Versäumnis mehr zu verzeichnen war.

Der Anteil der Vorwürfe wegen eines vom Patienten als unbefriedigend empfundenen Verlaufs nach Bandscheibenoperation ist wohl auch deswegen so hoch, weil die Erwartungshaltung in die operative Therapie oft zu hoch ist; dies wird insbesondere deutlich im Vergleich zum Anteil der Vorwürfe bei Behandlung wegen

Hirntumoren und gilt in gleichem Maß bei der Behandlung peripherer Nerven und der Trigeminusneuralgie.

Die Ergebnisse der Umfrage bei den im Kammerbereich Nordrhein tätigen neurochirurgischen Kliniken scheint zu bestätigen, daß das von der GAK in ihren Statuten gesetzte Ziel einer schnellen und unbürokratischen Aufklärung bei Vorwurf ärztlicher Behandlungsfehler erreicht wird, da ungefähr die Hälfte der Vorwürfe durch die GAK geklärt werden konnten und nur ein Teil durch gerichtliche Verfahren weiterer Entscheidung zugeführt wurde. Die gegenüber gerichtlichen Auseinandersetzungen schnellere Abwicklung dürfte wesentlich dazu beitragen, da auch bei Einschaltung der Gesamtkommission, d.h. der „nächsten Instanz", die meisten Anträge innerhalb von 2 Jahren beschieden werden können.

Als Konsequenz aus den Erfahrungen der GAK kann zusammenfassend gesagt werden, daß genaue Indikationsstellung und Aufklärung, sorgfältige Durchführung und Dokumentation des operativen Eingriffs und gut durchgeführte und gut dokumentierte Nachsorge es ermöglichen, dem Vorwurf ärztlicher Behandlungsfehler zu begegnen. Die hierfür aufgewandte Zeit wird eingespart durch ausbleibende Gerichtsverfahren.

Dank. Herrn Schumacher, Geschäftsführer der Gutachterkommission für Ärztliche Behandlungsfehler bei der Ärztekammer Nordrhein sei für seine Hilfe und Anregungen gedankt. Den Kliniken, die die ausgesandten Fragebögen so sorgfältig und offen beantwortet haben, sei an dieser Stelle herzlich gedankt: Neurochirurgische Klinik der RWTH Aachen; der Universität Düsseldorf; der Städt. Krankenanstalten Duisburg; der Evangelischen Krankenanstalten Duisburg-Nord; der Universität Essen; der Universität Köln; der Städt. Krankenanstalten Köln-Merheim; und des Bethesda-Krankenhauses Wuppertal.

Literatur

1. Statuten der Gutachterkommission für Ärztliche Behandlungsfehler bei der Ärztekammer Nordrhein. Rhein Ärztebl 4/81, S 94

Sorgfaltsmangel oder typische Komplikation – Auswertung der Begutachtungen für die Schlichtungsstelle für Arzthaftpflichtfragen unter besonderer Berücksichtigung der Bandscheibenoperation

D. Stolke, V. Seifert und H. Dietz[1]

Einleitung

Einerseits möchte der heutige Patient aufgeklärt und informiert sein. Er möchte als aufgeklärter Patient, obwohl er Laie ist, als Partner des Arztes gelten.

Andererseits hat die Ärzteschaft durch Erfolgsmeldungen und falsch genutzte und falsch verstandene Plublicity den Patienten und dem staunenden Publikum klargemacht, daß es in der Medizin keine Probleme mehr gäbe, daß sozusagen eine „Allmachbarkeit" vorläge, daß es nach der Behandlung durch den Arzt besser sei als je vorher. Diese falschen, diese geweckten und nicht zu erfüllenden Hoffnungen führen dazu, daß der Patient nach der Behandlung häufig unzufrieden ist, weil er eben nicht besser geworden ist als vorher, sondern nur gebessert ist.

Diese ungesunde Allianz zwischen dem sog. aufgeklärten Patienten und dem sog. allmächtigen Arzt hat zu einer Zunahme der Regreßansprüche geführt.

Anliegen der Schlichtungsstelle

Um dieser Flut von Haftpflichtansprüchen gerecht zu werden und (a) um eine außergerichtliche Regelung zu ermöglichen, (b) sowie ein nachprüfbares Gutachten zu erreichen, (c) zu prüfen, ob die Ansprüche begründet sind, (d) um dem Antragsteller dies kostenlos zu ermöglichen und (e) den Arzt vor unbegründeten Prozessen und Anzeigen zu schützen, ist in Niedersachsen am 1.11.76 die Schlichtungsstelle der 5 norddeutschen Ärztekammern mit Sitz in Hannover eingerichtet worden [1].

[1] Neurochirurgische Klinik, Medizinische Hochschule Hannover, D–3000 Hannover 61

Advances in Neurosurgery, Vol. 18
K.-A. Bushe, M. Brock, and M. Klinger (Eds.)

Tabelle 1. Gründe für die Antragstellung (BSV-Patienten)

	Anzahl
Discitis	5
Caudalähmung pOP	4
EDH pOP mit Daudalähmung	3
Perforation der abdom. Gefäße	2
kons. Behandlung	2
pers. Beschwerden pOP	5
Duraleck	1
Caudalähmung präOP verkannt	1
Wurzelschädigung pOP	1
	24

Anzahl der Anträge

Im Jahr 1986 haben 1556 Anträge die Schlichtungsstelle erreicht, im Jahr 1987 1357 [3]. Betreffend das neurochirurgische Fachgebiet war der Anteil der Anträge gering, 2,6% bzw. 2,75%. Ein ähnlich günstiges Bild ergibt sich, wenn wir die Regreßansprüche der Gesamtzahl der Patienten der Medizinischen Hochschule Hannover mit denen der Neurochirurgischen Klinik vergleichen. Unter ca. 35 000 stationären und ca. 95 000 ambulanten Patienten sind 43 bzw. 56 Anträge auf Regreß eingereicht worden. Die ca. 3 000 stationären und 4 000 ambulanten Patienten pro Jahr der Neurochirurgischen Klinik stellen in den beiden Jahren je 4 Anträge.

Gründe für die Antragstellung

Wenn die 35 Gutachten, die in diesem Zeitraum ausgearbeitet worden sind, aufgeschlüsselt werden, so treten folgende Antragsgründe zutage (Tabelle 1): 24mal handelt es sich um Antragssteller, die wegen einer, wie sie meinen, mißlungenen lumbalen oder cervikalen Bandscheibenoperation Regreß fordern, 4 Antragssteller fordern nach einer Schädeloperation (3mal nach einer Subarachnoidalblutung und Aneurysmaoperation und einmal nach einer Janetta-Operation) Entschädigung. In 3 Fällen haben Patienten nach einer Carpaltunneloperation bzw. nach einer Akzessoriusschädigung nach Lymphknotenexstirpation den Antrag gestellt. In je einem Fall geht es um Schäden nach einer DREZ-Lesion, um einen epiduralen Abszeß nach Myelographie, ein Geburtstrauma und eine postoperative Armvenenthrombose.

Die Bandscheibenpatienten hatten folgende Gründe für die Antragstellung: Bei 5 Patienten war postoperativ eine Discitis intervertebralis aufgetreten, bei 7 eine Cau-

Tabelle 2. Sorgfaltsmangel anerkannt (BSV-Patienten)

1.	Cauda – mangelnde Dokumentation (7 Blutkonserven ohne Op-Bericht – Begründung)
2.	Liquorleck – Redondrainage – Hirninfarkt
3. + 4.	enger Spinalkanal – nur Flavektomie – pOP Cauda
5.	Cauda – falsche Höhe – Blutung
6.	Cauda pro-op – kons. Behandlung
7.	Cauda – EDH – 4 Tage bis zur Revision
zusätzlich bei typischer Komplikation mangelnde Aufklärung festgestellt 2×	

dalähmung, bei 2 Patienten war es zu einer Perforation der abdominellen Gefäße gekommen, einmal kam es zu einer Liquorfistel und zwei Patienten stellten den Antrag, weil sie „nur" konservativ behandelt worden waren.

Gründe für die Anerkennung des Anspruchs

Von den 35 bearbeiteten Anträgen wurde in 9 Fällen ein Sorgfaltsmangel bejaht (Tabelle 2) und in 26 Fällen verneint. Die Bejahung des Antrages hatte in operativen Ursachen, in der Dokumentation und in der Aufklärung ihre Begründung. Welche Sorgfaltsmängel sind nun anerkannt worden:

Natürlich muß ein Sorgfaltsmangel anerkannt werden, wenn von einer komplikationslosen Operation im Operationsbericht die Rede ist, der Patient hinterher eine Caudalähmung hat und im Narkoseprotokoll 7 transfundierte Blutkonserven protokolliert sind und darüberhinaus die Operation mehrere Stunden gedauert hat.

Sobald die falsche Höhe operiert worden ist, muß selbstverständlich ein Sorgfaltsmangel anerkannt werden, wobei besonders gravierend ins Gewicht fällt, wenn dann auch noch eine Caudalähmung postoperativ auftritt.

Bei 2 nicht die Bandscheibenoperationen betreffenden Fällen mußte ein Sorgfaltsmangel anerkannt werden, weil eine suffiziente Aufklärung nicht dokumentiert war und der Patient behauptete, nicht über die erheblichen postoperativ erlittenen Komplikationen aufgeklärt worden zu sein. Dieses betrifft in den hier vorliegenden Fällen einmal eine Akzessoriusparese nach Lymphknotenbiopsie und eine DREZ-Lesion mit konsekutiver Beinparese und Hemikauda.

Konsequenzen [1–3]

Welche Konsequenzen sind nun aus den Anträgen zu ziehen?

Als Gutachter ist man immer wieder erschüttert, wie wenig in den Krankenunterlagen dokumentiert ist, wie lückenhaft die Aufklärung ist, aber auch wie unsinnig sie sein können, wenn darin schon auf die Operation in der falschen Höhe hingewiesen wird, da ein solches operatives Fehlverhalten selbstverständlich nicht durch die Aufklärung gedeckt werden kann.

Als weitere Konsequenz aus den Anträgen an die Schlichtungsstelle muß gezogen werden, daß die Indikation zur Operation enger zu stellen ist und daß die Operation tatsächlich sauber und korrekt durchgeführt wird. Besteht zum Beispiel klinisch, laut vorliegender Dokumentation, der Verdacht auf einen engen Spinalkanal und alle radiologischen Untersuchungen (CT, Myelographie) bestätigen diese Diagnose und im Operationsbericht wird dann nur eine Flavektomie durchgeführt und postoperativ ist ein schweres Caudasyndrom mit Beinparesen und Sphinkterstörungen aufgetreten, dann kann hier nicht von einer korrekten Indikation und sauberen operativen Ausführung gesprochen werden.

Ebenso sollte sich kein Operateur aus falsch verstandenem Mitleid zu einer Operation bewegen lassen. Insbesondere, wenn bereits in allen präoperativen Arztbriefen eine Indikation zur Operation verneint wird und zum Ausdruck gebracht wird, daß der Patient psychisch erheblich überlagert sei. Auch dann kann der Gutachter nicht von einer indizierten Operation sprechen, besonders wenn auch noch alle radiologischen Zusatzuntersuchungen keinen Hinweis auf einen Bandscheibenvorfall zeigen.

Als besonders wichtig, jedoch, wäre zu nennen, daß wir Operateure uns unserer Fähigkeiten und Möglichkeiten bewußt bleiben und das in angemessener Bescheidenheit unseren Patienten im Aufklärungsgespräch vermitteln.

Literatur

1. Berner W, Trostdorf E, Vogel R (1981) Schlichtung in Arzthaftpflichtfragen. Dtsch Ärztebl 78:183–191
2. Hirthammer C (1989) Erfahrungen mit Gutachterkommissionsverfahren bei Arzt-Haftpflichtansprüchen. Versicherungsmedizin 2:64–69
3. Löblich HJ (1989) Ärztliche Behandlungsfehler aus der Sicht der Schlichtungsstelle. Arzt u. Krhs 4:108–110

Zur Begutachtung von traumatischen Hirnschlagaderverletzungen

H. Wassmann und L. Solymosi[1]

Einleitung

Ein gesunder 34jähriger Polizeibeamter verunfallte mit seinem Motorrad während des Dienstes als ein Taxifahrer die Vorfahrt mißachtet. Er war nicht bewußtlos und wurde ambulant behandelt. Es wurde diagnostiziert: eine Commotio cerebri, Schnittverletzungen der linken Schädelhälfte und ein Schleudertrauma der Halswirbelsäule.

Nach zwei Monaten wird er wieder für voll dienstfähig gehalten, wobei die anfänglich starken Nackenschmerzen in leichter Form fortbestehen. Nach 3 Jahren treten linkshirnige ischämische transitorische Attacken auf und nach einem weiteren Jahr ein kompletter Schlaganfall mit Aphasie und Hemiplegie rechts, wovon sich der Patient nur gering erholt und an dessen mittelbaren Folgen er 10 Jahre nach dem erlittenen Unfall stirbt.

Die Sektion und neuropathologische Untersuchung ergab eine ausgedehnte Erweichung der linken Großhirnhemisphäre bei Thrombosierung der linken A. carotis interna bis in die gleichseitige A. cerebri media hinein. Die A. carotis interna zeigte im Siphonbereich einen unruhigen Verlauf der Muskelfasern, kleine Narbenfelder, vernarbte Intimazellproliferationen und Ablagerungen von Eisenpigment als Zeichen einer alten stattgefundenen Blutung in die Gefäßwand, also insgesamt die Zeichen einer traumatischen Wandschädigung. Von diesem Gebiet aus hat sich ein Thrombus bis in die A. cerebri media hinein fortentwickelt bis es zu einer Apposition von frischem thrombotischen Material gekommen ist, welches das Gefäßlumen der A. cerebri media verschloß und den bis dahin bestehenden Kollateralkreislauf unterbrach.

Andere Ursachen für die Thrombose wie Arteriosklerose oder entzündliche Gefäßerkrankungen wurden ausgeschlossen.

Die Problematik bei der Beurteilung der Zusammenhangsfrage zwischen Trauma und erlittenem Schlaganfall mit Todesfolge bei einerseits symptomfreiem Intervall von mehr als 3 Jahren und andererseits eindeutigem pathologischen Befund ver-

[1] Neurochirurgische Universitätsklinik, Sigmund-Freud-Str. 25, D–5300 Bonn

Advances in Neurosurgery, Vol. 18
K.-A. Bushe, M. Brock, and M. Klinger (Eds.)

Tabelle 1. Liste der Patienten, bei denen obstruktive Veränderungen eines hirnversorgenden Gefäßes mit einem zuvor erlittenen kraniozervikalen Trauma in Verbindung gebracht wurden

Fall – Alter Geschlecht	Seite	Trauma	Neurolog. Status	Intervall	Angio.	Therapie	GOS
1 – 40 – w	R	Verkehr	IV	2 h	ICA	Konserv.	IV
2 – 33 – m	R + L	Verkehr	IV	7 Tage	VA	Fib. lys.	III
3 – 20 – m	L	Verkehr	IV	2 h	ICA	Konserv.	III
4 – 43 – w	L	Sport	II	20 min	ICA	OA++	I
5 – 45 – m	L	Sport	III	6 h	ICA	OA++	I
6 – 40 – w	R	Wohnung	II	3 Tage	ICA	Lig.	I
7 – 34 – m	R	Verkehr	II	7 Tage	ICA	Lig.	I
8 – 36 – w	L	Wohnung	II	2 Tage	ICA	Lig. + EIAB	I

Neurologischer Status: II = transitorisch ischämische Attacke; III = prolongiertes ischämisches neurologisches Defizit, IV = kompletter Schlaganfall. GOS = Glasgow Outcome Skala: I = gute Erholung, II = mäßige Behinderung, III = schwere Behinderung, IV = vegetativer Status. ICA = A. carotis interna, VA = A. vertebralis, OA++ = erfolgreiche Durchführung einer offenen Angioplastie, Lig. = Karotisligatur, EIAB = extra-intrakranielle Bypass-Operation

anlaßte uns, ähnliche zu begutachtende oder zu behandelnde Fälle der letzten Jahre zusammenzustellen.

Fallanalysen

In diesem Zeitraum führten weitere 8 Patienten unserer Klinik (4 weibl., 4 männl.) mit einem Durchschnittsalter von 36,4 Jahren, die Obstruktion eines hirnversorgenden Gefäßes mit neurologischen Ausfällen auf ein Trauma zurück (Tabelle 1).

In 3 Fällen handelte es sich um Patienten, die ein schweres Schädelhirntrauma erlitten hatten, bis dahin gesund waren und bei denen die neurologische Ausfälle in einem unmittelbaren zeitlichen Intervall von 20 Minuten bis 48 Stunden nach dem Trauma auftraten, so daß in der gutachterlichen Beurteilung das Trauma mit überwiegender Wahrscheinlichkeit als Ursache der Hirngefäßdissektion angesehen wurde.

So wurde der PKW eines 20jährigen Patienten (Fall 3) auf einer Dienstfahrt wegen überhöhter Geschwindigkeit aus der Kurve geschleudert.

Hierbei erlitt der Patient eine rechtshirnig betonte Contusio cerebri und 2 Stunden nach dem Unfall eine Hemiplegie links als Folge einer Dissektion mit Verschlusses der A. cerebri media im Ursprungsgebiet (Abb. 1). Die Therapie erfolgte konservativ, worunter sich die Halbseitensymptomatik nur gering besserte.

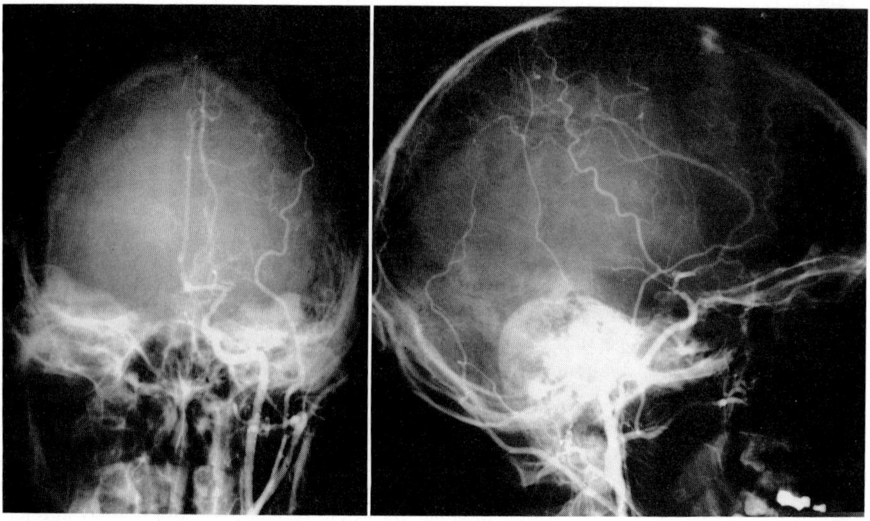

Abb. 1. Die Angiographie bei Fall 3 nach erlittenem zerviko-kranialem Trauma zeigt eine Dissektion der A. carotis interna mit Verschluß der A. cerebri media in ihrem Ursprungsbereich

Problematischer in der Beurteilung ist der Verlauf bei einem 33jährigen Patienten (Fall 2), der einen Auffahrunfall erlitt und 7 Tage abgesehen von Kopfschmerzen symptomlos blieb bis es zum Auftreten der Symptomatik eines Basilarisverschlusses kam. Hier fanden sich angiographisch die Zeichen einer Dissektion der Vertebralarterien mit Verschluß der A. basilaris (Abb. 2), so daß unverzüglich in unserer Klinik eine Fibrinolyse mit Urokinase (30 000 IE) und anschließend eine Liqueminisierung mit 800 IE/h durchgeführt wurde, worauf es zu einer inkompletten Rückbildung der schweren Ausfälle kam.

Auch hier haben wir bei fehlenden Zeichen einer andersartigen Gefäßerkrankung den Zusammenhang mit dem Trauma als mit überwiegender Wahrscheinlichkeit gegeben angesehen.

In den übrigen 4 Fällen wurde ein sogenanntes Bagatelltrauma wie z.B. Sprung ins Schwimmbecken, Sportverletzungen mit leichten Prellungen oder ein leichter Autounfall mit der Hirngefäßdissektion in Verbindung gebracht, wobei das zeitliche Intervall in 3 Fällen mehrere Tage betrug, so daß in der Beurteilung das Trauma mit überwiegender Wahrscheinlichkeit im Sinne einer Gelegenheitsursache gedeutet wurde.

Als Beispiel sei über einen 45jährigen Soldaten berichtet (Fall 5), der bei einem Fußballspiel zusammenstieß und 6 Stunden später eine linkshirnige TIA erlitt. Da sich trotz Antikoagulation die neurologische Symptomatik im Sinne eines PRIND wiederholte entschlossen wir uns zur Durchführung einer offenen Angioplastie der

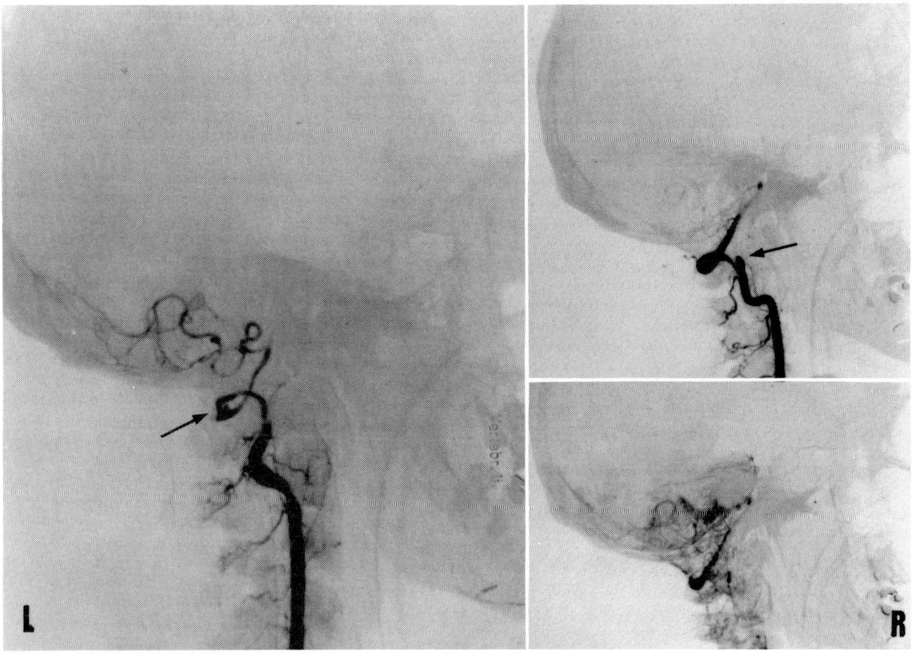

Abb. 2. Die selektive Angiographie beider Vertebralarterien bei Fall 2 zeigt Dissektionen im Gebiet der Atlasschlinge mit einem sackförmigen Aneurysma (*Pfeile*) und einem kompletten Verschluß der A. basilaris an der Spitze. Die linke A. vertebralis versorgt alleine die A. cerebelli inferior posterior

langstreckig dissezierten linken A. carotis interna. Nach gelungener Rekanalisation (Abb. 3) kam es zu einer kompletten Erholung des Patienten.

Ein ursächlicher Zusammenhang mit dem leichten Trauma wurde nicht mit hinreichender Sicherheit gesehen, zumal es sich im Bifurkationsbereich ausgedehnte atheromatöse Plaques fanden.

Therapie

Um ein systematisches Vorgehen zu erreichen, halten wir derzeit folgendes Therapieschema für sinnvoll:

Alle Patienten mit einer traumatischen Dissektion einer hirnversorgenden Arterie werden unverzüglich nach computertomographischer und angiographischer Diagnosestellung mit *Heparin antikoaguliert*. Bei Dissektionen im vertebrobasilären

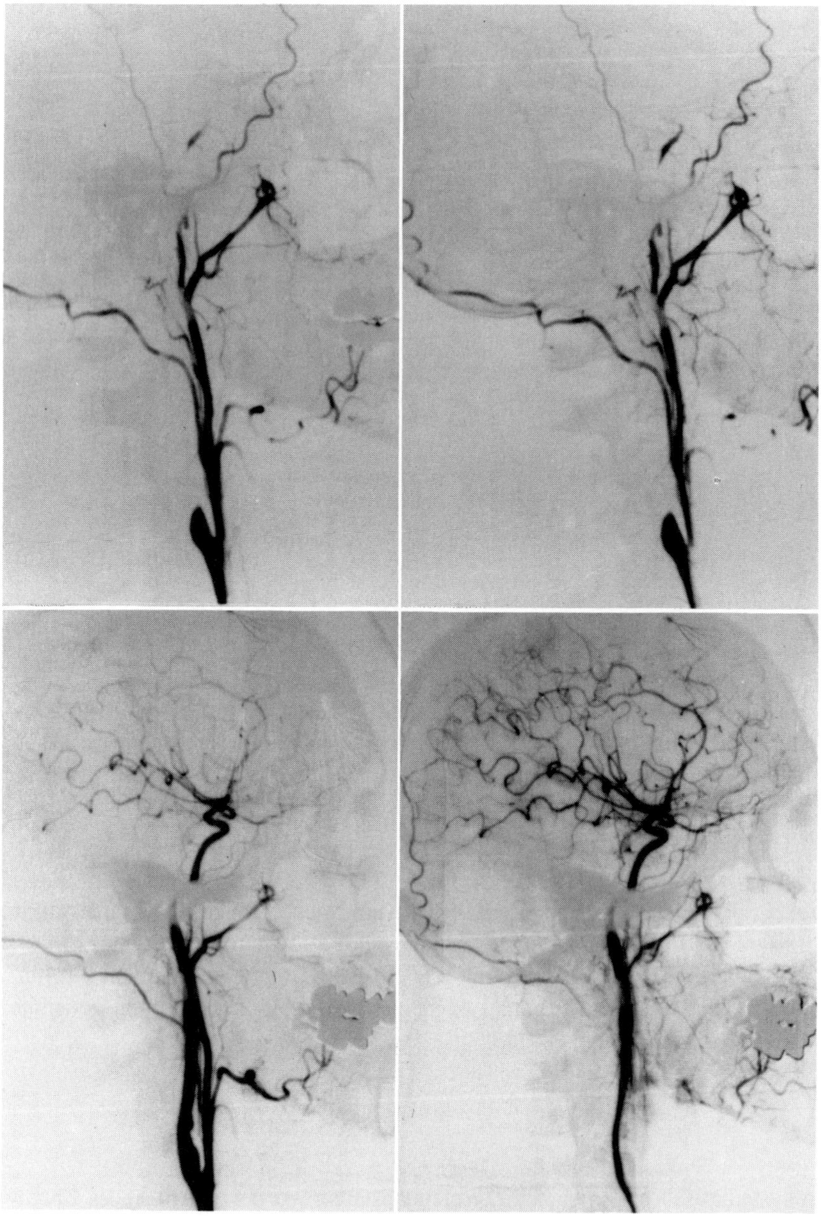

Abb. 3. Die Hirngefäßdarstellung bei Fall 5 zeigt eine ausgedehnte traumatische Dissektion der linken A. carotis interna vor (*oben*) und nach (*unten*) Durchführung einer offenen Angioplastie

Kreislauf erfolgt zunächst eine *lokale Fibrinolyse* des Embolus innerhalb der ersten 6 Stunden nach Beginn der Symptomatik.

Zeigt der Patient die therapieresistente Symptomatik einer zerebralen Ischämie im Stadium II oder III, sei es embolisch oder hämodynamisch bedingt, so schlagen wir je nach angiographischem Bild die elektive Durchführung einer *offenen Angioplastie* oder die *Karotisligatur* in Abhängigkeit von der zerebrovaskulären Reservekapazität mit gleichzeitiger extra-intrakranieller arterieller Bypass(EIAB)-Operation vor.

Unter Beachtung unseres Therapieschemas führten wir bei 2 der 4 Fällen mit rezidivierenden transitorisch ischämischen Attacken eine offene Angioplastie durch, die zu einer vollständigen Rekanalisation der Karotis führte [6]. In 3 weiteren Fällen wurde eine Kartosiligatur durchgeführt, einmal in Kombination mit einer EIAB-Operation bei unzureichender zerebrovaskulärer Reserve.

Alle diese Patienten zeigten eine komplette neurologische Erholung ohne erneute TIA's.

Begutachtung

Für die Anerkennung einer überwiegenden Wahrscheinlichkeit der Kausalität des Traumas für die Hirngefäßdissektion sehen wir derzeit folgende Punkte als wesentlich an:

1. ein stattgefundenes adäquates kranio-zervikales Trauma,
2. einen typischen angiographischen Befund,
3. das Fehlen einer systemischen Gefäßerkrankung und
4. ein plausibler zeitlicher Zusammenhang zwischen Trauma und dem Auftreten der neurologischen Symptomatik.

Diskussion

Schon der Punkt 4 unserer Begutachtungskriterien zeigt die Problematik, da hier vermutlich ein Intervall von Minuten bis zu Jahren vorkommen kann, wenn wir uns an den ersten Beispielfall erinnern, der nach einem freien Intervall von mehr als 3 Jahren erst durch die differenzierte pathologische Untersuchung als Traumafolge anzuerkennen war.

Abgesehen von Peters [4], der über ein 25jähriges freies Intervall berichtet, wird dies von Krauland [3] mit einem Zeitraum bis zu mehreren Monaten angegeben. Er stellt 9 ausführlich untersuchte Patienten vor, bei denen an dem ursächlichen Zusammenhang der Thrombose einer großen Hirnschlagader mit einem stumpfen Schädelhirntrauma nicht zu zweifeln ist.

Um eine Hirngefäßdissektion überhaupt als traumatisch bedingt anzuerkennen und von der spontanen Dissektion zu unterscheiden, die durch fibromuskuläre Dysplasie oder systischer Medianekrose bedingt sein kann, ist das adäquate Trauma eine unbedingte Voraussetzung, welches z.B. in der Lage ist durch Dehnung oder Zerrung der Arterie die Ruptur der Intima oder Media auszulösen und ein intramurales Hämatom mit Entwicklung einer Thrombose zu verursachen [2]. Hierbei erinnern wir uns, daß die früher durchgeführte Direktpunktion der A. carotis meistens folgenlos blieb. Die traumatisch bedingte Obstruktion einer Hirnarterie wurde bisher mit einem Anteil von 2%–3% bei kranio-zervikalen Traumen als eine relativ seltene Begleitkomplikation dargestellt.

Steiger [5] beobachtete in einer Periode von 3 Jahren 12 traumatische und 8 spontane Dissektionen der A. carotis. Bollinger [1] beschrieb 1891 eine Form der Spätapoplexie, die erst Stunden bis Monate nach einem Trauma auftrat, wobei er eine verzögert aufgetretene Blutung vermutete. Denkbar ist es jedoch auch, daß in einigen dieser Fälle wie oben beschrieben eine traumatische Hirngefäßdissektion die Ursache der Symptomatik war.

Die modernen wenig invasiven diagnostischen Möglichkeiten wie transkranielle Doppler-Sonographie, intravenöse DSA ermöglichen eine unmittelbare Untersuchung jedes Patienten mit Verdacht auf traumatische Hirngefäßdissektion, so daß diese Komplikation möglicherweise in der Zukunft häufiger diagnostiziert werden wird und die Zusammenhangsfrage besser zu beurteilen sein wird. Hierbei soll unser systematisches Therapieschema zeigen, ob die Mortalitätsrate der traumatischen Karotisdissektion von derzeit 40% gesenkt werden kann.

Literatur

1. Bollinger O (1891) Über traumatische Spät-Apoplexie. Rudolf Virchow Festschrift Bd. 2. v. August Hirschwald, Berlin, S 459–470
2. Födisch HJ, Kloss K (1966) Thrombotische Verschlüsse im Stromgebiet der Arteria carotis nach stumpfen Schädel-Hals-Traumen. Hefte zur Unfallheilkunde 88. Springer, Berlin Heidelberg New York, S 1–48
3. Krauland W (1982) Verletzungen der intrakraniellen Schlagadern. Springer, Berlin Heidelberg New York
4. Peters G (1959) Die Veränderungen an Gehirn und Hirnhäuten bei chronischen traumatischen Störungen. Verh Dtsch Ges Pathol 43:103
5. Steiger HJ (1988) Zur Behandlung der traumatischen Karotisdissektion. Neurochirurgia 31:128–133
6. Wassmann H, Solymosi L (1985) Revaskularisation langstreckig stenosierter oder nahezu verschlossener Karotiden. Neurochirurgia 28:131–133

Kann ein Schädelhirntrauma ein Akustikusneurinom auslösen? Rechtsmedizinische Überlegungen anhand eines Falles

C. Plangger, K. Twerdy, V. Grunert und H. Kostron[1]

Der 20jährige Patient J.M. wurde mit den klinischen Zeichen eines erhöhten intrakraniellen Druckes der Universitätsklinik für Neurochirurgie Innsbruck zugewiesen. Ein Schädel-Nativröntgen, das einen Wolkenschädel, Sprengung der Schädelnähte und eine Erosion der Sella gezeigt hatte (Abb. 1 u. 2), hatte zu einer computertomographischen Abklärung und zur Entdeckung eines Neurinoms im rechten Kleinhirnbrückenwinkel geführt (Abb. 3 u. 4). Dieses verlegte den Aquaeductus Sylvii und führte so zu einem Verschlußhydrocephalus. Da der Patient 7 Jahre vorher ein Schädelhirntrauma mit Fraktur der Schädelbasis erlitten hatte, war eine rechtsseitige Facialisparese und Taubheit am rechten Ohr nicht beachtet worden. Der Patient hatte nämlich erklärt, er könne seit dem Unfall nichts mehr hören und hatte sogar auf Schadenersatz geklagt.

Ein Jahr nach Exstirpation des Neurinoms im rechten Kleinhirnbrückenwinkel entwickelte der Patient eine Rhinoliquorrhoe. Die chirurgische Exploration der vorderen Schädelgrube zeigte einen knöchernen Defekt in der Lamina cribrosa mit Zerreißung der Dura.

In einem 2. Prozeß gegen den Unfallgegner, wurde die Frage aufgeworfen, ob das Trauma mit der Schädelfraktur für die Entwicklung des Neurinoms im Kleinhirnbrückenwinkel und die Rhinoliquorrhoe verantwortlich gemacht werden kann. Der Tumor, der die Taubheit am rechten Ohr, die rechtsseitige Facialisparese und die Ataxie verursachte, steht sicherlich in keinem kausalen Zusammenhang mit dem Trauma. Jedoch wurde wegen der Vorgeschichte einer Schädelbasisfraktur mit anschließendem Hörverlust und Facialisparese, der Tumor im Kleinhirnbrückenwinkel lange übersehen. Die Rhinoliquorrhoe mit der Fraktur der Lamina cribrosa, der Durazerreißung und dem Hirnprolaps in den knöchernen Defekt sind Folge eines Traumas. Auch nach so langer Zeit, kann diese Rhinoliquorrhoe durch den 8 Jahre vorher erfolgten Unfall verursacht sein. Der Patient könnte jedoch auch später auf Grund seiner ataktischen Gangunsicherheit gefallen sein.

---

[1] Universitätsklinik für Neurochirurgie, Anichstr. 35, A–6020 Innsbruck

Advances in Neurosurgery, Vol. 18
K.-A. Bushe, M. Brock, and M. Klinger (Eds.)

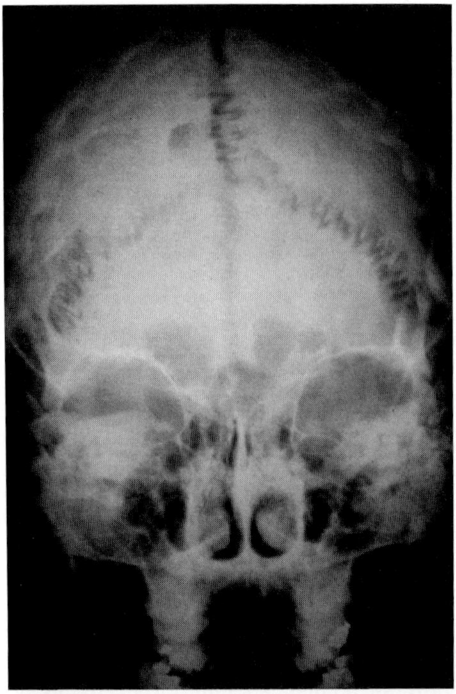

Abb. 1

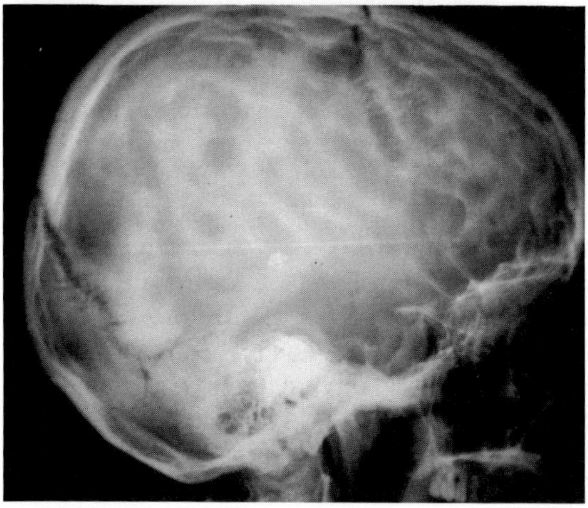

Abb. 2

Abb. 1. u. 2. Das Schädelröntgen des 20jährigen Patienten J.M. zeigt einen Wolkenschädel, Sprengung der Schädelnähte und eine Erosion der Sella durch den erhöhten intrakraniellen Druck verursacht durch die Abflußbehinderung des Liquors im Aquaeductus Sylvii

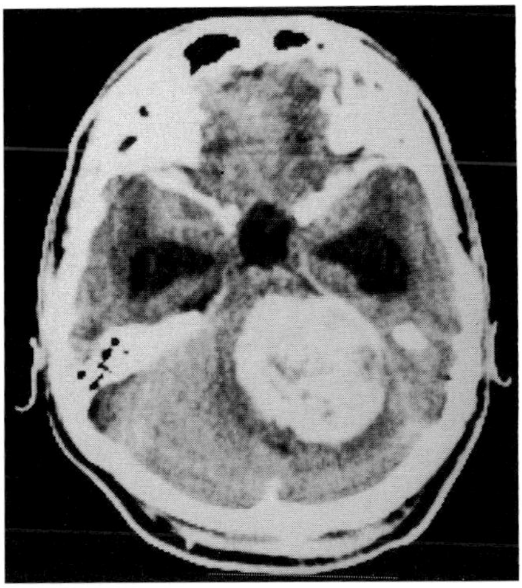

Abb. 3

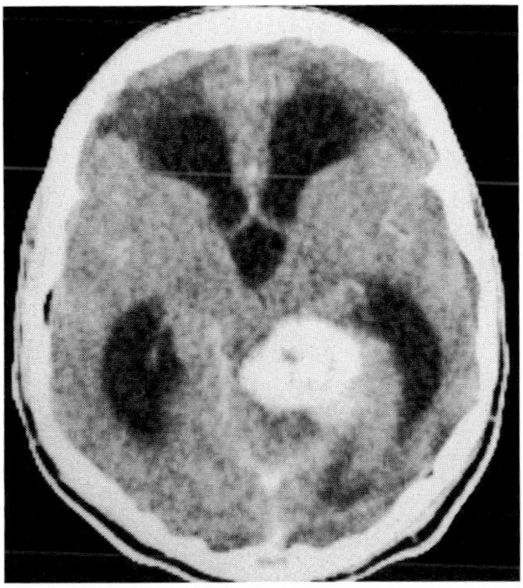

Abb. 4

Abb. 3 u. 4. Die Computertomographie des Schädels des 20jährigen Patienten J.M. zeigt ein Neurinom im rechten Kleinhirnbrückenwinkel mit Verlagerung des Hirnstamms, mit Verschluß des Aquaeductus Sylvii und mit dadurch verursachtem supratentoriellen Hydrocephalus mit periventrikulärer Dichteminderung

Gibt es histologische Beweise für eine traumatisch bedingte Bandscheibenschädigung?

L. Gerhard, H.-E. Nau, V. Reinhardt, F. Rauhut und G. Kötz[1]

Von vielen Patienten wird beim Auftreten akuter radikulärer Beschwerden bei Bandscheibenvorfall ein vorangegangenes traumatisches Ereignis angegeben. Dies trifft für Patienten aller Altersgruppen und für alle Abschnitte der Wirbelsäule zu. Neben dieser großen allgemeinen Gruppe von Bandscheibenschäden gibt es jedoch Patienten, wo das traumatische Ereignis offenkundig ist. Die klinische Erfahrung zeigt, daß die Häufigkeit traumatischer und degenerativer Bandscheibenschädigungen sich innerhalb der Abschnitte der Wirbelsäule und hinsichtlich des Alters des Patienten deutlich unterscheiden. Im Gegensatz zu lumbalen Anteilen der Wirbelsäule ist innerhalb des cervikalen Abschnittes eine eindeutige traumatische Ursache in den ersten Lebensjahrzehnten besonders häufig. Daher erscheint es von Bedeutung, die morphologischen Möglichkeiten zu überprüfen, die eine klinische Annahme einer traumatischen oder degenerativen Ätiologie bei Bandscheibenvorfällen unterscheiden oder stützen können.

Der allgemeine Ablauf von Reaktionen am Gewebe ausgelöst durch ein Trauma oder andere Schädigungsfaktoren kann in folgende Schritte unterteilt werden.:

Zu Beginn findet sich eine erste areaktive Phase mit Schädigung der Gefäßschranke (Ödem, Austritt von Blutzellen) und mechanischer Gewebsdestruktionen.

Danach folgt die Phase der Mobilisation von Makrophagen, Fibroblasten und Gefäßendothel bzw. Vermehrung von Gefäßlumina, was in einer 3. Phase zur Bildung von Organisationsgewebe mit Resorption des zerstörten Gewebes bei Aufbau neuer Gewebsstrukturen führt. Dies schließlich mündet in die Endphase, die Ausbildung einer ruhenden Narbe.

Es gibt grobe Markierungen für die einzelnen Phasen und für das zeitliche Alter einiger histologischer Reaktionen. Ein Beispiel ist die Bildung neuer kollagener Fasern innerhalb eines Organisationsgewebes, die nicht früher als 6 Wochen nach Einwirken der Schädigung eintritt.

Wenn wir diese allgemeine Regeln über Gewebsreaktionen auf die Bandscheiben anwenden, müssen wir allerdings berücksichtigen, daß Bandscheibenknorpel und Nucleus pulposus nicht vaskularisiert sind mit Ausnahme von frühen Ent-

[1] Neurochirurgische Klinik, Institut für Neuropathologie, Universitätsklinikum Essen, D-4300 Essen 1

Advances in Neurosurgery, Vol. 18
K.-A. Bushe, M. Brock, and M. Klinger (Eds.)

wicklungsstadien. Beim Erwachsenen reichen die Gefäße vom Knochenmark des Wirbelkörpers bis in den Deckplattenknochen und an den Deckplattenknorpel. Dorsal und ventral enden die Gefäße an der äußeren Oberfläche des Annulus fibrosus und der inneren Grenze des Periosts. Daher wird eine Zellreaktion innerhalb des Nucleus pulposus und das Penetrieren von Organisationsgewebe bzw. einer Organisationsreaktion im Nucleus pulposus mehr Zeit benötigen, als die allgemeinen Regeln für ein vaskularisierteres Gewebe wie oben ausgeführt benötigen. Zellen und Gefäße brauchen eine zusätzliche Zeit, um von den äußeren Grenzen her Bandscheibenknorpel und Nucleus pulposus zu durchdringen. Die Tatsache der Nichtvaskularisation ist bei der Bandscheibe auch verantwortlich für das Fehlen einer Blutung bei traumatischer Einwirkung. Andererseits führen auch degenerative Veränderungen nicht selten zu einer Revaskularisation infolge Schädigung des Annulus fibrosus oder der Deckplatte. Dies konnten wir in unterschiedlichen Graden und Ausbildungsmustern bei allen unseren Patienten mit nicht traumatischen Bandscheibenerkrankungen beobachten. Daher bleibt das Vorliegen einer Vaskularisation und das Bestehen von Organisationsreaktionen innerhalb des Nucleus pulposus als solche ein unspezifischer Befund. Das gleiche gilt für Eisenablagerungen als Markierungen für resorbierte Blutungen bei vorhandener Revaskularisation.

Die sogenannten „degenerativen" Veränderungen des Nucleus pulposus sind vorzugsweise durch die Verflüssigung zentraler Knorpelportionen bedingt, was als solches ein physiologischer Vorgang ist. Mit nur wenigen Ausnahmen können diese „degenerativen Veränderungen" im Zentrum des Bandscheibenknorpels bei Kindern und Erwachsenen gefunden werden. Die entscheidende Gewebspartie, um Reaktionen und Veränderungen zu beurteilen, ist daher die Grenzzone der Deckplatte und die äußere Oberfläche des Annulus fibrosus. Von hier aus beginnt die Revaskularisation und die Invasion von Organisationsgewebe. Bei traumatischer Ursache kann man hier zum Beispiel Rißfolgen oder das Einreißen kollagener Fasern nachweisen mit anschließender Knorpelneubildung und zusätzlich Osteoidformation, die dem histologischen Bild der klassischen degenerativen Bandscheibenerkrankung fremd sind. Neu gebildetes nicht verkalktes Knochengewebe fanden wir in etwa 25% traumatischer cervikale Bandscheibenschäden bei Kindern aber auch bei sicherer traumatischer Einwirkung bei einzelnen Erwachsenen. Dieser neue Knochen läßt sich in seinem Aufbau und seiner Beziehung zum Umgebungsgewebe meist gut von osteochondrotischen Veränderungen bei degenerativen Bandscheibenschäden unterscheiden.

Ein weiterer wichtiger Faktor ist der Vergleich zwischen Einwirkung des Traumas, abgelaufener Zeit bis zur Operation und der möglichen Korrelation zu den histologischen Stadien der Reparationsreaktion. Wenn die Zeitkorrelation innerhalb der Phasen übereinstimmt, die oben ausgeführt wurden, ist der traumatische Ursprung entweder hochgradig wahrscheinlich oder sogar sicher. Wenn der Vergleich Gewebsreaktionen aufweist, die deutlich älter sind als der Zeit zwischen Trauma und Operation entsprechen, muß eine prätraumatische degenerative Veränderung der Bandscheibe angenommen werden. Daraus folgt auch, daß ein negatives Resultat hinsichtlich der Gewebsreaktionen, d.h. das Vorliegen keiner Zellreaktion, eine traumatische Ätiologie nicht ausschließt. Das negative Resultat kann auf zu geringem Zeitabstand (1. Phase), zu wenig Biopsiematerial oder auf dem Fehlen der äußeren Anteile der Bandscheibe wie Periost und Oberfläche des Annulus fibrosus beruhen. Wenn ein Patient mit dem Verdacht einer traumatischen Ätiologie operiert wird, sollte dies bei der Asservierung des Biopsiematerials berücksichtigt werden und das gesamte entfernte Gewebe zur histologischen Untersuchung eingesandt werden.

Zusammenfassend möchten wir folgende Voraussetzungen für die Beurteilung eines traumatischen Bandscheibenschadens aufführen: Vollständige histologische Untersuchung des gesamten beim operativen Eingriff entfernten Materials unter Einschluß von Anteilen des Annulus fibrosus, des Periosts und der Deckplatte. Wenn notwendig, sollten Serienschnitte durch das in toto eingebettete gesamte Biopsiematerial durchgeführt werden. Die Interpretation der Befunde sollte nur mit exakter Kenntnis des Zeitabstandes zwischen traumatischer Einwirkung und Operationstermin durchgeführt werden sowie unter Kenntnis aller konservativen Behandlungsversuche vor der Operation sowie der radiologischen Befunde.

Versicherungsrechtliche Bewertung von Traumen für die Entstehung und den Verlauf zervikaler und lumbaler Bandscheibenvorfälle

D. Terhaag und R.A. Frowein[1]

Bei der Jahrestagung 1979 in Berlin und bei ICRAN 1986 in Köln wurde über die Grundsätze der Annahme eines Zusammenhanges zwischen einem Trauma und einem zervikalen oder lumbalen Bandscheibenvorfall berichtet [1,4].

Die auch weiterhin in Praxis und Begutachtung kontroverse Diskussion gab zusammen mit dem 1987/88 beobachteten Krankheitsverlauf einer 42jährigen Hausfrau Veranlassung nochmals über eindeutige Verlaufszusammenhänge zu berichten.

Die bis dahin ohne Wirbelsäulenbeschwerden lebende Hausfrau prallte am 12.6.1987 mit dem Schädel gegen Holm und Türe ihres PKW. Sofort Kopf- und in die linke Schulter ausstrahlende Nackenschmerzen. Keine Bewußtseinsstörung. Sie verließ selbständig den Unfallort, legte sich aber daheim der Beschwerden wegen ins Bett. Am 4. Tag wegen zunehmender Schwäche im linken Arm und Gefühlsstörungen im Arm, Hand und Fuß Krankenhausaufnahme. Im Schädel-Computertomogramm im rechten Marklager zwei primär hypodense, nach Kontrastmittel isodense Herde mit geringem Randenhencement. Bei der Kontrolle nach 7 Tagen keine Befundänderung. Cortisontherapie. Neurologisch/Neurochirurgische Untersuchung. Vorschlag: Kernspintomographie des Schädels, Liquor-Diagnostik. Während der zweiwöchigen stationären Behandlung geringe Besserung der Parese. Liquor o.B. Im Kernspintomogramm stellten sich die Marklagerherde signalverändert, bei der Kontrolle nach 6 Monaten etwas kleiner, nach 12 Monaten gleichbleibend dar.

Exstirpation eines Weichteiltumors in der Nackenmuskulatur in Höhe C4, auch zur evtl. Klärung der intrakraniellen Prozesse. Histologisch: Cavernom. Auf dem Kernspintomogramm der Halsregion läßt sich im Nachhinein die Raumforderung bei C6/7 gerade noch erkennen.

Bei Fortbestehen radikulärer Schmerzen, Sensibilitätsstörungen und Schwäche der linken Hand wurde bei der Operation am 15.11.1988 – 18 Monate nach dem Unfall – ein perforierter Bandscheibenvorfall mit im Spinalkanal freien mit der Dura aber bereits bindegewebig verbackenen Sequestern entfernt.

[1] Neurochirurgische Universitätsklinik Köln, Joseph-Stelzmann-Str. 9, D–5000 Köln 41

Advances in Neurosurgery, Vol. 18
K.-A. Bushe, M. Brock, and M. Klinger (Eds.)

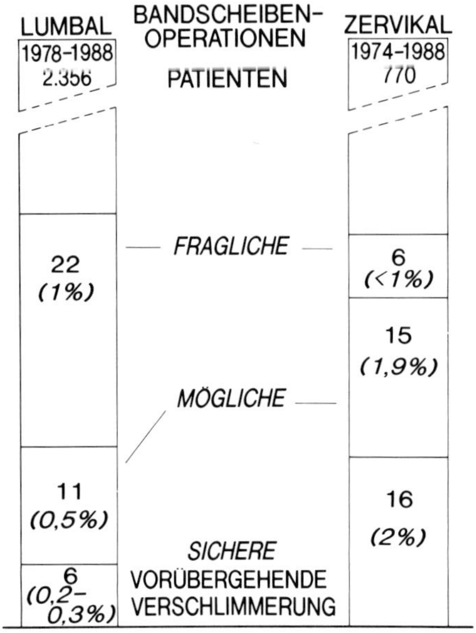

Abb. 1. Häufigkeit eines sicheren, möglichen oder fraglichen Einflusses eines Traumas im Sinne einer vorübergehenden Verschlimmerung eines Bandscheibenvorfalles bei 3 126 Patienten mit Bandscheiben-Operation

Ätiologisch ungeklärte Marklagerherde, Cavernom der Nackenregion und perforierter zervikaler Bandscheibenvorfall zusammentreffend, führten erst 18 Monate nach dem Trauma zur Diagnosenstellung und operativen Behandlung.

Traumatische Bandscheibenvorfälle sind zervikal wie insbesondere lumbal selten. Unter 770 operativ behandelten Patienten mit zervikalem Bandscheibenvorfall gaben 37 Patienten (ca. 5%) bei der Vorgeschichte ein Trauma. an. Unter Berücksichtigung von Unfallart und -hergang und insbesondere der Syndromentwicklung nach dem Trauma war ein sicherer Zusammenhang bei 16 Patienten (2%), ein möglicher bei 15 Patienten (ca. 2%) anzunehmen. Bei 6 Patienten blieb der Zusammenhang fraglich (Abb. 1).

Das Syndrom-Zeitdiagramm (Abb. 2) läßt erkennen, daß bei allen Patienten, bei denen ein sicherer oder möglicher Zusammenhang anzunehmen war, sich radikuläre Schmerzen unmittelbar nach dem Trauma und Gefühlsstörungen und/oder Paresen fast ausnahmslos innerhalb von Stunden bis wenigen Tagen entwickelten.

Beispiel 1: Bei dem 45jährigen bisher beschwerdefreien Patienten, der beim LKW-Entladen mit einer Rinderhälfte stürzte, traten sofort Schmerzen im Nacken auf, die in den linken Arm ausstrahlten. Erst nach der Feststellung von Paresen wurde das Syndrom gedeutet. Bei der Operation – 8 Wochen nach dem Unfall –

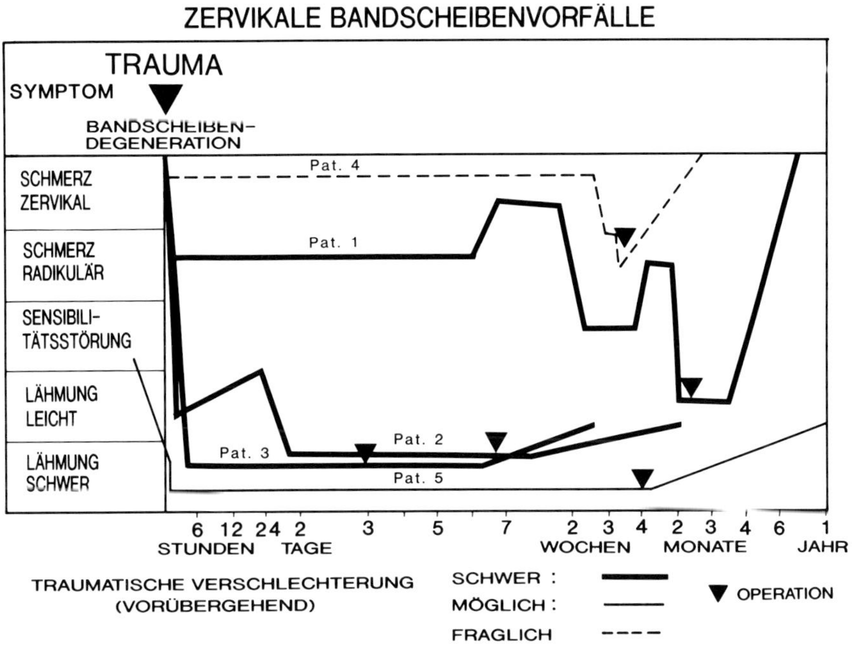

Fig. 2. Syndrom-Zeitdiagramm typischer Krankheitsverläufe im Sinne einer sicheren, möglichen oder fraglichen vorübergehenden Verschlimmerung bei Trauma und zervikalem Bandscheibenvorfall

wurde ein perforierter Bandscheibenvorfall mit freien Sequestern im Spinalkanal entfernt. Rasche Erholung. Es wurde eine vorübergehende Verschlimmerung für die Dauer von 3 Monaten angenommen.

Beispiel 2: Nur fraglich der Zusammenhang zwischen Trauma und Bandscheiben-Vorfall bei der 37jährigen Patientin, die bei einem Spaziergang stürzte. Kopfschiefhaltung. Nach 3 Wochen radikuläre Schmerzen. Bei der Operation wurde eine Bandscheiben-Protrusion C5/6 und ein perforierter Vorfall C6/7 entfernt. Rasche Erholung.

Wegen der engen räumlichen Nachbarschaft im oberen Abschnitt des Wirbelkanals sind bei zervikalen Bandscheibenvorfällen sowohl radikuläre wie ausch radikulär-medulläre Syndrome möglich. Letztere wurden posttraumatisch auch hinter stärkeren – als Drucklager wirkenden – osteochondrotischen/spondylotischen Randwulstungen beobachtet, wie dies auch von Saternus beschrieben worden ist [2, 3].

Von 2 356 Patienten, bei denen ein lumbaler Bandscheibenvorfall operiert werden mußte, gaben 39 (1,8%) anamnestisch ein Trauma an, das subjektiv zum Beginn bzw. zur wesentlichen Verstärkung der Symptomatik geführt hatte.

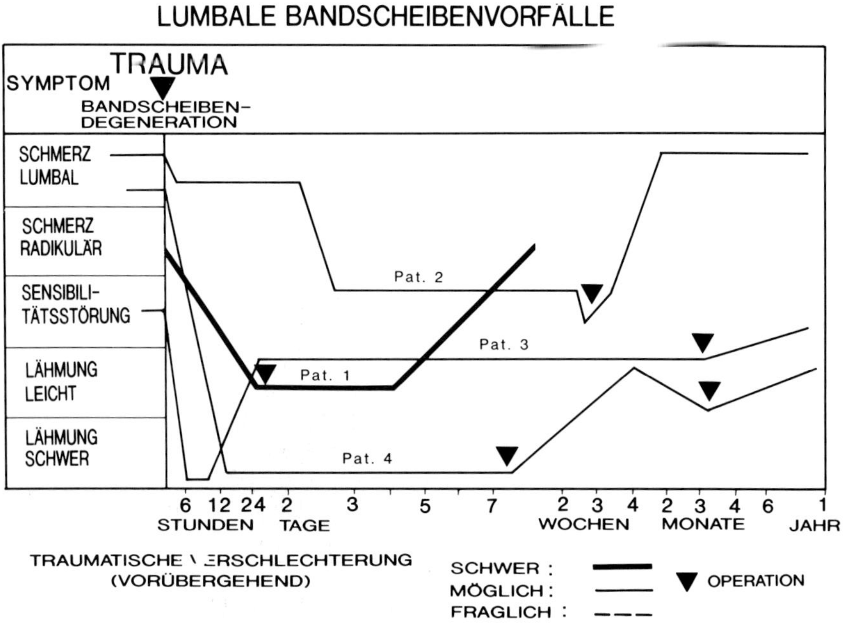

Abb. 3. Syndrom-Zeitdiagramm typischer Krankheitsverläufe im Sinne einer sicheren, möglichen oder fraglichen vorübergehenden Verschlimmerung bei Trauma und lumbalem Bandscheibenvorfall

Neben der Bewertung von Art und Hergang des Ereignisses war wiederum der zeitliche Ablauf der Syndromausbildung die Hilfe für die Bewertung des sicheren, möglichen oder fraglichen Unfallzusammenhanges (Abb. 3).

Beispiel 1: Der 36jährige Rohrreiniger stürzte mit einer Schlauchrolle beladen die Kellertreppe hinunter. Die Rolle schlug ihm ins Kreuz. Sofort radikuläre Schmerzen, noch am Unfalltag Fußheberparese. Bei der Operation Entfernung eines perforierten Bandscheibenvorfalles L4/5. Gute Rückbildung der Parese. Trotz des Fehlens von Vorbeschwerden wurde eine vorübergehende Verschlimmerung für die Dauer der Arbeitsfähigkeit – 3 Monate – angenommen.

Beispiel 2: Der 56jährige Patient, der bereits seit mehreren Jahren Rückenschmerzen und ein Taubheitsgefühl an der Außenseite des rechten Unterschenkels hatte, stürzte. Zunächst Gehunfähigkeit, dann radikuläre Schmerzen und Gefühlsstörung auch im linken Bein. Fußheberparese. Bei der Operation 3 Monate nach dem Sturz Entfernung einer Bandscheibenprotrusion L4/5. Postoperativ nur geringe Funktionsbesserung. Es wurde eine unfallbedingte, vorübergehende Verschlimmerung für die Dauer eines Jahres angenommen.

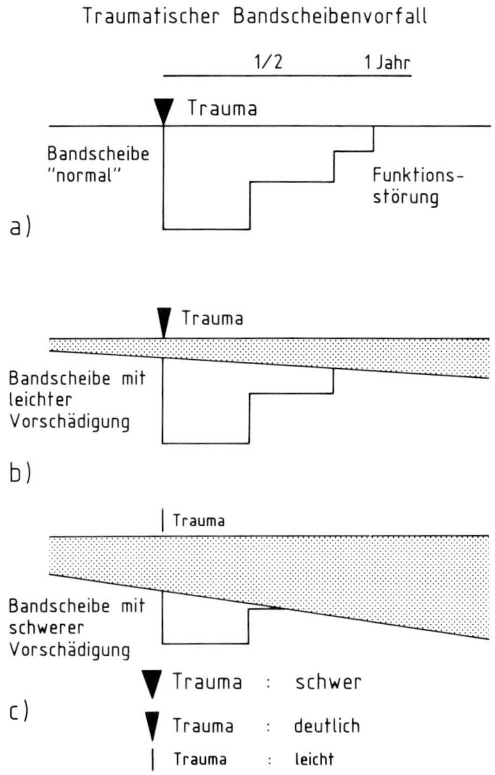

Abb. 4a–c. Einfluß eines Traumas auf Entstehung und Verlauf des akuten Syndroms eines Bandscheibenvorfalls bei jungen Patienten (**a**) mit „normalem" Bandscheibengewebe und bei Erwachsenen mit (**b**) leichter oder (**c**) schwerer Bandscheibendegeneration

Wegen der bei Erwachsenen im Gegensatz zu Jugendlichen auch bei subjektiv bis zum Unfall bestehender Beschwerdefreiheit stets anzunehmenden – individuell jedoch unterschiedlich starken – Vorschädigung durch Osteochondrose usw., kann das Trauma immer nur die Bedeutung haben, zu einer vorübergehenden, nicht richtunggebenden Verschlimmerung zu führen. Diese Vorschädigung muß individuell beurteilt werden. Die vorübergehende Verschlimmerung ist auf einen Zeitraum von maximal 12 Monaten in der Regel zu begrenzen (Abb. 4).

Auch die weiter gestiegene Zahl der Beobachtungen bestätigt, daß es den „traumatischen Bandscheibenvorfall" gibt, aber selten. Ein Zusammenhang im Sinne der vorübergehenden Verschlimmerung konnte bei Patienten mit zervikalem Bandscheibenvorfall sicher bzw. möglich jeweils mit 2% festgestellt werden. Bei lumbalen Bandscheibenvorfällen war ein solcher Zusammenhang mit sicher 0,3% und möglich mit 0,5% noch seltener.

Literatur

1. Frowein RA, Terhaag D (1977) Traumatische Bandscheibenvorfälle. Advances in Neurosurgery 4:82–89
2. Saternus K-S (1979) Die Verletzungen von Halswirbelsäule und Halsweichteilen. In: Junghanns H (Hrsg) Die Wirbelsäule in Forschung und Praxis, Bd 84. Stuttgart, Hippokrates
3. Saternus K-S (1982) Die Begutachtung des Schleudertraumas der Halswirbelsäule. Aktuelle Traumatologie 12:4–11
4. Terhaag D, Frowein RA (1988) Traumatic disc prolapses. Neurosurgery Rev

Eine seltene Komplikation nach Bandscheibenoperation in Knie-Ellenbogenlage

H.-E. Nau, H. Gerhard, H. Wiedemayer und V. Reinhard[1]

Einleitung

Komplikationen, die mit einem operativen Eingriff zusammenhängen, können durch die operative Technik selbst, die Narkose und durch die Lagerung des Patienten hervorgerufen werden. Bei lumbalen Bandscheibenoperationen werden entzündliche, mechanische und neurogene Komplikationen häufig berichtet [1–3]. Wir möchten eine seltene Komplikation hinzufügen, die durch die Lagerung des Patienten in Knie-Ellenbogenlage (Häschenstellung, Mekka-Stellung) bedingt ist, die wohl in vielen, sehr wahrscheinlich sogar in der Mehrzahl der neurochirurgischen und orthopädischen Kliniken vorgenommen wird; wobei diese Lagerung von Klinik zu Klinik etwas individuell modifiziert wird. Der Kopf ist in einer Kopfschale gelagert, der Körper des Patienten wird einerseits hier sowie auf den unterlagerten Ellbogen und andererseits durch die zusammengeknickten Beine gehalten und so die Wirbelsäule nach dorsal aufgebogen.

Nach dieser Lagerung haben wir Bewegungsstörungen der Arme beobachten können. Der erste von uns beobachtete Fall solch einer beidseitigen „Plexusparese" ging als Zivilprozeß bis an den Bundesgerichtshof und führte zu höchstrichterlichen Entscheidungen, wer für die Lagerung zuständig sei, ob Operateur oder Anästhesist.

Kasuistik

Fall 1

Die 38jährige Patientin wird wegen eines S1-Syndroms bei myelographisch gesichertem lumbosakralen Bandscheibenprolaps in Knie-Ellenbogenlage operiert. Postoperativ fallen Paresen der kleinen Handmuskulatur links mehr als rechts und Paresen der Handbeuger und -strecker beidseits auf, eine Hypästhesie und Hypalgesie ist im Bereich der volaren Handseiten beidseits zu finden. Lediglich der direkt nach der Operation untersuchende Arzt findet außer diesen Befunden auch ein

[1] Neurochirurgische Klinik, Universitätsklinikum Essen, Hufelandstr. 55, D-4300 Essen 1

Advances in Neurosurgery, Vol. 18
K.-A. Bushe, M. Brock, and M. Klinger (Eds.)

beidseitiges positives Babinski-Zeichen und eine Fußheberschwäche links. Diese Befunde können von allen nachfolgenden Untersuchern nicht mehr festgestellt werden.

Fall 2

Die 52jährige Patientin wurde einige Jahre vor der Bandscheibenoperation wegen HWS-Beschwerden in der Ambulanz vorstellig. Später wird sie wegen eines L5-Syndroms mit Fußheberparese, radikulären Sensibilitätsstörungen und persistierendem Schmerzbild an dem computertomographisch gesicherten Bandscheibenprolaps in Höhe LWK 4/5 in Häschenstellung operiert. Postoperativ fällt eine beidseitige hochgradige Schwäche der proximalen und auch distalen Armmuskulatur auf, ferner eine Herabsetzung der Sensibilität bei C7 beidseits. Gleichzeitig bemerkt der Erstuntersucher eine proximal und linksbetonte Schwäche der Beinmuskulatur, was noch am Abend nachweisbar ist, jedoch am nächsten Tag sich vollständig zurückgebildet hat.

Diskussion

Für unser Patientengut mit insgesamt drei solcher Beobachtungen ergibt sich bei rund 5 000 Bandscheibenoperationen in dieser Lagerung eine Häufigkeit von 0,6 Promille oder eine solche Läsion auf rund 1 700 Operationen in dieser Lagerung. Zu diesen eigenen Fällen kamen weitere vier durch Anfragen und Gutachtertätigkeit hinzu. Daher waren wir erstaunt, daß auch durch Literaturrecherchen kein anderer Fall zu eruieren war.

Gemeinsam ist allen uns bekannten Fällen die klinische Symptomatik: stets eine deutliche Lähmung der Muskeln an Armen oder Händen direkt postoperativ und nur bei direkt postoperativ stattfindender Untersuchung auch eine Schwäche von Beinmuskeln und Reflexstörungen. So wundert es nicht, wenn in den meisten Fällen an eine Plexusschädigung gedacht wird. Die neurologischen Defizite an den oberen Extremitäten bilden sich im Gegensatz zu den an den unteren, wenn überhaupt, dann nur schlecht zurück. Die spinale Läsion kann elektrophysiologisch durch verlängerte Latenzen der Tibialis-evozierten Potentiale objektiviert werden.

Als Erklärung, wie diese Symptome entstehen könnten, bietet sich als mechanische Lösung an, daß zum einen durch Zerren der sehr variablen Zuflüsse der A. spinalis anterior eine Durchblutungsstörung hervorgerufen wird, zum anderen (im Fall 2 mit bereits vor Jahren auftretenden Halswirbelsäulenbeschwerden) die individuelle Disposition durch z.B. osteochondrotische Randzacken eine Rolle spielt.

Zusammenfassung

Aufgrund der dargelegten Fälle muß davon ausgegangen werden, daß es eine durch Knie-Ellenbogenlagerung ausgelöste Plexusläsion nicht gibt, sondern daß es sich um eine Durchblutungsstörung des Zervikalmarks handeln muß, zu der es erst kommen kann, wenn noch weitere individuelle Faktoren hinzu kommen. Für die spinale Störung sprechen die Bilateralität der Paresen, die zumindest passager wohl stets nachweisbaren Zeichen eine Störung der langen Bahnen und die Latenzverzögerung der Tibialis-evozierten Potentiale.

Literatur

1. Horwitz NH, Rizzoli HV (1987) Postoperative complications of extracranial neurological surgery. Williams and Wilkins, Baltimore
2. Oppel F, Schramm J, Schirmer M (1977) Results and complicated course after surgery for lumbar disc herniation. Adv Neurosurg 4:36–51
3. Schüwer U, Roosen K (1988) Komplikationen bei lumbalen Bandscheibenoperationen. Neurochirurgia 31:192–195

Gesprächsführung in der Neurochirurgie

H. Poimann[1]

Marco Mummenthaler läßt sich wie folgt zitieren: „Ein blinder Neurologe ist besser als ein tauber" [15]. Wie Sie mir zugeben, läßt sich diese Aussage nicht zwanglos auf die Neurochirurgie übertragen, wenngleich sie die Bedeutung des ärztlichen Gespräches und des Zuhörens in den Neurologischen Fachgebieten hervorhebt und unterstreicht. Wie aus der umfangreichen Untersuchung von Schunemann und Mitarbeitern über 110 Chirurgen bekannt, sind die Vorbedingungen für eine gute operative Tätigkeit nicht Wissen und Intelligenz, sondern ein hohes visuell-räumliches Auflösungsvermögen, eine ausgebildete Augen-Hand-Koordination und eine relativ hohe Streßtoleranz [19]. Der Neurochirurg verbringt aber einen Großteil seiner Arbeitszeit nicht im Operationssaal, sondern im Gespräch mit anderen [17], nämlich mit Patienten, Angehörigen, Pflegekräften, Kollegen, Pharmareferenten und führt unzählige Telefonate. Leider gibt es keine Untersuchungen, die den Arzt als sehr kompetent in Sachen Kommunikation und Gesprächsführung ausweisen, Belege für das Gegenteil gibt es dagegen sehr viele [3, 6, 10, 13]. Dies ist auch nicht verwunderlich, denn der Arzt ist für die Gesprächsführung schlecht oder gar nicht ausgebildet [10, 13], obwohl ein professionelles ärztliches Gespräch, ebenso eine ärztliche und damit auch eine justiziable Handlung ist, wie jede andere Behandlung mit Medikamenten, Strahlen oder dem Skalpell [4, 20]. Ebensowenig wie jedoch der in der Kindheit erlernte und gekonnte Umgang mit Messer und Gabel, ausreichende Fertigkeiten, oder die Befähigung für einen chirurgischen Eingriff zur Verfügung stellen, sind die im Alltag erworbenen Kenntnisse und Fertigkeiten mit der Sprache ausreichend, um sie als medizinisches Instrument einzusetzen. Aber auch das Wort ist in der Lage, tiefgreifende Verletzungen und womöglich Schäden anzurichten, vergleichbar dem Messer, schon Hamlet sagte: „Nicht brauchen will ich Dolche, sondern reden" [18]. Das heißt, das ärztliche Gespräch in all seinen Facetten und Variantenreichtum hat vergleichbar mit anderen ärztlichen Tätigkeiten, z.B. einer Operation, analoge Voraussetzungen, Grundbedingungen, Ziele, Wirkungen und Folgen, d.h. der Arzt muß ebenso verantwortungsvoll damit umgehen. Für Operation, Medikation und Gespräch gilt also gleichermaßen:

[1] Neurochirurgische Klinik und Poliklinik der Universität Würzburg, Josef-Schneider-Str. 11, D–8700 Würzburg

Advances in Neurosurgery, Vol. 18
K.-A. Bushe, M. Brock, and M. Klinger (Eds.)

1. Der Arzt sollte für durchzuführende Tätigkeiten (OP/Gespräch), entsprechendes Wissen, Können und Erfahrung besitzen und seine persönliche Grenze respektieren.
2. Damit versetzt er sich in die Lage Indikationen adäquat zu stellen, erreichbare Ziele einzuschätzen und die verfügbaren Mittel einer bestimmten Tätigkeit kunstgerecht zum Nutzen der Patienten einzusetzen.
3. Er ist sich der Wirkungen, Folge und möglichen Komplikationen seines Eingriffes bewußt und besitzt Mittel, um mit diesen Folgen umgehen zu können, um evtl. Schäden zu vermeiden.
4. Der Arzt sollte die Rahmenbedingungen so gestalten, daß das Ergebnis, seiner ansonsten kunstgerechten Handlung, nicht unnötig beeinträchtigt wird.

Ich bitte Sie in Anbetracht dieser Dinge einmal sich folgendes Szenario vorzustellen: ein unerfahrener Chirurg, gleichwohl ein gestandener Mann, oder eine Frau, mit Lebenserfahrung und einer gewissen Lebenspraxis, lediglich assistiert von einem Studenten, operiert im Op-Vorraum ohne Beachtung der Asepsis mit bereits benutzten Instrumenten ein Meningeom!

Bei Ihnen dürfte er das natürlich nicht, dies widerspricht allen Regeln des operativen Vorgehens; doch wie wäre das vor 100 Jahren gewesen, zur Zeit von Victor Horsley? Wie hoch wäre die Komplikationsrate, die sie beim operativen Eingriff heutzutage tolerierten: 20%? 10%? 5%? 1%? – Und wie ist es beim ärztlichen Gespräch?

Der Patient hat nicht nur das Recht auf ein ausführliches Aufklärungsgespräch, sondern nach einem Urteil des Landesgerichtes Köln [12] präoperativ auch das Recht auf Zuspruch und Vermittlung von Zuversicht. Und es wäre nicht das erste Mal, daß Justitia eine Anpassung alltäglicher Praxis an bestehende ethische Forderungen beschert.

Doch abgesehen von juristischen Konsequenzen ist das ärztliche Gespräch ein Mittel, das nicht nur das für die Behandlung unersetzliche Vertrauen schaffen kann; mit den unzähligen psychologischen Untersuchungsergebnissen möchte ich Sie in keiner Form langweilen (wogleich diese auch den von uns zu behandelnden Patienten betreffen). Durch das präoperative, professionelle Aufklären ist es nicht nur möglich die postoperative Liegezeit zu verkürzen [7, 14], wobei bekanntermaßen Liegezeiten nicht unbedingt von der Schwere eines Eingriffes und oder der Verletzungsfolgen abhängen [5], oder der intraoperative Blutverlust ist bei Eingriffen am Rücken bei entsprechendem Verhalten des Arztes im präoperativen Gespräch geringer [1, 2].

Diese Fakten mögen Ihnen vielleicht ein kleines Schmunzeln entlocken und Ihnen utopisch erscheinen – sie sind jedoch belegte Fakten. Und bevor Sie eine abschließende Bewertung vornehmen, möchte ich Sie doch bitten, kurz zu überlegen, inwieweit Ihre Kompetenz tatsächlich reicht, um aufgrund Ihres Wissens über Kommunikationstechniken und Gesprächsführung darüber fundiert urteilen zu können.

Wenn Sie nur die Zeit berücksichtigen mögen, die Sie dafür aufgewendet haben, im Vergleich zum Erwerb von chirurgisch technischen Fertigkeiten und Wissensmöglichkeiten.

Waren im Handbuch der Neurochirurgie von Olivecrona und Tönnis, 1960 nur ein paar Absätze auf Seite 25 über dem Gespräch mit dem Kranken und seinen Angehörigen daniedergeschrieben worden [9], sind im Wilkens schon 3 Seiten über "The Art of History Taking" von B.J. Hurwitz [8], ich will nicht sagen geopfert worden. Und auch die Tatsache, daß an einer Neurochirurgischen Tagung ein neurochirurgisch Tätiger über ein so unwichtiges Thema wie Gesprächsführung 7 Minuten Ihrer Zeit in Anspruch nehmen darf, möchte ich als Indiz für die zunehmende Bedeutung der Gesprächsführung auch für den Chirurgen sehen.

Kein Neurochirurg wird heute auf eine adäquate, medikamentöse Begleitbehandlung verzichten wollen und ich bin der festen Überzeugung, in nicht allzu ferner Zukunft wird auch der Neurochirurg auf die Wirkung des Wortes nicht gerne verzichten wollen. Nicht nur bei der Anamnese und im Aufklärungsgespräch, auch im Bereich der Patientenführung, der Prognoseeröffnungen und der postoperativen Betreuung sollte der Chirurg, dem im Alltag nur wenige Minuten pro Patient zur Verfügung stehen [17], dieses Instrument so effektiv wie möglich zur Informationsgewinnung, Vertrauensbildung und Instruierung des Patienten nutzen. Besondere Probleme stellen sich bei der Prognoseeröffnung Schwerkranker und von Tumorpatienten sowie im Umgang mit Angehörigen. In der Betreuung schädelhirnverletzter und tumoroperierter Patienten, bei der Reintegrierung in Familie und Gesellschaft können eine kompetente Beratung und kurzpsychotherapeutische Führung durch den Neurochirurgen eine entscheidende Überbrückungsmaßnahme darstellen. Durch gezielte, bewußt eingesetzte Gesprächsintervention und Kommunikationstechniken, z.B. bei Aufklärungsgesprächen oder während der Visite (s. Tabellen 1 u. 2) lassen sich auch in der Neurochirurgie neben einer für die Therapie unerläßlichen Vertrauensbildung häufig kraft- und zeitkonsumierende postoperative Auseinandersetzungen mit dem Patienten und den Angehörigen vorab vermeiden: dabei ist die Grundlage aller effektiven Gesprächsführung eine adäquate ärztliche Haltung in Kombination mit einer entsprechenden Gesprächstechnik. Aufgrund dieser Bedeutung der ärztlichen Gesprächsführung wird an der Universität Würzburg in der Neurochirurgie seit 5 Jahren ein entsprechender Unterricht angeboten [13, 16].

„Wer die Arztkommunikation beherrscht, beherrscht seine Zukunft" [21]. Mit diesem Zitat, entnommen einer Pharmamarketing-Zeitung aus dem Jahr 1982, möchte ich schließen, nicht ohne Sie auf die Vieldeutigkeit oder Vielschichtigkeit des Wörtchens „seine" aufmerksam gemacht zu haben. Vielleicht ist es mir gelungen, Sie etwas auf die Art und Weise neugierig zu machen, wie man Gespräche führen und was man damit erreichen kann – und es wird Ihnen damit so ergehen, wie mit dem operativen Tätigsein – das ist eine weitere Analogie – je weiter man sich damit beschäftigt und eindringt, umso spannender wird es, und

Tabelle 1. Anhaltspunkte für das Visitengespräch nach L. Geisler

1.	Die Visite muß dem Kommunikationsbedürfnis des Patienten entgegenkommen und systematisch angelegt sein.
2.	Der Patient soll zum Fragen angehalten werden, der Arzt soll Abweisungsstrategien vermeiden.
3.	Die Visite ist auf ein patientenzentriertes 2-Personen-Gespräch hin orientiert.
4.	Patientenorientierter und organisations- bzw. teamorientierter Teil der Visite sollen deutlich getrennt werden.

Tabelle 2. Anhaltspunkte für das präoperative Gespräch nach L. Geisler

1.	Aufklärung über die Operation um eine rechtskräftige Einwilligung des Patienten zu erhalten.
1.1.	Maßnahme und Ziel darstellen.
1.2.	Die Menge der Information auf den individuellen Patienten anpassen.
1.3.	Aufklärung im juristisch erforderlichen Umfang durchführen.
2.	Psychologische Stabilisierung des Patienten.
2.1.	Vertrauen und Motivation aufbauen.
2.2.	Ängste herausfinden und soweit unbegründet abbauen helfen.
2.3	Zeitlichen Ablauf benennen können.
2.4.	Möglicherweise mit Patienten, die eine Operation erlebt haben, sprechen lassen.

umso mehr macht es Spaß, zum Nutzen des Patienten, zum Nutzen des Arztes; nur der Jurist wird ausnahmsweise in diesem Fall nicht gewinnen.

Literatur

1. Bennet HL (1985) Behavioral anaesthesia, advances. Institut of the Advancement of Health 2 (4):11–21
2. Bennet HL, Benson DR, Saigo SH (1985) Preoperative suggestions of decreased bleeding during spine surgery. Paper presented to the Scoliosis Research Society Annual Meeting, September 20
3. Bliesener T (1982) Die Visite – ein verhinderter Dialog. Narr, Tübingen
4. Bochnik HJ, Gärtner-Hutz C, Richtbeig W (1986) Schwierige Ärzte – Schwierige Patienten. Deutscher Ärzte Verlag, Köln, S 16–22
5. Frey D (1987) Psychologische Variable für Verweildauer und Arbeitsbeginn nach schweren Verletzungen. Psychol Heute 2:10
6. Geisler L (1987) Arzt und Patient – Begegnung im Gespräch. Pharma-Verlag, Frankfurt a.M.
7. Hofling S (1988) Psychologische Vorbereitung auf chirurgische Operationen. Springer, Berlin Heidelberg New York, S 136–143
8. Hurwitz BJ (1985) The art of history taking. In: Wilkins RH, Rengachary SS (eds) Neurosurgery, vol 1. McGraw-Hill, New York

9. Irsigler FJ (1960) Die Unterredung mit dem Kranken und seinen Angehörigen. In: Olivecrona H, Tönnis W (Hrsg) Handbuch der Neurochirurgie, Bd 4 (1):25
10. Korsch BM, Gozzi EK, Francis V (1968) Gaps in doctor-patient communications. Paediatr 42:855–871
11. Ley P, Spelman S (1967) Communication with the patient. Staples, Chicago
12. LG Köln (31.07.1979) 5 0 32/11. In: Giese B (1988) Medizinschaden und Arzthaftpflicht. Jüngling, Tübingen
13. Maquire P, Rutter D (1976) Training medical students to communicate. In: Bennet AE (ed) Communication between doctors and patients. Oxford University Press, London, pp 47–73
14. Mumford E, Schlesinger HJ, Glas GV (1982) The effects of psychological intervention on recovery from surgery and heart attacks: an analysis of the literature. Am J Public Health 72:141–151
15. Mummenthaler M (1973) Neurologie. Thieme, Stuttgart
16. Poimann H (1985–1989) Einführung in die Ärztliche Gesprächsführung und praktische Übungen zur idiolektischen Gesprächsführung. Vorlesungsverzeichnis der Universität Würzburg
17. Poimann H, Lohr W (1988) Arbeitsplatzanalyse in der Neurochirurgie mit dem FAA (unveröffentlicht)
18. Shakespeare W (1984) The tragedy of Hamlet, Prince of Danemark. Cambridge University Press, London, Act 3, Scene 2:852
19. Schunemann AL, Pickleman J, Hesslein R, Freeark RJ (1984) Neuropsychologic predictors of operative skill among general surgery residents. Surgery 96:288–295
20. Wesiack W (1971) Das ärztliche Gespräch. Hippokrates 42:207–217
21. Wilkes MW (1982) Wer die Arzt-Kommunikation beherrscht, beherrscht seine Zukunft. Pharma-Marketing J 4/82:115

Gutachtertätigkeit in Sozialgerichtsverfahren

Y. Altay[1]

In den von den Sozialgerichten angeforderten Sachverständigengutachten muß die Frage über den Gesundheitszustand des Versicherten und dessen berufliches Leistungsvermögen ausführlich erörtert werden.

Die Aufgabe, sachverständiger Berater des Gerichts zu sein, kann der Arzt nur dann erfüllen, wenn es ihm gelingt, juristischen Gedankengängen zu folgen. Für den Juristen ist der Weg zur Entscheidung durch die Rechtsnorm vorgezeichnet. Sein Denken geht vom allgemeinen ins einzelne, vom abstrakten Gesetz zum konkreten Fall, wobei es ein vorwiegend deduktives Denken ist. Anders als der Jurist geht der Arzt vom Einzelfall aus und bedient sich der induktiven Methode [1,2].

Im Sozialrecht ist die Leistungsfähigkeit immer nur im Hinblick auf die Arbeit als Mittel zum Erwerb rechtserheblich. Erwerbsfähigkeit bedeutet die Fähigkeit, Arbeitskraft im Erwerbsleben wirtschaftlich zu verwerten.

Die Kläger begehren die Gewährung einer Rente wegen Berufs- und/oder Erwerbsunfähigkeit. Der Neurochirurg wird öfter als Sachverständiger ernannt und mit der Erstellung eines Gutachtens beauftragt, wenn die dem Antrag zugrunde liegende Erkrankung auf degenerative Veränderungen im Wirbelsäulenbereich beruht, das heißt, wenn es sich um den Zustand nach Bandscheibenvorfall-Operation handelt.

In den letzten zwei Jahren haben wir in den im Auftrage der Sozialgerichte erstellten Gutachten in allen Fällen die Berufsunfähigkeit bejaht und Erwerbsunfähigkeit abgelehnt. Das Durchschnittsalter der Kläger betrug 52 Jahre. Das Gerichtsverfahren dauerte im Schnitt 1 1/2 Jahre. Die Verfahren wurden in einem Drittel der Fälle durch Vergleich, in zwei Dritteln der Fälle durch Rücknahme der Klage abgeschlossen.

Für die Beurteilung der Frage, ob einem Versicherten eine Rente wegen Berufs- oder Erwerbsunfähigkeit zu gewähren ist, kommt es in der Regel nicht darauf an, ob der leistungsgeminderte Versicherte vermittelbar ist, also einen seiner körperlichen und geistigen Leistungsfähigkeit entsprechenden und sozial zumutbaren Arbeitsplatz erhält, oder ob für ihn ein offener Arbeitsmarkt besteht. Nach der Rechts-

[1] Neurochirurgische Universitätsklinik, Knappschaftskrankenhaus, In der Schornau 23, D-4630 Bochum 7

Advances in Neurosurgery, Vol. 18
K.-A. Bushe, M. Brock, and M. Klinger (Eds.)

sprechung des Bundessozialgerichtes ist aber eine andere Betrachtung im Einzelfall dann notwendig, wenn dem Versicherten aus gesundheitlichen Gründen der Zugang zu den Arbeitsplätzen besonders stark erschwert ist. In diesen Fällen ist dem Versicherten der Arbeitsmarkt praktisch verschlossen (vgl. Bundessozialgericht, SozR 2200 § 1247 Nr. 33).

Die Frage, wie die Fähigkeit, Arbeitsplätze von der Wohung aus zu erreichen, zu beurteilen ist, ist in der Rechtssprechung noch nicht eindeutig geklärt. Der 5. Senat des Bundessozialgerichts geht davon aus, daß eine Gehfähigkeit von bis zu 500 Metern in der Regel nicht ausreicht, um einen Arbeitsplatz zu erreichen (Urteil vom 11.09.1979, AZ: 5 RJ 86/78; Urteil vom 10.03.1982, AZ: 5 b RJ 70/81; Urteil vom 06.06.1986, AZ: 5 b RJ 52/85 und Urteil vom 05.02.1987, AZ: 5 b RJ 22/86).

Der 5. Senat des Bundessozialgerichtes stellt dabei angesichts der Zumutbarkeit des Wohnsitzwechsels nicht darauf ab, welche Wegstrecken zum öffentlichen Verkehrsmittel am Wohnort des Klägers zurückzulegen sind, sondern sieht es als maßgeblich an, welcher Weg zu einem öffentlichen Verkehrsmittel als üblich angesehen werden kann. Dabei stützt sich der 5. Senat des Bundessozialgerichts hinsichtlich der Feststellung, daß eine Gehfähigkeit von bis zu 500 Metern nicht ausreicht, um ein öffentliches Verkehrsmittel zu erreichen, auf Feststellungen des Landessozialgerichts Rheinland-Pfalz und des Schleswig-Holsteinischen Landessozialgerichtes (vgl. Urteil vom 11.09.1979, AZ: 5 RJ 86/78 und Urteil vom 10.03.1982, AZ: 5b RJ 70/81).

Demgegenüber stellt der 4. Senat des Bundessozialgerichts (Urteil vom 26.05. 1987, AZ: 4a RJ 21/86) in dem Fall, daß ein Versicherter in einem durch öffentlichen Personennahverkehr erschlossenen industriellen Ballungsraum wohnt, auf die örtlichen Verhältnisse und die Üblichkeit der Wegstrecken in diesem Ballungsraum ab. Nach der Rechtssprechung des 4. Senats des Bundessozialgerichtes ist dabei zur Ermittlung der in einem Ballungsraum üblichen Fußwege zwischen der Wohnung des Versicherten und einem Arbeitsplatz aufzuklären, welche durchschnittlichen Wegstrecken ein Versicherter von einem nicht ungewöhlich liegenden Wohngebiet aus, zu einer Haltestelle eines öffentlichen Verkehrsmittels und sodann von der Zielhaltestelle aus zu einem nach Lage, Art und Größe nicht untypischen Betrieb des Ballungsraumes üblicherweise zurücklegen muß. Soweit dabei nach den Gegebenheiten des Ballungsraumes Betriebe mit weiteren Betriebswegen in Betracht zu ziehen sind, ist nach Auffassung des 4. Senats des Bundessozialgerichtes zu klären, ob Werksverkehr mit Personenbeförderung auf dem Betriebsgelände, insbesondere für Behinderte üblich oder möglich ist.

Der 13. Senat des Landessozialgerichts für das Land Nordrhein-Westfalen (Urteil vom 16.08.1987, AZ: L 13 J 144/86) geht dagegen davon aus, daß zwar für die Beurteilung der Frage, ob für einen Versicherten der Arbeitsmarkt wegen Einschränkung der Gehfähigkeit als verschlossen anzusehen ist, auf die Verhältnisse am Wohnort des Versicherten abzustellen ist, und hält demgemäß die Ermittlung der Entfernung von der Wohnung des Versicherten zur Haltestelle und die Ermittlung der üblichen Entfernung von einer Haltestelle bis zum Eingang der Betriebsstätten für erforderlich.

Entgegen der Rechtssprechung des 4. Senats des Bundessozialgerichts hält der 13. Senat des Landessozialgerichts für das Land Nordrhein-Westfalen Ermittlungen über die Verhältnisse auf dem Betriebsgelände für nicht erforderlich, da es nach Auffassung des 13. Senats des Landessozialgerichts für das Land Nordrhein-Westfalen offenkundig ist, daß es für alle Verweisungstätigkeiten Arbeitsplätze in der Nähe des Betriebseinganges gibt (vgl. Urteil des Sozialgerichts Dortmund vom 18.12.1987, AZ: S 34 J 216/86).

Damit führt die medizinische Einschätzung einer Gehfähigkeit bis zu 500 Metern aus juristischer Sicht zu der Annahme einer Erwerbsunfähigkeit.

Literatur

1. Engisch L (1964) Einführung in das juristische Denken, 3. Aufl. Kohlhammer, Stuttgart
2. Hennies GG (1968) Unterschiede zwischen juristischem und medizinischem Denken. Med Sachverst 64:214

Notwendigkeit und Voraussetzungen einer Gebrechlichkeitspflegschaft bei neurochirurgischen Patienten

M. Bettag, C. Bettag, J. Römer, J.C.W. Kiwit und W.J Bock[1]

Jede in die körperliche Unversehrtheit eingreifende Behandlungsmaßnahme erfüllt objektiv den Tatbestand der Körperverletzung gemäß § 223 StGB.

Ein ärztlicher Eingriff bedarf daher einer besonderen Rechtfertigung, die grundsätzlich durch die rechtswirksame Einwilligung des Patienten erfolgt [2]. Eine Einwilligung ist dann rechtswirksam, wenn der Patient das Wesen, die Bedeutung und die Tragweite eines erforderlichen ärztlichen Eingriffs in seinen Grundzügen erkennen und abwägen kann [3,4]. Verschiedenartige neurochirurgische Krankheitsbilder wie Hirntumoren, Hirnblutungen oder Schädel-Hirn-Verletzungen können zu derart ausgeprägten psychopathologischen Störungen führen, daß der Patient nach den genannten maßgeblichen Kriterien keine rechtswirksame Einwilligung mehr erteilen kann. Der Gesetzgeber spricht bei diesen Krankheitsbildern von sog. „geistige Gebrechen" und differenziert zwischen Geistesschwäche und Geisteskrankheit, wobei sämtliche Definitionen zeigen, daß es hierbei erhebliche Abgrenzungsschwierigkeiten gibt [1,4,6,8].

Speziell in der Neurochirurgie muß bei allen Krankheitsbildern, die zu einer Bewußtseinsstörung oder Bewußtlosigkeit führen oder aber Störungen der Wahrnehmung und des Erlebens bewirken, ärztlicherseits geprüft werden, ob bei dem betreffenden Patienten eine freie und von der vorliegenden Störung unbeeinflußte Willens- und Entscheidungsfähigkeit gegeben ist.

Liegt diese Fähigkeit nicht vor und besteht ein entsprechendes Fürsorgebedürfnis, so sollte der Arzt ein Pflegschaftsverfahren in Hinblick auf die durchzuführende Behandlungsmaßnahme einleiten. Hierdurch wird die Respektierung des Selbstbestimmungsrechts des Patienten gewährleistet [5].

Es kann nicht pauschal von einem bestimmten Krankheitsbild auf die Notwendigkeit einer Pflegschaftsanordnung geschlossen werden, da gleichartige Läsionen bekanntlich zu unterschiedlichen Funktionsstörungen führen können. Jeder Fall ist daher individuell zu beurteilen.

Grundsätzlich sollte bei nicht mehr einwilligungsfähigen Patienten, bei denen ein Wahleingriff vorgenommen werden soll – also keine Notfallsituation vorliegt – der Weg des Pflegschaftsverfahrens angestrebt werden. Dies ist nicht notwendig

[1] Neurochirurgische Klinik der Universität Düsseldorf, Moorenstr. 5, D–4000 Düsseldorf

Advances in Neurosurgery, Vol. 18
K.-A. Bushe, M. Brock, and M. Klinger (Eds.)

bei denjenigen Krankheitsbildern, die zu einer vitalen Gefährdung oder erheblichen Verschlechterung des Gesundheitszustandes führen, wenn nicht ein sofortiger ärztlicher Eingriff vorgenommen wird. Die Rechtfertigung erfolgt in diesen Fällen über die sog. mutmaßliche Einwilligung [2, 5].

Es wird z.B. bei einem Patienten mit einem bifrontalen Meningeom und dem klinischen Befund eines ausgeprägten hirnorganischen Psychosyndroms für einen geplanten elektiven Operationseingriff eine Behandlungspflegschaft erforderlich sein, wenn dieser Patient keine rechtswirksame Einwilligung erteilen kann. Kommt derselbe Patient aber mit einer progredienten Bewußtseinstrübung und beiderseitigen Stauungspapillen zur stationären Aufnahme, so wird für den sofortigen operativen Eingriff aufgrund der Notfallsituation die mutmaßliche Einwilligung angenommen.

Sollte die Anordnung einer Gebrechlichkeitspflegschaft erforderlich sein, so kann sich die Pflegschaft allein auf ärztliche Behandlungsmaßnahmen beschränken, im Einzelfall auch auf eine bestimmte Operation [3, 8]. Mit Erledigung der besonderen Angelegenheit endet auch die Pflegschaft, vgl. § 1918 Abs. III BGB. Gemäß § 1910 Abs. III BGB darf die Pflegschaft nur mit Einwilligung des Gebrechlichen angeordnet werden, es sei denn, daß eine Verständigung mit ihm nicht möglich ist. Da bei den vorher genannten neurochirurgischen Krankheitsbildern eine Verständigungsmöglichkeit mit dem Patienten gerade nicht vorhanden ist, handelt es sich dabei fast immer um eine sog. Zwangspflegschaft [4, 6, 8].

Das Verfahren wird in der Regel aufgrund einer entsprechenden Anregung des Arztes eingeleitet [3]. Es handelt sich um ein Amtsverfahren der freiwilligen Gerichtsbarkeit und nicht um ein Antragsverfahren, d.h. eine formlose Benachrichtigung des zuständigen Vormundschaftsgerichts genügt [1, 6].

Der Vormundschaftsrichter ist funktionell zuständig für die Anordnung einer Pflegschaft, vgl. § 14 Nr. 4 RPflG. Die örtliche Zuständigkeit des Gerichts richtet sich nach dem Wohnort des Gebrechlichen, vgl. § 38 FGG i.V. mit § 36 FGG.

Der Arzt als Sachverständiger hat über den Gesundheitszustand des betroffenen Patienten Auskunft zu erteilen und die Krankheitsbefunde darzulegen, die seines Erachtens eine freie Willensentscheidung unmöglich machen. Die Benennung einer Krankheitsdiagnose allein reicht nicht aus, die psychopathologischen Befunde mit ihren Folgewirkungen müssen konkret aufgeführt werden.

Im Einzelfall kann zuzsätzlich ein psychiatrisches Fachgutachten oder eine amtsärztliche Untersuchung angeordnet werden [3, 4, 8]. Trotz der Zuhilfenahme von ärztlichen Sachverständigengutachten kann sich das zur Entscheidung berufene Gericht nur durch den zusätzlichen eigenen persönlichen Eindruck vom Betroffenen eine verläßliche Entscheidungsgrundlage verschaffen, der zuständige Vormundschaftsrichter hat den Gebrechlichen also grundsätzlich – möglichst im Beisein des betreuenden Arztes – anzuhören [3, 4, 7].

Die Geschäftsfähigkeit des Pflegebefohlenen wird durch die Gebrechlichkeitspflegschaft im Unterschied zur Vormundschaft grundsätzlich nicht berührt, wenn auch bei den betroffenen neurochirurgischen Patienten häufig eine zumindest partielle Geschäftsunfähigkeit besteht.

Im Vergleich zur Vormundschaft stellt die Pflegschaft daher einen weit weniger einschneidenden Eingriff in das Persönlichkeitsrecht des Pfleglings dar [7].

Literatur

1. Brand/Hensel (1963) Die Vormundschaft-, Familienrechts- und Fürsorgeerziehungssachen in der gerichtlichen Praxis, 2. Aufl. Springer, Berlin Göttingen Heidelberg
2. Dreher/Tröndle (1986) StGB-Kommentar, 43. Aufl. München, Vor§ 32, § 223 StGB
3. Göppinger (1980) Betrachtungen zur Unterbringung psychisch Kranker. FamRZ 856–864
4. Kraiß H (1980–81) Probleme der Gebrechlichkeitspflegschaft. BW Not Z 59–62
5. Münchener Kommentar zum BGB (1978) Mertens, Bd 3 2. Hb., 1. Aufl. § 823 Rdnr. 448ff
6. Münchener Kommentar zum BGB (1987) Goerke, Bd 5, 2. Hb., 2. Aufl. § 1910 Rdnr. 1ff
7. OLG Koblenz in FamRZ (1989) 213–214
8. Palandt (1986) BGB-Kommentar, 45. Aufl., München, § 1910 Anm. 2ff

Subject Index